ASTRO-NAVIGATION

-from Square One to Ocean-Master

2011 Edition

Alan Murray

Visit us online at www.authorsonline.co.uk

An Authors OnLine Book

Copyright © Alan Murray 2011

Cover design by James Fitt ©

All rights reserved. No part of this publication may be reproduced, stored in a retrieval system, or transmitted in any form or by any means, electronic, mechanical, photocopy, recording or otherwise, without prior written permission of the copyright owner. Nor can it be circulated in any form of binding or cover other than that in which it is published and without similar condition including this condition being imposed on a subsequent purchaser.

The moral rights of the authors have been asserted.

ISBN 978-07552-0648-3

Authors On Line Ltd
19 The Cinques
Gamlingay
Sandy
Bedfordshire
SG19 3NU
England

This book is also available in e-book format, details of which are available at
www.authorsonline.co.uk

This book is dedicated to the memory of Paddy Murray

Acknowledgements

My thanks go to the following:-

Peter Webb and the members of Leigh on Sea Sailing Club for teaching me all I know about sailing.

Richard Fitt my Editor, James Fitt the designer and AuthorsOnLine.co.uk for the preparation of the manuscript.

Gillie Macadam, for reading the manuscript and for his general support and encouragement.

Carmen Farrugia for her support during the editing of the 2011 Edition.

Captain James P. Hughes, Lecturer in Nautical Studies, South Tyneside College and Mr G. P. Luker, Senior Lecturer, Warsash Maritime Academy, Southampton, for their invaluable comments on the textual content and style.

HM Nautical Almanac Office for permission to use data from HM Nautical Almanac NP 314, NP 303, NP 401 and to reproduce the World Map of Time Zones. '© Crown Copyright and/or database rights. Reproduced by permission of the Controller of Her Majesty's Stationery Office and the UK Hydrographic Office (www.ukho.gov.uk).'

Weems and Plath, Annapolis, MD, USA, for the photograph of the Star finder 2102-D.

Please note that while all reasonable care has been taken in the preparation of this book, both in the data contained and the worked examples presented, no responsibility can be taken by either the author or publisher for the further use of the methods described. The author welcomes suggestions for future improvements and any reference attribution.

To contact the author: astronavsq1@gmail.com or astronav-square-one.com

Astro-navigation from Square One: Contents:

Those sections paragraphs immediately relevant to the current RYA Yacht Master Ocean course syllabus are identified in the contents pages and section paragraphs ***headed in italics***.

Introduction		**viii**
Glossary		**ix-xxi**
Section 1: The basic terms and concepts		**1**
1.1	*The celestial sphere*	2
1.2	*Fixing position on the Earth*	4
1.3	The Nautical Mile	7
1.4	*Positioning on the Celestial Sphere*	8
1.5	*Celestial latitude: Declination*	9
1.6	*Celestial longitude - Greenwich Hour Angle*	10
1.7	*Celestial longitude - Aries and Sidereal Hour Angle*	11
1.8	*Local Hour Angle*	13
1.9	*Celestial movement*	15
1.10	The celestial sphere and the view from the deck	16
1.11	The strength of the Sun	19
1.12	The Earth's tilt and the seasons	20
1.13	Converting to the geocentric view of the Sun	22
1.14	Why does the Sun rise and fall	24
1.15	*Local apparent noon*	26
1.16	*Linking Angle to Time*	26
1.17	*Linking GMT/UT to local apparent noon*	30
1.18	*The Sun to measure latitude*	31
1.19	*The importance of the Sun at local noon*	34
1.20	Summarising Section 1	34
Section 2: Astro-navigation: Down to the details		**36**
2.1	*The horizon when measuring altitude*	37
2.2	*The "visible sky" view of the horizon*	38
2.3	*Parallax*	40
2.4	*Semi-diameter*	42
2.5	*Refraction*	43
2.6	*Altitude corrections for parallax, semi-diameter and refraction*	44
2.7	*Summarising the manipulation of a sextant sight*	44
2.8	*What can you deduce from an altitude?*	45
2.9	*Relating an observed altitude to an observer's position - the PZX triangle*	47
2.10	*Local Hour Angle and the PZX triangle*	53
2.11	*Azimuth Angle and the PZX triangle*	54
2.12	*Finding the Azimuth from Azimuth Angle*	55
2.13	*Summarizing the PZX Triangle*	56
2.14	The difficulty with Zenith Distance and Azimuth	56
2.15	*Calculated altitude – how and why do you work it out?*	57

2.16	Explaining spherical trigonometry	57
2.17	*Comparing calculated and true altitude*	60
2.18	*Working out the azimuth*	62
2.19	*Converting a quadrantal notation azimuth to 360° notation*	63
2.20	*The relative sizes of observed and calculated altitudes*	64
2.21	The distance between true and calculated positions	64
2.22	*The intercept*	66
2.23	*Summarizing the plotting of a sight*	68
2.24	*Limitations in the plotting of a single sight*	70
2.25	*Using transferred position lines*	71
2.26	Summarising Section 2.	72

Section 3: The Sextant and its use — 73

3.1	The predecessor of the Sextant – the seaman's quadrant.	73
3.2	*The development of the modern sextant*	74
3.3	*Basic sextant structure and optics*	75
3.4	*The identification and correction of potential and actual sextant errors*	78
3.5	*Setting the heavenly body on the horizon*	82
3.6	Acquiring a sextant	83

Section 4: Tables, tables, tables — 84

4.1	*The Nautical Almanac*	85
4.2	*The Ephemeris Tables*	86
4.3	*Tables of Conversion*	90
4.4	Tables of Increments and Corrections	91
4.5	*Tables of Increments*	91
4.6	*Tables of Corrections*	93
4.7	*Tables for Calculated Altitude*	96
4.8	*Tables of Azimuth*	101
4.9	Tables of Mathematical functions	101
4.10	Lists and Indexes	102
4.11	Interpolation	102
4.12	ABC tables and interpolation	105
4.13	The problems of interpolating Azimuths by eye	110
4.14	*Calculator methods - practical alternatives to ABC Tables*	110

Section 5: The Sun and Time – their important link — 112

5.1	Our Sun	113
5.2	*Apparent and Mean Solar Time*	114
5.3	*Mean Solar Time*	114
5.4	*The Equation of Time*	115
5.5	*Apparent and Mean solar time*	116
5.6	How do planets orbit our Sun?	117
5.7	*Describing orbits by their Eccentricity*	118
5.8	*Explaining planetary motion*	119
5.9	*How earth's orbit affects apparent solar time*	120

5.10		How Earth's axial obliquity affects apparent solar time	123
5.11		Visualizing the effect of the changing speed and obliquity	128
5.12		Other causes of orbital variation	130
5.13		The relevance of each day's equation of time	132
5.14		*GMT/UT and Time zones*	132
5.15		*Keeping up with the times*	134
5.16		Summary of Section 5	135

Section 6: Navigating with the Sun — **136**

6.1	*The Greenwich Hour Angle*	137
6.2	*Determining Local Hour Angle*	140
6.3	*The Sun's changing declination*	144
6.4	Navigating with the Sun	146
6.5	The Sun's motion across the sky	146
6.6	*The Sun at meridian passage*	147
6.7	*Timing non-meridian Sun sights*	149
6.8	Sun Sight Reduction	150
6.9	*The Noon Sight*	152
6.10	*The 3 equations used to produce latitude*	154
6.11	*Using a proforma*	160
6.12	Summing up the meridian passage	162
6.13	Taking the meridian sight	162
6.14	*The horizon and height of eye*	163
6.15	*Non-meridian Sun sights and their handling*	164
6.16	*Sun-run-Sun position fixing*	172

Section 7: Navigating with the Moon — **179**

7.1	The Moon in general	180
7.2	The Moon in navigation	180
7.3	The Moon we see	181
7.4	The Moon's varying orbit	181
7.5	The phases of the Moon	184
7.6	The time of day and the Moon's phase	186
7.7	Apparent Moon size, its orbit and Augmentation	187
7.8	The Moon's parallax	188
7.9	The Moon and eclipses	189
7.10	Moonrise and Moonset	191
7.11	The Moon's meridian passage	193
7.12	*The Moon's GHA and Declination*	194
7.13	*Sighting the Moon*	196

Section 8: Navigating with the Planets — **206**

8.1	The Planets	206
8.2	*The navigational planets*	207
8.3	Planetary inclinations and eccentricities	207
8.4	The brightness of planets	208
8.5	The Planets and the celestial sphere	210

8.6	The special orbit of Venus	211
8.7	A word about Mercury	213
8.8	Planets with orbits larger than Earth's	213
8.9	*The NP 314 Planet diagram summary chart*	215
8.10	*Planetary Greenwich hour angle and Declination*	217
8.11	*Twilight at varying latitude and longitude*	218
8.12	*Working up a Planet sight*	222

Section 9: The stars in general and *Polaris* in particular — 231

9.1	The stars in general	232
9.2	The celestial sphere	232
9.3	*The characteristics of navigationally useful stars*	233
9.4	Brightness or Magnitude	235
9.5	The celestial sphere's movement	237
9.6	The changing night sky	238
9.7	Movement of the stars on the celestial sphere	239
9.8	Leading and trailing stars in constellations	241
9.9	Recognizing patterns in the sky	241
9.10	Invisible stars	244
9.11	Circumpolar stars	245
9.12	Non-circumpolar stars	247
9.13	*Sight planning and choosing*	248
9.14	*Reducing a star sight*	251
9.15	*Rapid sight reduction table*	259
9.16	*Polaris-the North star*	259
9.17	Finding Polaris in the sky	266
9.18	*The special position of Polaris*	269
9.19	*Polaris altitude equals latitude*	270
9.20	*Adjusting for the off-centre position of Polaris*	271
9.21	Summarizing navigation using the stars.	276

Appendix - Understanding spherical trigonometry – from square one — 277

A.1	Relating angles to the sides of triangles	277
A.2	The cosine ratio of the angles of a right-angled triangle	280
A.3	The tangent ratio of the sides of a right-angled triangle	281
A.4	The mathematics of describing 3 dimensional space	283
A.5	Defining the cross product of two vectors	284
A.6	Describing 3 vectors in terms of dot and cross products	285
A.7	Describing the full spherical triangle	288
A.8	Describing the radian as a unit of angular measure	289
A.9	Deriving the equations to solve the spherical triangle	290
A.10	Transforming the cosine formula to an easily usable form for a calculator	293
A.11	*The use of plotting sheets and defining Departure*	294

Bibliography — 296
Index — 301

Astro-navigation from Square One: Introduction

Why learn astro-navigation? It appeals to all sailors to use the Earth's natural elements such as wind, current and tide to propel them across seas and oceans and a natural extension of this must be to use the multitude of heavenly bodies as a guide. It matters not that the means of electronic navigation have proved, so far, to be extremely accurate and reliable. There is little more satisfying for a sailor than to gaze in to the heavens and use the Sun, Moon, planets or stars, together with a little maths, to produce lines on a chart indicating a likely position. For the training of professional sailors, astro-navigation remains a mandatory subject.

Many who try fail to finish a book on learning astro as the subject is relatively complex. It requires knowledge of an entirely new language of several dozen terms and the use of seemingly impenetrable maths. Others simply stick to proformas and sight reduction tables, content not to understand the underlying science.

Many astro books leave much to the imagination, explaining little from first principles with scant attention to the language and detail. This book endeavours to leave nothing unexplained, no word undefined and aims to teach the subject by UNDERSTANDING and not by ROTE. Those section paragraphs immediately relevant to the current RYA Yacht Master Ocean course syllabus are identified in the contents pages and section paragraph in *italics*.

Emphasis is needed that the astro-navigator does not need to know a myriad of stars and that most of the work in astro-navigation has been done for you. The precise predicted positions of all bodies at all times are recorded in the astro-navigational almanac. Essentially all that needs to be done is to compare the height in the sky of a body from your unknown position to the known height of the body from a known position. It is even easier using the Sun for latitude at mid-day.

The subject requires an ability to think in three-dimensional space, problematic for some, less so for others and to ease this potential pitfall I have used diagrams freely with every explained concept. If background information helps in the understanding, then I make no apology for including what might seem to be superfluous detail. Astro-navigation is a wonderful mix of art and science. The art is in the use of the sextant, in its care and adjustment and the taking and reading of a sight. The maths and physics may appear complex but can be rendered less fearsome.

Do not become over-anxious at the thought of spherical trigonometry, intimate knowledge is unnecessary and most techniques minimize its use on the chart table, reducing the maths to careful addition and subtraction. Of course the modern wizardry of computer programs and internet access can provide useful help and take some of the work out of "astro", but all sailors will recognise the importance of self-reliance in navigation as they do in boat-handling; a £250 sextant, a scientific calculator and a set of annual tables make a good place to start, for an outlay of £300.

Many when tackling this book may find the early pages too simplified but as with swimming; it is better to start gracefully in the shallow end rather than to immerse oneself in the middle, only to drown. The glossary is meant to be comprehensive and at first sight is daunting. Although it is at the beginning of this book, it is probably best not read at the beginning but referred to as necessary from within the text. The book's divisions into sections will be self-explanatory, addressing the basic vocabulary, the details of the theory, the sextant and the tables used before considering in turn, time and the heavenly bodies. The examples shed light and explain the available options in handling the data. The appendix is a digest of the relevant plane and spherical trigonometry and explains the maths used in detail; an understanding being useful but not imperative.

Glossary

Glossary of astro-navigational terms:

This list of explanations of words relating to the various aspects of astro-navigation is included here, not necessarily to be immediately read and understood in isolation. As each new or significant term arises in the text of this volume it will appear in **bold type**, indicating that its definition appears in the Glossary. Where a specific explanation in this glossary uses a further astronomical term which also appears independently in the glossary, then this too appears in **bold type**.

- **ABC formula** – a mathematical formula from which to derive the **quadrantal azimuth** of a heavenly body, given its **local hour angle** and **declination** together with the **latitude** from which it is observed.

- **ABC tables** – tables to enable the computation of a **quadrantal azimuth** of a heavenly body, given its **local hour angle, declination** and the **latitude** of the observation.

- **albedo** – the measure of the light reflected to that incident upon a heavenly body e.g. the Moon has an albedo of 0.07, meaning that 7% of the Sun's incident light is reflected from its surface.

- **altitude** – the angular distance of a body above an **horizon**. The **sextant altitude** (H_s) is that measured by a sextant before any corrections are made for instrumental error. **Apparent altitude** is the angular distance above an horizon after corrections for both **sextant error** and **angle of dip** have been applied. The **observed altitude** (H_o - and called by some authors: **true altitude**), is the altitude that would have been found if the observation had been made at the centre of the Earth. It is found by applying a correction to **apparent altitude** (where appropriate) for **semi-diameter, parallax** and **refraction**. The **calculated altitude** (H_c) is the altitude of an observed body found mathematically from the body's **local hour angle** and **declination** together with the **latitude** from which the observation is made.

- **angle of dip** – the angle at an observer's position between the **visible** and **sensible horizons** i.e. the angle between the visible horizon and the intersection of the tangential plane to the Earth's curvature at the level of the observer's eye, projected out to meet the **celestial sphere**.

- **angle of parallax** – the angle at an observed body between straight lines drawn to the observer and the centre of the Earth and from which is derived a correction to be applied to the **apparent altitude** of the Sun, Moon and closer planets.

- **aphelion** – the point in its solar orbit when Earth is farthest from the Sun, at present in early July, but changing very slowly.

- **apogee** – the point in its orbit when the Moon is farthest from the Earth.

- **apparent altitude** – the angle of altitude of a heavenly body above the **sensible horizon**, derived by correcting the sextant altitude for both **index error** and the **angle of dip**. Abbreviated to H_a.

apparent magnitude – the **magnitude** or brightness of a heavenly body when viewed from Earth.

apparent solar time – the passage of time as experienced on Earth which depends upon the movements of the Sun. It is the time that would be depicted on a sundial. The variations in this time are primarily due to the Earth's fixed axial **obliquity** and speed changes due to its elliptical orbit. Apparent solar time minus **mean solar time** gives rise to a difference called the **equation of time**.

apparent Sun – the Sun and its movements as seen from Earth, apparent meaning: "that which is seen".

Aries – see **first point of Aries**

assumed position – a **chosen, dead-reckoning** or **estimated position** on a chart and through which is drawn an **azimuth** of an observed heavenly body and for which a **calculated altitude** is derived for comparison with an **observed altitude** in order to produce an **intercept** and corresponding **position line**.

augmentation – the name given to the change in apparent diameter of the Moon with changing latitude of the observer on Earth, caused by the significant ratio of the Earth's diameter and the distance between the Moon and the Earth.

axial obliquity – see **obliquity**

azimuth – the direction from an observer to an observed body, measured in true degrees from North. Calculated azimuth tables and calculated azimuth formulae generally produce a **quadrantal azimuth**; a direction measured in degrees East or West of North or South. Abbreviated to Z_n.

azimuth angle – the internal angle of the **PZX triangle** at the corner marked by the **zenith** of the observer. In the northern hemisphere, if the **local hour angle** is less than 180°, then the **azimuth** = 360° – azimuth angle. If the **local hour angle** is more than 180° then the **azimuth** = azimuth angle. Abbreviated to Z.

azimuth distance – the angular distance over the surface of the Earth from an observer to a heavenly body's **geographical position**.

calculated altitude – the altitude of a heavenly body derived mathematically or from tables using the body's declination, local hour angle and a chosen latitude. The calculated altitude is compared to the body's **observed altitude** in order to derive an **intercept** on its **azimuth**. Abbreviated to H_c.

celestial equator – the horizontal projection of the plane of the Earth's equator on to the celestial sphere, producing the datum line from which is measured **declination**.

celestial horizon – a plane projected from the Earth's centre to form an arc on the **celestial sphere** at right angles to a line from the Earth's centre through the position of the

Glossary

observer up to the observer's **zenith**. It is from the celestial horizon that the **observed altitude** (also called the **true altitude** by traditionalists) is calculated by correcting a **sextant altitude**.

celestial latitude – properly called **declination**: the angle subtended at the Earth's centre by a line from a heavenly body on the celestial sphere and the plane of the **celestial equator**.

celestial longitude – properly called **Greenwich Hour Angle**: the angle subtended at the North Celestial Pole between the **Greenwich meridian** and the meridian on which lies a particular heavenly body.

celestial poles – **North and South** - the projection of the North and South earthly poles directly above and below the Earth to meet the **celestial sphere**, forming an axis around which the sphere is presumed to rotate.

celestial sphere – a theoretical sphere of infinite dimensions surrounding the Earth, on which lie the heavenly bodies, some being fixed (stars) and others moving across it (Sun, Moon and planets). The Earth lays centrally, its own centre being coincident with that of the celestial sphere. The sphere rotates about an axis through the **celestial poles**.

chosen position – an **assumed**, **dead-reckoning** or **estimated position** on a chart through which is drawn an **azimuth** of an observed heavenly body and for which a **calculated altitude** is derived for comparison with an **observed altitude** in order to produce an **intercept** and corresponding **position line**.

circumpolar – a term applied to a star which when viewed from a particular **latitude** stays above the horizon during its 24 hour journey around the celestial pole, neither rising nor setting.

co-latitude – an angular measurement, akin to **latitude**, but measured, not from the equatorial plane to the position of the observer, but from the North Pole to the position of the observer. Co-latitude = 90° - latitude for an observer in the northern hemisphere.

collimation error – an error in the transmission of light received from the **horizon mirror** of the **sextant** as it passes through the optics of the telescope, if used. It implies that the telescope alignment is not parallel to the plane of the instrument.

conjunction – a term applied when the positions of heavenly bodies are in line with one another when viewed from Earth. Venus, having an orbit inside that of Earth, can be in conjunction with the Earth and Sun either between the Sun and Earth, or when it lies beyond the Sun. These positions are known respectively as an inferior and superior conjunction.

constellation – a term applied to a group of stars which when viewed from Earth form a recognizable and named pattern.

correction – an amendment to a figure or measurement, in order to adjust it because of a known variation or potential error e.g. a **sextant altitude** (H_s) is corrected for **index error** and **angle of dip** to give the apparent altitude (H). Correction values may be in the form of a table or arrived at by the use of a mathematical formula.

cosine – a mathematical ratio of the side adjacent to the angle and the hypotenuse (long side) of a right-angled triangle. The ratio is used to determine the size of the angle or if the angle and one side is known, then to determine length of the other side. The cosine function features prominently in the solution of the navigational **PZX triangle**.

cross product – a mathematical method of describing two **vectors** from the same point in space, in which both vectors are described by a third vector in a plane at right angles to them. Its value is given by their product and the sine of the angle between them: for vectors **a** and **b** the cross product **a**x**b** = $ab\sin\theta$. Used in the trigonometric description of a **spherical triangle**.

datum line – a real or assumed line from which a measurement is taken e.g. the **Greenwich meridian** for longitude, or the **visible horizon** for the **sextant altitude** of a heavenly body.

dead-reckoning position – an assumed position for navigational purposes arrived at by noting course, time and distance run from a known position, but <u>without</u> allowances for tide and current.

departure – the distance in nautical miles of one meridian of longitude from the next at a particular latitude. This distance varies as the cosine of latitude, being 60nm at the equator and 0nm at the pole.

dot product – a mathematical method of describing two vectors from the same point in space, allowing one to be defined in terms of the other's magnitude and the cosine of the angle between them. For vectors **a** and **b** the dot product **a·b** = $ab\cos\theta$. The result is a scalar in the same plane as the two vectors, having magnitude but not direction. Used in the trigonometric description of a **spherical triangle**.

eccentricity – a mathematical description of the elliptical extent of an orbit of one heavenly body to another e.g. the Moon's orbit around Earth or a planet around the Sun.

eclipse – a condition in which the earthly view of a heavenly body is altered temporarily due to a specific spatial alignment such that light from the Sun is either obscured from reaching Earth (an eclipse of the Sun by the Moon) or the shadow of the Earth is cast over the Moon (an eclipse of the Moon by the Earth).

ecliptic – the apparent annual path of the Sun around the **geocentric** Earth, lying at an angle of 23° 27´ to the plane of Earth's equator and giving rise to the changing seasonal pattern.

Glossary

elongation – a term applied to the solar orbit of a planet, e.g. Venus as seen from Earth, describing the horizontal angle between the planet, the Earth and the Sun.

ephemeris – tables that give the daily positions of the navigationally useful heavenly bodies: the Sun, Moon, planets and stars. The word ephemeris, comes from the Greek *ephemeros*, meaning living a day and from which we derive the word ephemeral.

equation of time – the difference between the **apparent solar time** kept by the Sun's movements when observed from Earth and **mean solar time**, based on an artificially fixed and exact 24 hour day. The difference is due primarily to the changing speed of the Earth's movements around the Sun and the effect of the **obliquity** of the Earth's axis.

equatorial view – a view of the geocentric Earth and **celestial sphere** from a point distant from but level with the plane of the **equator**; useful for the appreciation of the fundamental angular measurements in order to calculate **latitude** from the **noon sight** of the Sun.

equinoctial points – see **equinox**.

equinox – the periods of equal night and day, around March 21st and September 23rd coinciding with the two **equinoctial points**, at which the Sun's apparent path round the Earth, the ecliptic, intersects with the plane of Earth's equator and at which point its **declination** is 0°.

estimated position – an assumed position for navigational purposes arrived at by noting course, time and distance run from a known position, with allowances for tide and current.

first point of Aries (Aries) – the point at which the Sun's annual orbital path, the **ecliptic**, crosses the **celestial equator** in March and is used as the hour angle point of reference for all stars. **Greenwich hour angle** of a star = **Greenwich hour angle of Aries** + **sidereal hour angle** of the star.

full Moon cycle – a period of approximately 14 **lunations** or **synodic months**, between the dates and times at which the full Moon coincides with its **perigee** (the point in its orbit when it is nearest point to Earth). It is caused by a form of **precession**, a gravitational effect producing cyclical alignment variations in the Moon's axis of rotation.

geocentric – the artificial Earth-centred arrangement of the **celestial sphere** upon which all heavenly bodies are situated and on which astro-navigational calculations are based.

geographical position – the point of intersection with the Earth's surface, of a straight line from the Earth's centre to a heavenly body.

great circle – a circle on the surface of the Earth's sphere that has its centre coinciding with that of the Earth. **Meridians of longitude** are all great circles, as is the equator. An arc of a great circle passing through two points on the surface of the Earth represents the shortest distance between those points.

Greenwich Hour Angle – of a heavenly body, is the angle measured at the North celestial pole between the **Prime Meridian** at Greenwich and the meridian on which lays a particular heavenly body. The angle is measured in degrees West of Greenwich from 0-360°.

Greenwich Hour Angle of Aries – the angle measured at the North celestial pole between the **Prime meridian** at Greenwich and the representative of the stars, the **first point of Aries**. The angle is measured in degrees West of Greenwich from 0-360°.

Greenwich Mean Time – is **Universal Time** + 1200hrs 00s. The term GMT is technically obsolete but widely used when accuracy to less than 1 second is required, as in much nautical navigation

haversine – a mathematical ratio of half a versed sine: (1–cosine)/2, now largely historical. Haversines can be used in a longhand method of finding a **calculated altitude** (H) and increases the accuracy over logarithmic cosines when considering small angles.

heliocentric – the true Sun-centred arrangement of the solar system.

horizon mirror – the mirror attached to the body of a **sextant** which reflects light received from the **index mirror**, towards the eye-piece or telescope. The mirror is either half-silvered or has special coatings such that the horizon is simultaneously viewed with the image of the observed body.

horizon – natural/visible – the line seen by an observer where the curvature of the Earth appears to meet the visible heavens. Its distance from the observer is dependent upon the observer's height above the surface of the sea.

horizon – sensible – the horizontal tangential plane at the level of the observer's eye which meets the **celestial sphere** just above the visible horizon

inclination – an angular description of one plane with respect to another e.g. the plane of the Moon's orbit around the Earth to that of the Earth's orbit around the Sun.

increment – a finite increase in a variable quantity. In astro-navigation it usually refers to the necessary additions to a whole hour figure so as to adjust it for both minutes and seconds of time e.g. the tabulated **Greenwich Hour Angle** of the Sun or **Aries** may be at one hourly intervals. A second table is required to look up the addition for the minutes and seconds for a specific moment between whole hours.

index error – an error in alignment in the **horizon mirror** (not index mirror) of a **sextant**, around its horizontal axis. It is adjusted for after correction of the **perpendicularity** of the **index mirror** and **side error** of the horizon mirror. Any residual error is then the result of the two mirrors not being perfectly parallel when the sextant reading is exactly zero.

Glossary

The value of this error can be applied to the sextant altitude as a correction, or taken out with the horizon mirror adjustment until it is as small as possible.

index mirror – the mirror attached to the index arm of a **sextant** which reflects light from an observed body to the **horizon mirror. Perpendicularity** of the plane of the mirror to the arm must be carefully adjusted.

intercept – given an **azimuth** to a heavenly body drawn on a chart through a **chosen** or **dead-reckoning position**, the point at which the **position line** is drawn across it at right angles is the intercept. The exact positioning of the intercept depends upon the relative size of the **calculated** and **observed altitudes**. Their angular difference is equal numerically in **nautical miles** to the distance from the chosen/dead-reckoning position to the intercept. If the **observed altitude** is larger than the calculated then the intercept is placed towards the observed body with respect to the **chosen position**. If calculated altitude is greater than the true altitude then the intercept on the azimuth is away from observed body with respect to the chosen position.

intercept method – the method introduced by Capt. Marcq St Hilaire in which a **calculated altitude** from a chosen position is compared to **observed altitude** and their difference used to arrive at an **intercept** on the **azimuth** of the observed body.

interpolation – a mathematical process done by eye or calculation to obtain intermediate values between ones given, usually in a table. It may be used in the interpretation of tables such as **ABC tables** that can be used for the production of an **azimuth** from an observer to a heavenly body, using **latitude**, **local hour angle** and **declination**.

latitude – the name given to the system of measurement used to fix position on the Earth in a North-South direction. The latitude of a point is the angle made at the centre of the Earth between the plane of the **equator** and a line from the point. Latitude is measured in degrees and minutes North or South of the equator from 0° at the equator to 90° at the poles.

libration – the apparent and marginal change in the face of the Moon visible to an observer on Earth due to the changing orbital position of the Moon relative to the Earth. It is a consequence of the changing angle between the Earth, Moon and Sun.

local hour angle (LHA) – of a heavenly body is the angle measured at the North **celestial pole** between the meridian of **longitude** of an observer and the meridian on which lays an observed heavenly body. It is measured in degrees West of the observer from 0-360°. Local Hour Angle = **Greenwich Hour Angle** – observer's **longitude** West, or Local Hour Angle = **Greenwich Hour Angle** + observer's **longitude** East.

local apparent noon – the moment in time at which the Sun crosses the meridian of **longitude** of an observer; it is the moment of local **meridian passage**.

logarithm – a method of computation using exponents. A logarithm (log) is the power (exponent) to which one number, the base, must be raised to give another number e.g. $10^2 = 100$, so that the logarithm to the base 10 of 100 is 2. Base 10 is used in "common logarithms". The logarithm to base 10 of 1000 = 3. The sum 100 x 1000 in log form is 2 + 3 = 5, and the antilog of 5 is 10^5 = 100000. For complicated numbers, tables of logs to four or more figures are provided, converting difficult multiplications and divisions into simpler additions and subtractions. Prior to the invention of the slide rule, calculator and computer, logarithmic tables were commonplace and a vital mathematician's tool.

longitude – the name given to the system of measurement used to fix position on the Earth in an East-West direction. The longitude of a point on the Earth's surface is the angle made at the centre of the Earth by a line from that point and a datum plane of a North-South line, the **Greenwich Meridian**. Longitude is measured in degrees and minutes East or West of the datum from 0° -180° East or West.

lunars – tables used before the introduction of the chronometer, in which angular distances between certain stars and the Moon were allied to **Greenwich Mean Time** to give the GMT/UT of a particular observation. This when compared to local time, governed by the meridian passage of the Sun, enabled **longitude** to be estimated.

lunation or **synodic month** – a period of 29.3 days in which the Moon completes an orbit of the Earth relative to the Sun and at which point it attains the same angular relationship with Earth and the Sun and hence the same appearance from Earth. Greek *synodus*: a meeting or conjunction.

magnitude – a term used to indicate the brightness of a heavenly body relative to others. The brightness as seen from Earth is the body's **apparent magnitude**. The magnitude of a number in mathematics is a measure of the greatness of its size.

mean solar time – an artificial adjustment of apparent solar time that smoothes out and averages the variation seen when the Sun is closely observed, these variations being due to the axial **obliquity** of the Earth and its cyclically changing speed due to the effect of its elliptical orbit. **Greenwich Mean Time** is the foremost mean solar time, also more formally known as Universal Time (UT).

meridian passage – the point in time at which a heavenly body crosses a defined meridian of longitude e.g. at Greenwich or the individual meridian of an observer. It may be applied to any heavenly body.

meridians of longitude – **great circles** orientated in the North-South plane on the Earth and passing through both poles. They are used to fix position in an East-West direction from 0-180º East or West of the **Prime meridian** (0º) at Greenwich.

micrometer – a device mounted on a sextant to increase the accuracy of the altitude reading, typically to the nearest 0.2 minutes.

nautical almanac – a general term for an annual publication giving tabulated data on the heavenly bodies useful for astro-navigation; the Sun, Moon, 4 major planets and the

Glossary

navigationally useful stars. It should perhaps strictly be termed an astro-navigational almanac as some nautical almanacs do not give astro-navigational data. Astronomical almanacs give tabulated astronomical data on bodies in the heavens. These data are above and beyond that needed for navigation.

nautical mile – exactly 1852 metres (6076 ft), by international agreement in 1929. A nautical mile was originally agreed to be 6080 feet in length and called an "Admiralty mile"; being the distance of arc on the Earth's surface subtended by an angle of one minute (1′) at the Earth's centre.

noon sight – the altitude sighting taken of the Sun as it crosses the observer's **meridian of longitude**, reaching its highest point in the sky on that day and used in order to establish **latitude**.

NP 314 – the code for the Nautical Almanac, the annual publication by HM Nautical Almanac Office containing all the data necessary for astro-navigation in that year.

nutation – a nodding oscillation seen as part of the movement called **precession** of the Earth's axis, due to slight variation in the Sun's gravitational attraction of the Earth, consequent upon the marginally aspherical shape of the Earth, it being slightly fatter in the equatorial regions when compared to the slightly more flat poles.

obliquity – the fixed angle of the Earth's equatorial plane to the plane of its orbital rotation around the Sun. The angle is 23° 27′ and has the effect, as the Earth orbits the Sun, of slowly changing the aspect of the Earth that most directly faces the Sun and consequent upon this producing the cyclical seasonal change.

observed altitude (H_o) – the altitude of a heavenly body as if it had been observed from the centre of the Earth. It is derived from the **apparent altitude (H_a)**, corrected where necessary for semi-diameter, parallax and refraction.

observer's view – a helpful view of the **geocentric** Earth used to depict what an observer sees of the celestial sphere above and the sea around. It illustrates the observer being "on top" of a particular part of the navigator's world.

opposition – a term applied to the position of a planet (Mars, Jupiter or Saturn) to the Earth in which the Earth is positioned directly on the line from the particular planet to the Sun. Venus, with its orbit inside that of Earth's cannot ever be in opposition but only in superior or inferior **conjunction**.

parallax – the apparent change in the position of an observed body caused by a change in the position from which it is viewed. It is used to describe one of the corrections of the altitude of a body from the observer's position on the surface of the Earth to the theoretical altitude measured from the centre of the Earth. The angle between lines from the observer and centre of the Earth to the body is called the **angle of parallax**. Parallax is zero for distant bodies such as stars but significant for near bodies such as the Moon and greater for a body when viewed relatively horizontally as compared to one viewed vertically.

parallel of latitude – **small circles** orientated parallel to the **equator** on the Earth's surface that are used to fix position in a North-South direction. The parallel of latitude that is the **equator** is a **great circle** as it has its centre coincident with that of the Earth's centre.

perigee – the point in its orbit when the Moon is nearest the Earth.

perihelion – the point in its solar orbit when Earth is nearest the Sun, at present in early January.

perpendicularity – a condition used to describe the right-angle that the plane surface of a mirror must have to its supporting arm or **sextant** body in order to accurately reflect light useful for the altitude measurement of an observed heavenly body. The term is primarily applied to the **index mirror** of a sextant in relation to the instrument's index arm.

phases – applied to the Moon and its changes in shape apparent to an Earthly observer over the period of a **lunation** or **synodic month** of 29.3 days. It is due to the changing angle between the Earth, Moon and Sun, altering the degree to which the illuminated face is visible to an observer.

plotting sheet – a sheet used for plotting position when the scale of available charts is too small for accurate plotting and which takes account of changing **departure** with latitude.

polar distance – the angular distance on the **celestial sphere** from the pole (P) of the hemisphere in which the observation is being made, to the heavenly body (X) and forming one side of the **PZX triangle**.

polar view – a helpful view of the **geocentric** Earth from above the point of the celestial pole, useful for the appreciation and the calculation of **Greenwich hour angle** and **local hour angle** of heavenly bodies.

position circle – the circle on the surface of the Earth on the circumference of which an observer must lie at a given time to produce a specific measured altitude of an observed body. The centre of that circle being the **geographical position** of that body

position line – the line on a chart drawn across the **azimuth** of an observed body at the point of the **intercept**. It is in effect a very small part of the **position circle** for the **observed altitude** of the observed body, but so small that it is drawn as a straight line. The position of the observer at the time of the observation is assumed to be on or very close to the position line.

precession – a cyclical rotating motion of the axis of the Earth caused by non-uniform gravitational forces, primarily from the Sun, on the Earth's oblate spheroid shape; the Earth being slightly fatter in the equatorial regions. A similar motion

Glossary

occurs in the orbit of the Moon around the Earth and relates to the **full Moon cycle**.

Prime meridian – the 0° meridian of longitude at the Royal Observatory building in Greenwich London. From this meridian, **longitude** is measured in an East and West direction, reaching 180° East and West.

PZX triangle – the astro-navigational **spherical triangle** on the surface of the **celestial sphere**; PZX denoting the Pole(P), **zenith** of the observer (Z), and celestial position (X) of an observed body given by its **Greenwich hour Angle** and **declination**..

quadrantal azimuth – a direction of a heavenly body from an observer described in terms of its position East or West of North or South It is usually found from an **ABC table** or **ABC formula**. It is converted to 360° notation for plotting on a chart.

radian – a dimensionless measure of the angular length of the arc of a circle in terms of its radius. One radian is the angle subtended at the centre of a circle by an arc of circumference that is equal in length to the radius of the circle. 360° = 2π radians, making one radian equal to: 360/(2 x 3.142) ≈ 57.3°.

refraction – the process by which light rays passing through a medium other than the vacuum of space have their speed slightly slowed when entering a medium of a higher density, such as the atmosphere. This has the effect of changing the angle of the rays which then appear to come from a position other than their true origin, which without correction, would lead to an error in altitude estimation.

semi-diameter – a heavenly body by virtue of its large size (Sun) or close proximity (Moon) will have a discernable diameter. In an altitude assessment, part of the body's circumference is brought to the horizon. A figure for its semi-diameter is used to adjust the altitude such that the figure achieved would be that obtained if it had been possible to align the centre of that body with the horizon.

sextant – a precision instrument used to measure the angle of altitude of a heavenly body above the **visible horizon**.

sextant altitude (H_s) – the altitude of a heavenly body measured by a sextant before correction for any known **index error**.

side error – an error of alignment of the **horizon mirror** of a **sextant** around its vertical axis and which must be adequately corrected for by adjustment, such that it accurately reflects light useful for the altitude measurement of an observed heavenly body.

sidereal hour angle – the angle measured at a given moment at the North celestial pole between the meridian on which lies the **First Point of Aries** and the meridian on which lies a particular star or in some circumstances, planet. The angle is measured in degrees West of the **Aries** meridian from 0-360°.

sidereal month – a period of 27.3 days in which the Moon completes one orbit of the Earth beginning and ending at the same point in the background of stars, with identical celestial coordinates of GHA and Declination. Latin *sidus* = star.

sight reduction tables – tables that list a **calculated altitude** and **azimuth** for a body, given its **local hour angle**, **declination** and the **latitude** from which it is observed e.g. NP 303, NP 401, published by the UK Hydrographic Office.

sine – a trigonometric function – the ratio of the opposite side and the hypotenuse (long side) of a right-angled triangle linked to the size of the angle at the foot of the hypotenuse. The sine ratio is used in the solution of the spherical triangle.

small circle – a circle on the surface of the Earth's sphere whose centre does not coincide with that of the Earth. **Parallels of latitude**, other than the equator, are small circles.

spherical triangle – a three-sided geometric entity formed on the surface of a sphere by the intersection of three **great circles**, which by definition have their centres coincident with that of the sphere. The mathematical solution to the side lengths and angles is used to produce a **calculated altitude** (H_c) for an observed heavenly body and from which, when compared to the body's **observed altitude** (H_o), is derived an **intercept** on the **azimuth** of that body.

star finder – a mechanism using concentric discs to identify and locate the azimuth and altitude of navigational stars given the **local hour angle** of **Aries**.

sun-run-sun – a technique using transferred **position lines** whereby a navigational fix may be arrived at using forenoon and/or afternoon Sun sights with or without a midday **meridian passage** sight to establish latitude.

synchronous rotation – the name given to the peculiar situation of the Moon's orbit around Earth as the same aspect of the Moon is caused to constantly face the Earth, due to gravitational forces having stopped the independent rotation of the Moon on its own axis.

synodic month or **lunation** – a period of 29.3 days in which the Moon completes an orbit of the Earth relative to the Sun and at which point it attains the same angular relationship with Earth and the Sun and hence the same appearance from the Earth. Greek *synodus*: a meeting or conjunction.

syzygy – a term used to describe the condition in which to an observer on Earth, there is spatial alignment of the Earth, Moon and Sun e.g. at new Moon and full Moon. Greek *syzgia*: union.

tangent – a trigonometric function – the ratio of the opposite side and the adjacent side of a right-angled triangle linked to the size of the angle at the foot of the hypotenuse. The tangent ratio may be used in a solution of the spherical triangle to give the azimuth of a heavenly body.

Glossary

transferred position line – the technique of moving a derived position line forward or backward in time to coincide with one or more other position lines in order to produce a position fix.

trigonometry – the mathematical branch of geometry that describes the relationship between the sides and angles of a right-angled triangle.

twilight – the period between the horizon becoming visible and the Sun rising at dawn and that between the Sun setting and the horizon becoming obscured by darkness in the evening. It is the period during which star and planet sights are taken. Both civil and nautical twilight are defined by the Sun's angular relationship to the horizon, but it is the former, civil twilight, during which sights are made.

Universal time (UT) – properly called UT1, it is the time tracked at the Royal Greenwich Observatory and based upon the movements of the Sun. Note that Universal Coordinated time (UTC) is based on the atomic clock.

vector – a term given to a value with both magnitude (size) and direction in space, used for example in **spherical trigonometry** where vectors are used for the derivation of the mathematical equations for the solution of the spherical triangle.

versine – a mathematical ratio of 1–cosine of an angle used in one particular long-hand method of finding a **calculated altitude** (H_c)(calculated altitude = 90° – **zenith distance**). Calculations using logarithmic versines of small angles attain a higher accuracy than when using logarithmic **cosines**.

visible sea – the area of sea visible to an observer from the deck, varying with the height of the observer above the sea and limited by the horizon formed by the curvature of the Earth's surface and its apparent intersection with the heavens.

visible sky – the area of the **celestial sphere** visible to an observer on the deck, limited by the horizon formed at the intersection of the sea as it curves away over the Earth's spherical surface.

zenith – the point on the **celestial sphere** vertically above an observer, at the intersection of the **celestial sphere** and a line from the Earth's centre passing up through the Earth's surface at the position of the observer.

zenith distance – the angular distance over the surface of the **celestial sphere** from an observer's **zenith** to a heavenly body.

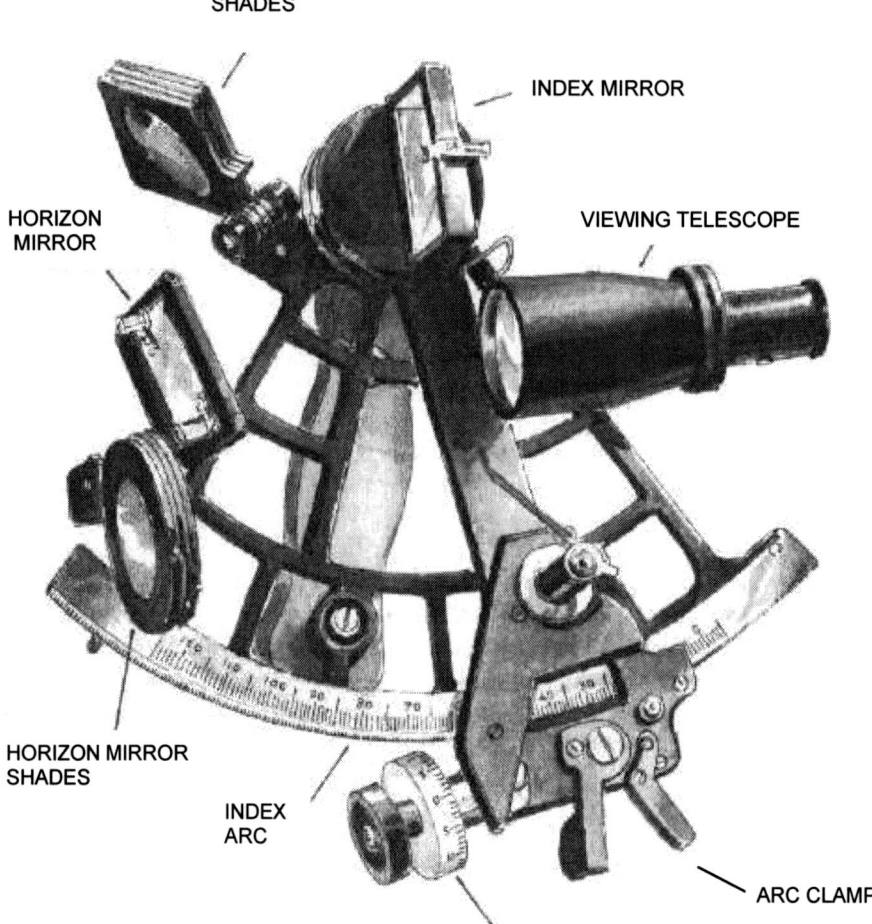

A typical sextant showing its component parts (Section 3 p. 75)

Section 1: The basic terms and concepts

This section explains the words, definitions and the basic theory involved in the observation of the heavenly bodies, with emphasis on the Sun. It begins the task of exploring the vital link between the Sun and time and their relationship to the finding of latitude and longitude. It constructs a model on which to build a working knowledge of the practice of astro-navigation.

By the end of Section 1, the following KEY FACTS should be understood:-

- **The astro-navigator's Earth lies central and still within the rotating celestial sphere.**
- **Latitude and longitude locate position on the Earth's surface.**
- **A nautical mile on the Earth's surface equates to 1′ of arc at the Earth's centre.**
- **The centres of great circles on the Earth coincide with the Earth's centre.**
- **The celestial sphere has an infinite and irrelevant radius.**
- **The stars are fixed relative to one another on the celestial sphere.**
- **The Sun, Moon and planets move predictably across the celestial sphere.**
- **The Sun, Moon and planets are located on the celestial sphere by their declination (celestial latitude) and Greenwich hour angle (celestial longitude).**
- **Sidereal hour angle relates the fixed stars to one point – the First point of Aries.**
- **The GHA of Aries added to the Sidereal hour angle of a star giving the Greenwich hour angle of the star.**
- **Local hour angle relates a body's Greenwich hour angle to the observer.**
- **The zenith is that theoretical point on the celestial sphere directly above the navigator's head.**
- **A heavenly body has its geographical point vertically below it on the Earth's surface.**
- **Earth's seasons are a consequence of the fixed tilt of the Earth's axis to its orbit.**
- **The rise and fall of the Sun in the sky are consequences of the Earth's fixed axial tilt and the position from which the Sun is observed.**
- **The Sun's westerly course across the sky defines the passage of time on Earth.**
- **The Sun's altitude at meridian passage is used to define the observer's latitude.**

1.1 *The celestial sphere* – the navigator's view of the heavens

In reality, the Moon rotates about our Earth and the Earth and planets rotate around the Sun and the stars we see are other Suns, uncountable billions of miles away. The situation in the heavens presumed by the astronavigator is very different and is based largely on the views of ancient observers, modified over the years. This difference is fundamental to understanding. The astronavigator needs to be able to see in his mind a condensed but important view in which the Earth lies centrally and is quite still. All the heavenly bodies of the universe are equidistant from the Earth, placed on the surface of a sphere – called the celestial sphere (**Figure 1**).

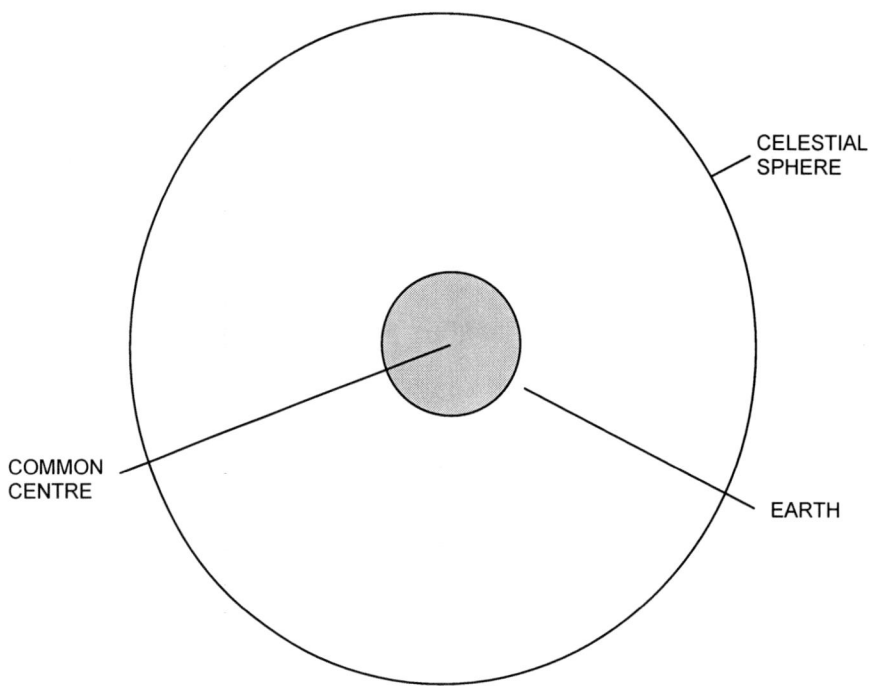

Figure 1. The basic model of the navigator's universe

It helps to think of two concentric transparent balls, the size of a tennis ball and a football, the smaller suspended within the larger, their centres coinciding <u>exactly</u> at a single point. Imagine an axis passing vertically from the North to the South Pole through our tennis ball Earth and projecting above and below to reach our large ball celestial sphere at its uppermost and lowermost points – called the **North** and **South celestial poles**. Project the circular equator of our tennis ball Earth to the horizontal periphery of the celestial sphere to produce the **celestial equator** on the larger sphere (**Figure 2**) mirroring that we use on the Earth.

– *the basics*

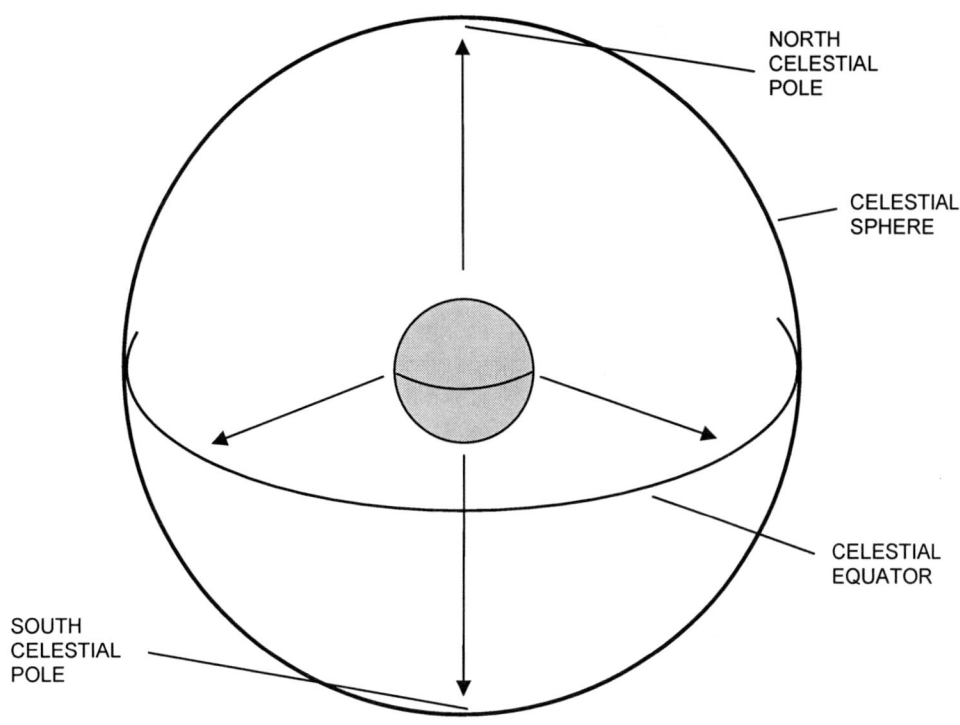

Figure 2. The celestial sphere with its own equator and poles

Now imagine the Sun, Moon, planets and stars all sitting on the surface of the larger ball, equidistant from the tennis ball Earth. Finally, imagine the outer celestial sphere rotating around the static Earth, once every 24 hours. This is the astronavigator's model of the arrangement of the heavens. The stars can be considered to be fixed in position on the celestial sphere, rotating as one, keeping the same positional relationship to one another. Observers as long ago as the Sumerian civilisation in the South of today's Iraq recorded and named certain prominent patterns of stars they saw, such as the signs of the Zodiac and the Great and Little Bear (*Ursa Major and Ursa Minor*)[1.1].

These patterns, still visible to us today form part of the basis of star identification for modern navigators. The Sun, Moon and planets however, do not stay in the same position on the celestial sphere, but move in predictable patterns across its surface. Knowing the expected positions of the stars and the predicted positions of the Sun, Moon and planets, the navigator is able to work out his own position. This is in essence all there is to astro-navigation.

Returning to reality to make a point; the current understanding of the cosmos is that the contents of the universe are part of an ever-increasing volume after an initial "big bang" formed the basis of all matter known to us from what is termed a single point. But no matter what the size and radius of our imagined celestial sphere and the distances of each particular star from the Earth (which are unimaginable), their angles relative to each other <u>and</u> their angles relative to the observer on Earth can be considered to remain constant and so are useful for navigational purposes.

1.2 *Fixing position on the Earth* – latitude and longitude

A navigator needs to be able to fix and describe his position on the surface of the Earth. To do this systems of measurement are needed so that a position can be related to fixed points of reference on the globe, such as the poles and equator. These systems of measurement are called **latitude** and **longitude**. Latitude is the name given to a measure of the position of an object North or South of the equator. Longitude is the name given to the measurement used to define a point East or West of an agreed datum line. If these terms are new to you it might help to think of **latitude** as the horizontal fix of position on a chart and **longitude** as the vertical.

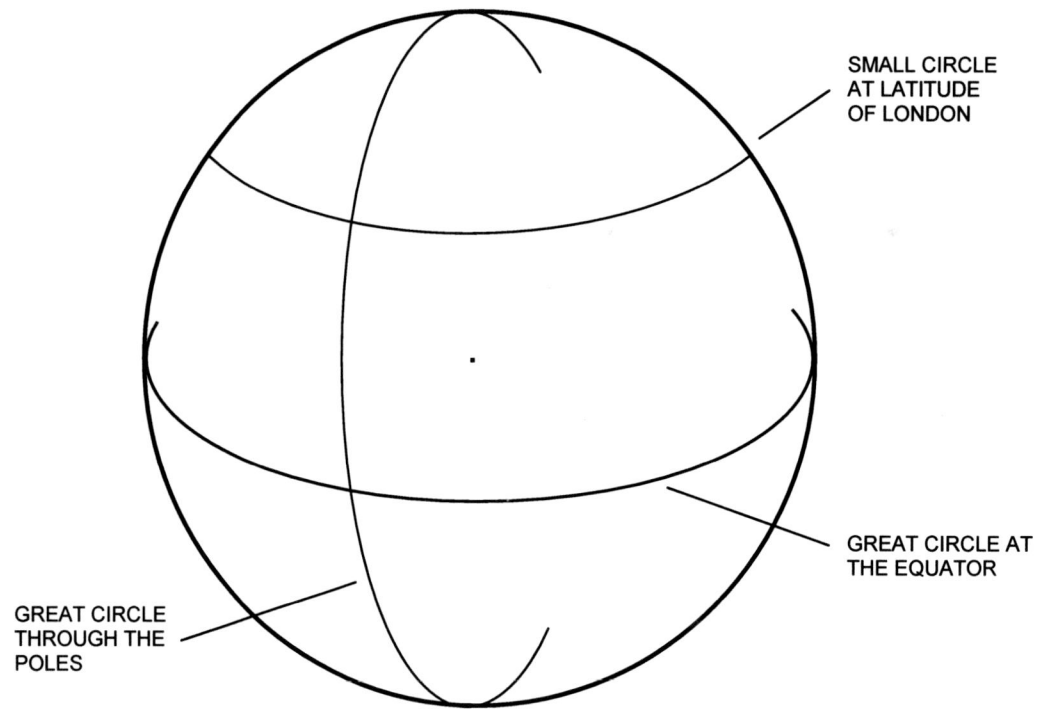

Figure 3. Great and small circles on the Earth

– the basics

To <u>understand</u> both, it is necessary to comprehend the conventions used in describing circles drawn on the surface of a sphere and how to measure distances on this surface expressed as angles at the sphere's centre. Consider the circle of the equator on the Earth. Where is the centre of the equatorial circle? Carefully consider a circle running from one pole crossing the equator to the opposite pole and then returning to the first pole. Where does the centre of this circle lie?

Then imagine a circle beginning at London in the northern hemisphere (or Auckland, New Zealand in the southern hemisphere) and running West around the Earth until it reaches its starting point. Where are the centres of these circles? Both the equatorial circle and the polar circle have their centres at the <u>centre</u> of the Earth and as such are called **great circles** (**Figure 3**).

A great circle on the surface of a sphere is one in which the centre of the circle coincides with the centre of the sphere. Of the circles based on London or Auckland and running West, neither has their centre at the centre of the Earth, they both lie elsewhere within their respective hemispheres and are called **small circles**. **Great circles** are of importance to the navigator as the shortest distance between two points on the surface of a sphere lies along the circumference of the great circle joining both points.

Returning to **latitude**; it is the name given to the system of measurement used to fix position on the globe in a North-South direction. The latitude of a point on the surface of the Earth is defined as the angle made at the centre of the Earth by a line from that surface point and the plane of the equator. It is measured in degrees and minutes (fractions of a degree) North or South of the equator from 0° at the equator to 90° at the poles.

There is only one **great circle** of latitude, that of the equator. All other latitudes are **small circles** lying on planes parallel to the equator, above it in the northern hemisphere and below it in the southern hemisphere. The centres of these circles do not coincide with the Earth's centre, but lie on the North-South polar axis. The latitude lines are called **parallels of latitude** (**Figure 4**), they neither converge nor diverge as they proceed around the globe when they are traced East or West, but remain parallel to one another.

The latitude of central London is 51° 30′ North and that of Auckland 36° 51′ South. **Longitude** is the name given to the system of measurement used to fix position on the globe in an East-West direction. The longitude of a point on the surface is defined as the angle made at the centre of the Earth by a line from that point and a datum plane of a North-South line, measured in degrees and minutes East and West of the datum, from 0° - 180° in both easterly and westerly directions. Each and every arc of longitude is a part of a **great circle** passing through both poles and having its centre coincident with the centre of the Earth. The longitude arcs are called **meridians of longitude** (**Figure 5**).

The meridians of longitude are not parallel to one another but are most widely spaced at the equator, converging to both the North and South and meeting at the geographical North and South poles. Unlike latitude, where the equator is a readily identifiable datum line, longitude

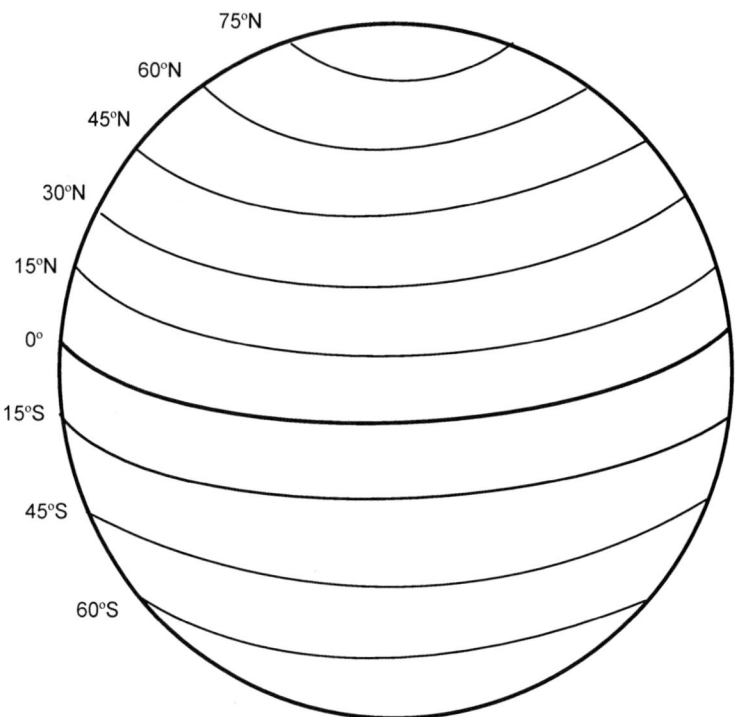

Figure 4. Parallels of latitude – all small circles except for the equator

has no natural East-West datum line and therefore one was chosen. The 0° meridian lies at the Royal Observatory building in Greenwich, London and is termed the **Prime Meridian,** named so by international agreement in 1880[1,2].

The term "meridian" is used to denote any and all longitudes, those of whole number or fractions of a degree. Wherever you are on the Earth you are on your own personal meridian. The result of carefully defining both latitude and longitude is that one can reliably pinpoint a position anywhere on the globe. The coordinates of Greenwich, London are 51° 30′ North 00° 00′ East and West and that of Auckland, New Zealand, 6° 51′ South 174° 47′ East.

– the basics

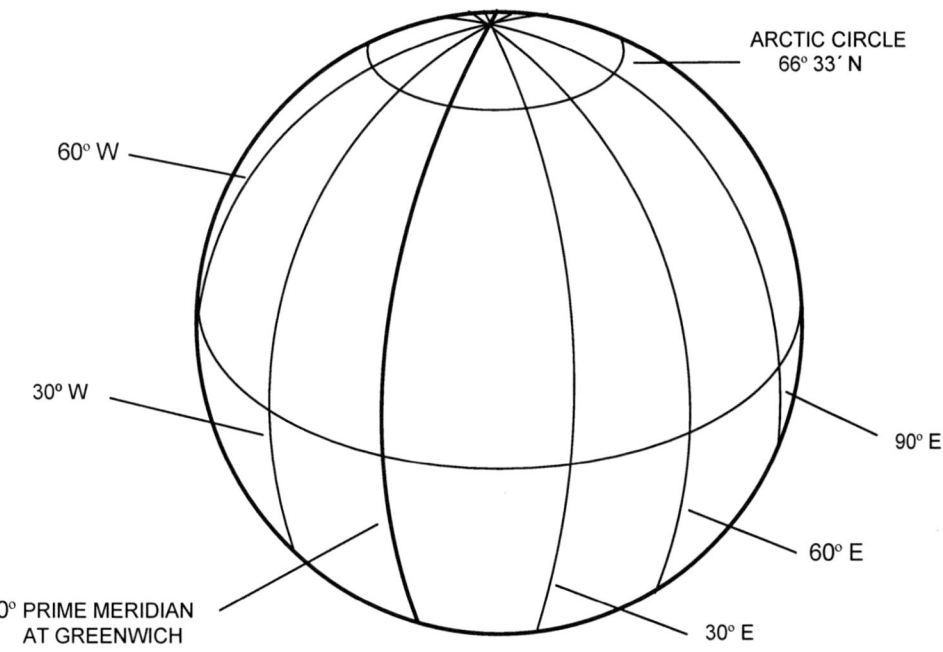

Figure 5. Meridians of longitude and the equator – all great circles

1.3 The Nautical Mile – defining angle, arc and distance

So far we have dealt with angular measurement for both latitude and longitude but our navigator needs some way to relate angular change at the Earth's centre to distance on the Earth's surface. These distances are often called arcs, because the line joining two points on the Earth's surface is part of the circumference of a circle and a segment of circumference is called an arc. Navigators measure distances in nautical miles, but why is this? By international agreement in 1929 the **nautical mile**[1.3] was defined as being exactly 1852 metres (6076.12 ft), but it was not always so. A nautical mile was originally agreed to be 6080 feet in length, a seemingly irregular number, but one with the title of an "Admiralty mile" (**Figure 6**).

The slightly odd length of 6080 feet was arrived at from the initial definition of a nautical mile, which is the distance of arc on the Earth's surface subtended by an angle of 1′ at the Earth's centre (1′ = one minute of latitude = $1/60^{th}$ of a degree).

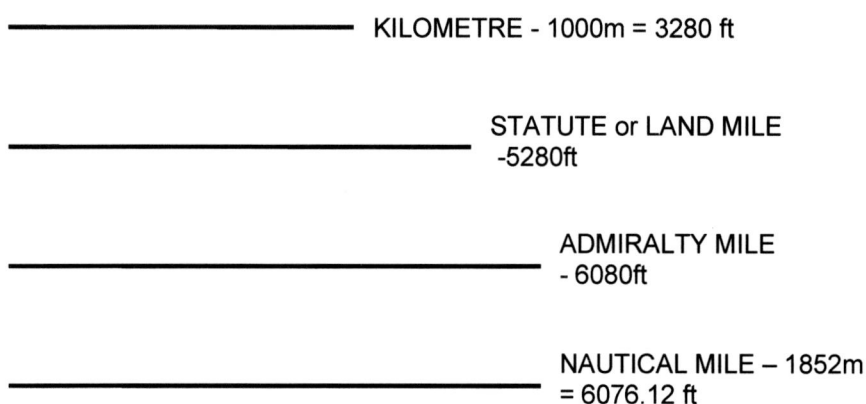

Figure 6. Comparing the "mile"-statute, admiralty and nautical

The Earth is not a perfect sphere, but a little flat at the poles and it bulges a little at the equator (called an oblate spheroid). The circumference at the equator is 24901 statute or land miles and that running through both poles is 24859 miles, giving an average circumference of 24880 statute miles. If our navigator travelled this distance around the world, he would move through 360° of angle measured at the Earth's centre.

Each degree of angle moved through would be equal to an arc on the surface of 69.11 statute miles (24880/360° = 69.11). Each degree is divided up into 60′ and therefore the length of arc travelled for each minute of angle moved through is 1.152 miles (69.11/60 = 1.152). Each statute mile is equal to 5280 feet and so each minute of angle moved through is equal to 1.512 x 5280 = 6080 feet on the Earth's surface. This is how the nautical mile was originally derived and explains its equivalence on the surface to an angle of 1′ at the Earth's centre. Because of the slightly aspherical shape of the Earth and imprecise measurement of the exact circumference of the Earth, several slightly differing measurements for the nautical mile were used by a variety of countries. Standardisation was deemed necessary and so followed the 1929 international agreement, basing the nautical mile on a whole number of metres – 1852m (6076ft).

1.4 *Positioning on the Celestial Sphere* – celestial latitude and longitude.

Our navigator has learned how a position on Earth can be fixed in terms of latitude and longitude. In a similar fashion the navigator needs some convention to describe the positions of the heavenly bodies on the celestial sphere, which for simplicity we will for the present call **celestial latitude** and **celestial longitude**.

– the basics

First however, it is important to review our description and understanding of the model we are using. Our model places all the heavenly bodies equidistant from Earth on the surface of the celestial sphere, when in reality stars are billions of miles away and much further away than our Sun and its planetary system. Surely this would introduce huge errors in navigation? In practice this is not the case and the model works very well as it is not the distance to a heavenly body that is used in astro-navigational calculations but it is, very importantly, their angular positions relative to one another and to fixed points introduced into the description of the celestial sphere. It is all about angles again, just as it is for describing latitude and longitude on Earth.

Remember, when using a hand-bearing compass on deck to fix a position from three land-bound features, you do not need to know how far away each feature lies. It is the angle that each bears from North that is important. Similarly when using the Sun, Moon, a planet or a star to help fix a position, it is an angle that is vital and you do not need to know how far away a body happens to be. Understand this and you will have mastered a fundamental of astro-navigation.

1.5 *Celestial Latitude: Declination* – what is it and how is it measured?

Just as on Earth, where the latitude of a point is determined in degrees North or South of the equator, so in our celestial sphere, the celestial latitude of a heavenly body is described as being a number of degrees North or South of the celestial equator. The angle in question is measured at the centre of the celestial sphere, which as we know, is coincident with that of the Earth. The slight complicating factor is that celestial latitude is given a new name, that of **declination** (from the Latin: *declinare* - to bend away). **Declination** is the angle subtended at the Earth's centre by a line from a heavenly body and the plane of the celestial equator.

For example and picking a day at random, on Friday 7th May 2011, the astro-nautical almanac lists the declination of the Sun at Greenwich at 1200 hours to be 16° 47.8´ North. This means that a line from the Sun to the centre of the Earth makes an angle with the plane of the celestial equator of 16° 47.8´ North of that equator (**Figure 7**).

Keep in your head the simple transparent tennis ball and football model, then project the image of the Sun onto it and see the angle made at the centre of the Earth between it and the celestial equator. The declination values for the Sun, Moon and planets change as they move over the celestial sphere and their values are tabulated in the astro-nautical almanac. Declination values of stars are fixed as they do not move on the celestial sphere.

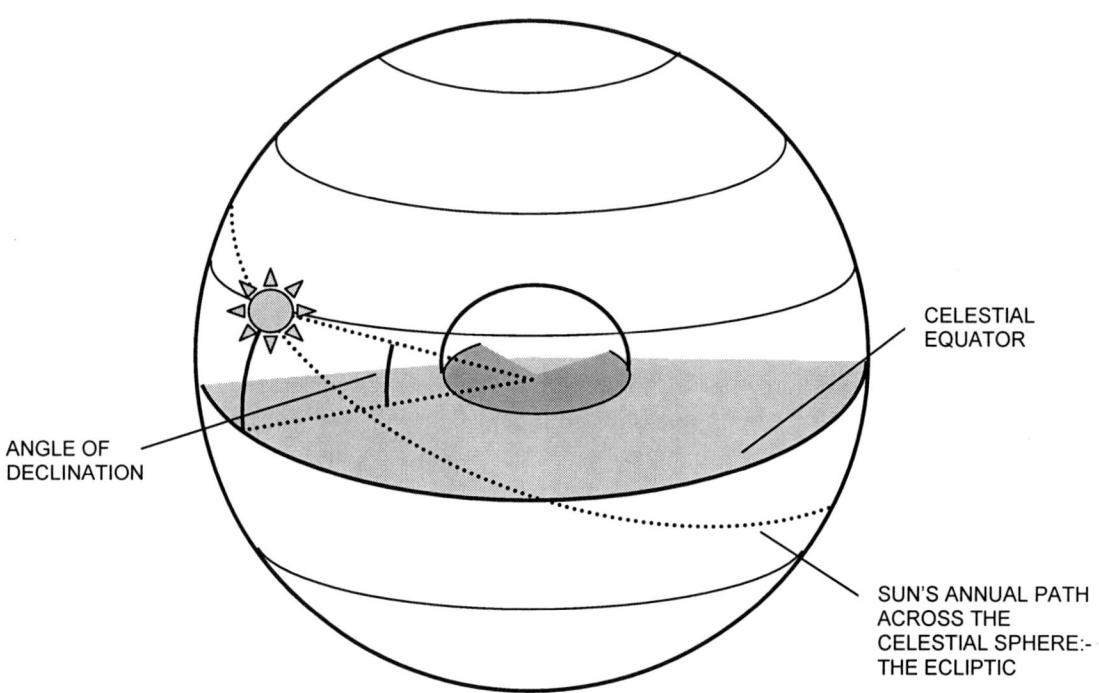

Figure 7. Declination: North-South position fixing on the celestial sphere

1.6 *Celestial longitude: Greenwich Hour Angle*

On Earth the longitude of a point is determined in degrees East or West of the Prime Meridian at Greenwich. On our celestial sphere, we also need a meridian in a North-South plane from which to measure the celestial longitude of a heavenly body. The Sun, Moon and planets move relatively independently on the celestial sphere and at any point in time, each will have its own measure of East-West position.

As previously described, the Greenwich meridian has been chosen as the datum line from which to measure terrestrial longitude and it is also the chosen meridian from which celestial longitude is measured. The term celestial longitude is perhaps unfortunately not generally used, but instead the term **Greenwich Hour Angle** is applied. It is the angle subtended at the North Celestial Pole between the **Greenwich meridian** and the meridian on which lays the heavenly body. Unlike longitude on Earth, which is measured in degrees East and West of Greenwich from 0-180°, Greenwich Hour Angle is measured in degrees West of Greenwich from 0-360°, as shown in **Figure 8**.

– the basics

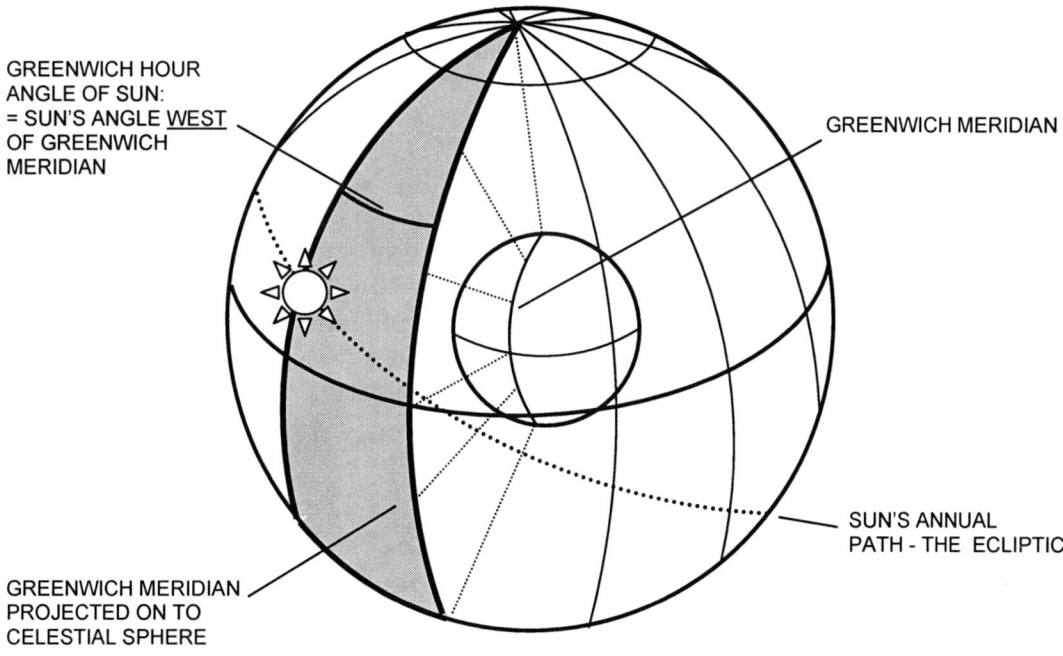

Figure 8. The celestial sphere showing the geometry of the Sun's Greenwich Hour Angle

In the astro-nautical almanac you will find listed the details of the Greenwich Hour Angle of the Sun, Moon and planets, for every hour of every day of the year. For example the **Greenwich Hour Angle** (GHA) of the Sun at exactly midday on March 21st 2011 is 358° 11.0′ almost, but not quite on the Greenwich meridian. You might question why the Sun is not exactly on the Greenwich meridian at midday every day, but a detailed explanation will have to wait until **Section 5**.

1.7 *Celestial longitude: Aries and Sidereal Hour Angle* – locating stars?

The Greenwich Hour Angle for the stars requires a little more explanation. The positions of the stars on the celestial sphere are fixed and they rotate as one, so although their declinations (North-South positions) are all different, their East-West positions can all be related to a single celestial meridian. The chosen meridian lies at what is termed the **First point of Aries** (abbreviated to **Aries**). The derivation of this term requires some knowledge of the Sun's movement on the celestial sphere; the annual path being called the **ecliptic** (**Figures 7, 8 & 9**). This path intersects the celestial equator in March and September and is discussed in detail in **Section 1.13**.

The ecliptic/celestial equator intersection is used as the point of reference for all stars and called the **First point of Aries**. The name derives from ancient observers having seen that the position of the Sun in March had as its background the constellation of Aries on the celestial sphere. The sphere rotates once every 24 hours and therefore the value for "**Aries**" changes every moment. The details of the changes in the astro-nautical almanac are listed as the **Greenwich Hour Angle of Aries**.

The definition of **GHA-Aries** is the angle subtended at the North Celestial Pole between the Greenwich meridian extended out to the celestial sphere and the meridian upon which lies Aries. Each individual star useful for navigation has listed in the astro-nautical almanac its longitudinal position on the celestial sphere. The angle is measured in degrees of angle West of **Aries**. The angle ranges from 0-360° and is called the **Sidereal Hour Angle (SHA)** and is fixed for each a star (Latin: *sidus* – star). The definition of **SHA** is the angle subtended at the North Celestial Pole between the meridians on which the star and **Aries** lie (**Figure 9**).

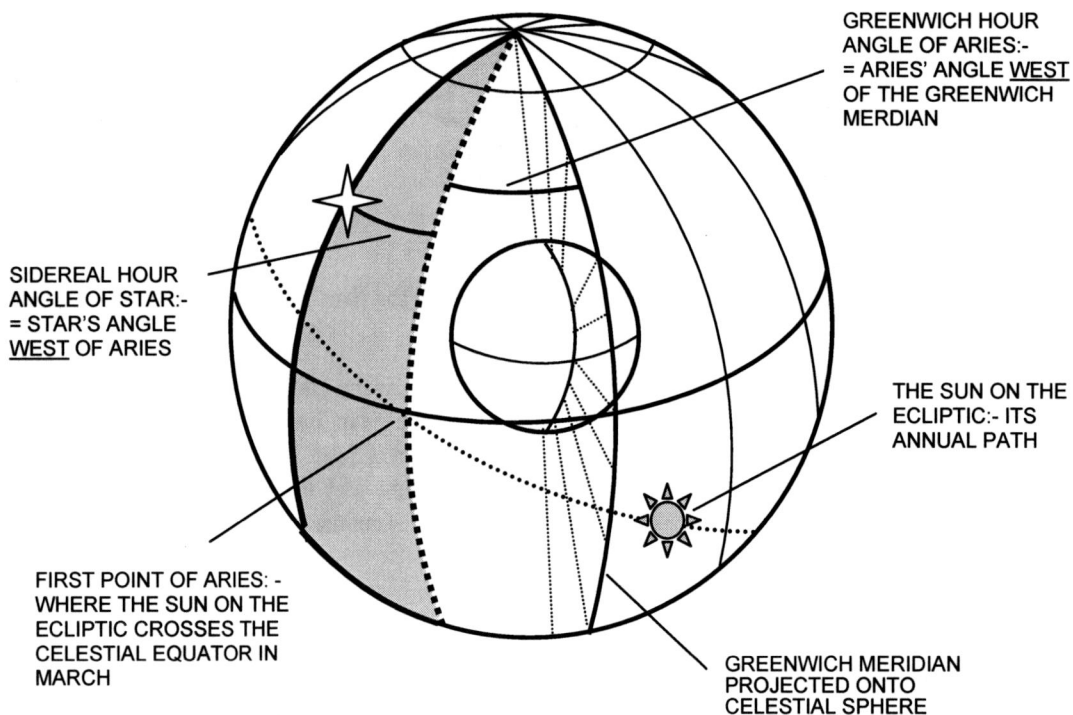

Figure 9. The celestial sphere showing Aries, GHA Aries and Sidereal Hour Angle

To find the Greenwich Hour Angle of a star, first the Greenwich Hour Angle of Aries is found in the astro-nautical almanac and then to this is added the star's Sidereal Hour Angle.

– the basics

Example 1. At 1200 on April 1st 2011, the Greenwich Hour Angle of Aries is tabled as 009° 44.6´. The Sidereal Hour Angle for the star Sirius is given as 258° 35.8´. They are added to give the Greenwich Hour angle of the Star:

GHA Aries:	009° 44.6´
SHA Sirius	258° 35.8´ +
GHA Sirius: =	**268° 20.4´** = the sum of the two.

The exact position of the Sun as it crosses the equator in March does not now correspond to the constellation of **Aries**. It has changed very slowly over the years but its name, the **First point of Aries**, has perpetuated. This slow change is called the precession of the equinoxes and need not unduly concern the navigator for the present, if at all.

1.8 *Local Hour Angle* – how the celestial sphere is related to an observer?

The values for **Greenwich Hour Angle** and **Declination** describe the point in the sky where a particular heavenly body can be found, in relation to the Greenwich meridian and celestial equator. In short, they give the celestial coordinates of the body. It is also necessary to have a means of knowing where a body lies in relation to an <u>observer</u> when placed at an earthly position other than at the Greenwich meridian.

The Declination of a body on the celestial sphere does not change if an observer's longitude changes. If the Sun has a declination of 10° North at a given time at the Greenwich meridian then this remains <u>unchanged</u> if observed from a meridian other than that of Greenwich, since declination describes the position North or South of the celestial equator. But most importantly, Greenwich Hour Angle is <u>highly</u> influenced by the East-West longitudinal position of an observer, because the celestial sphere rotates around its vertical axis.

Put simply, the sphere rotates around the vertical and not the horizontal aspect of the Earth. The angular relationship at the North celestial Pole, between the meridian on which lies a particular heavenly body and the meridian from which its sighting is made, is called the **Local Hour Angle**. If a heavenly body is sighted at the moment it crosses the observer's own meridian, its **Local Hour Angle** (LHA) is zero and this is a key point.

It is useful to remind yourself that GHA is a measure of celestial longitude, an angle that is measured from the projected Greenwich Meridian. It can help to think of **LHA** as the local celestial longitude of a heavenly body, measured in a westerly direction from the observer's meridian. Local Hour Angle for a body is calculated from the Greenwich Hour Angle and the observer's longitude.

Local Hour Angle = **Greenwich Hour Angle** <u>plus</u> observer's longitude <u>East</u>

Local Hour Angle = **Greenwich Hour Angle** <u>minus</u> observer's longitude <u>West</u>

The Local Hour Angles for the Sun, at two different Greenwich Hour Angles observed from different points, one with an easterly longitude and one with a westerly longitude, are illustrated in **Figures 10 & 11**. The view is from <u>above</u> the North celestial pole, usefully called the **polar view**. To help with orientating the image with respect to Greenwich, note the pole P is uppermost, towards the reader, with eastern longitudes to the left of the Earth's sphere and western longitudes to the right. All the heavenly bodies move from East to West, clockwise in the diagram. Greenwich Hour Angle and Local Hour Angle and how they are used are discussed in more detail in **Section 6.1** (p.137).

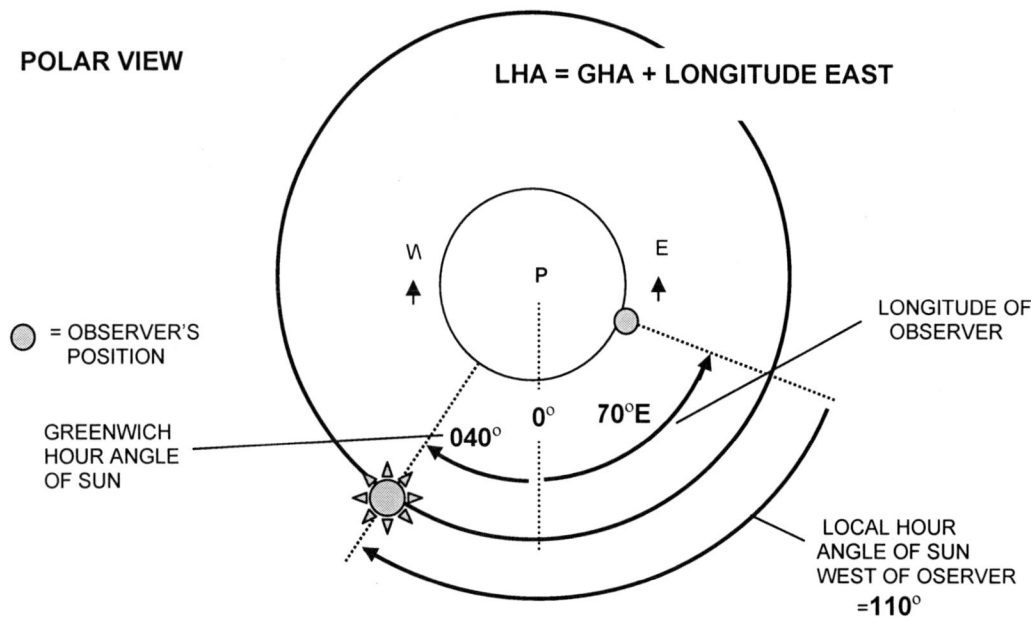

Figure 10. Measuring Local Hour Angle from an Easterly longitude

To summarise what we have learned in this section: for the Sun, Moon and planets, which move continuously across the celestial sphere, their position at a particular moment is given by their **Greenwich Hour Angle (GHA)** and **Declination (Dec)**. Stars are fixed on and do not independently move upon the celestial sphere. Their positions are given by their **Declination** (the equivalent of latitude) and **Sidereal Hour Angle**. To locate a particular star on the celestial sphere as the whole sphere rotates, the **First Point of Aries** is used as the locator on the celestial equator. The angle between **Aries** and the **Greenwich Meridian** projected on to the celestial sphere gives the **Greenwich Hour Angle of Aries**. The Greenwich Hour Angle of a star is arrived at by adding the GHA of Aries at the particular moment of observation to the SHA of the particular star. **Local Hour Angle**, the angle measured West of an observer, is calculated from the observer's longitude and the Greenwich Hour Angle of the observed body.

– the basics

The **Local hour Angle** of the Sun, Moon, planets and stars, for an observer at a particular moment, is given by the GHA of the body and the longitude of the observer: LHA = GHA minus west longitudes and LHA = GHA plus East longitudes.

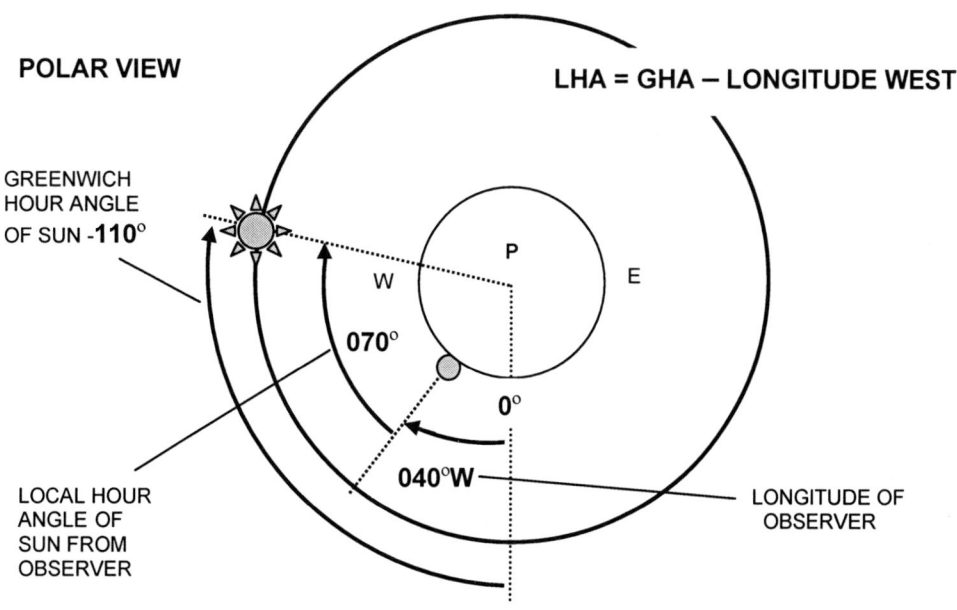

Figure 11. Measuring Local Hour Angle from a westerly longitude

1.9 *Celestial movement* – why are heavenly bodies seen to move as they do?

The stars are fixed in position on the celestial sphere and move as one, but the Sun, Moon and planets are seen to move. Why is this? Let us go back to one of the opening paragraphs: "In reality, the Moon rotates about our Earth and the Earth and planets rotate about the Sun and the stars we see are other Suns, uncountable millions of miles away". The celestial movement we see here on Earth is based on relative distance. Plainly the Moon is close to us, a mere 250,000 miles away and it reflects some of the light of the Sun in the direction of Earth, according to the Moon's relative position.

The visible planets useful for navigation; Venus, Mars, Jupiter, and Saturn, are visible because they reflect the Sun's light whilst circling around it and even they are relatively close to us when compared to the stars. Our own star the Sun, is 93 million miles away but the next nearest star, *Proxima centauri* is 4.2 light years away or 63,000 times the distance from Earth to our Sun; totalling 5.4×10^{14} miles. *Polaris*, the navigator's North-indicating star is unbelievably, another hundred times as far away as *Proxima centauri*.

The Voyager 2 spacecraft left Earth in 1977 to photograph the outer planets. It has now passed through our solar system, moving in the direction of the star Sirius at a speed of 10km/sec. It should reach its destination in 300 000 years[1.4]. This gives some idea of the immense distances of space. The stars are literally billions of miles away and our own Sun only millions. The stars, over the time periods relevant to navigation, stay constant in relation to one another's positions, so they can be represented by an agreed datum point, that of **Aries**. Within our own solar system however, the planets and Moon are close enough such that any movement of one body relative to another is discernable from hour to hour and day to day. This relative movement is also most importantly, measurable in terms of angular change.

1.10 *The celestial sphere and the view from on deck* – what do we really see?

The early part of this section has used a model of the Earth lying within the celestial sphere, both being on a common axis. It is now important to get to grips with the salient features of the visible segment of the Earth and sea from the deck of the observer's craft. The segments of the sea around the observer and the celestial sphere above are best termed the **visible sea** and **visible sky** (**Figure 12**). The area of **visible sea** varies directly with the height of the observer above it, the higher you are the more you see, limited by the horizon formed as the sea curves away over the Earth's spherical surface. The observer is placed on top of the visible world – this arrangement depicts an **observer's view**, seen from the side, level with the plane of the equator; the **equatorial view**.

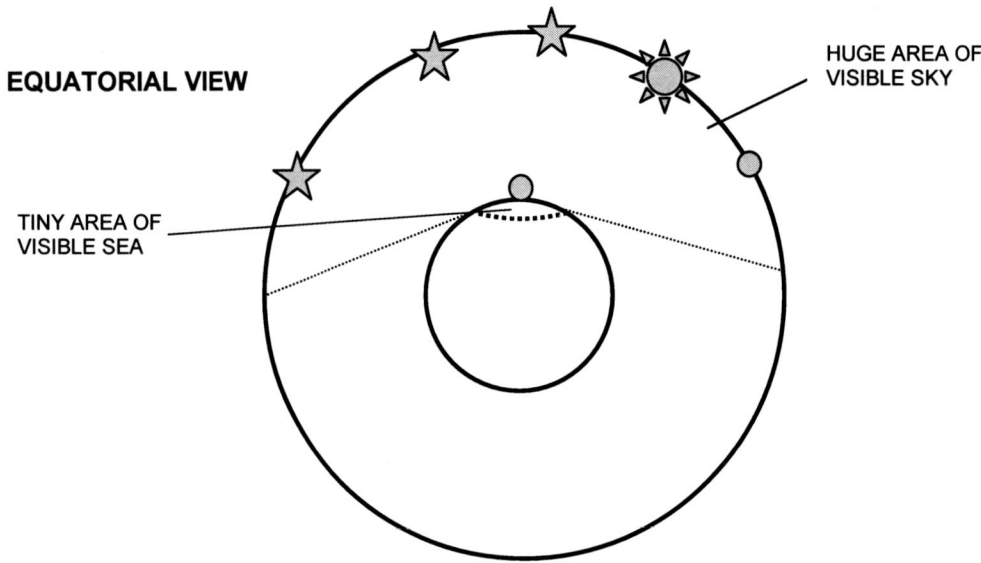

Figure 12. An observer's view of the visible sea and the visible sky

– the basics

From a small sailboat, the height of the observer's eye above the sea may be little more than 2m, giving a horizontal view of about 3 miles or so and a visible area of about 27 square miles. At an eye height of 4m, the horizon distance extends to 4 miles with a visible sea area of 53 sqM. Ship's bridges, from where professional mariners take their sights are plainly much higher and the eye height might be tens of metres above the sea. All astro-nautical almanacs have a height of eye-horizon distance tables and there is a mathematical formula for calculating it directly. This important variable will be considered further in subsequent sections.

A surprisingly small amount of the sea's surface is visible from the deck but contrastingly, a vast amount of the celestial sphere can be seen as over half of its surface area is visible, with the observer seemingly at its centre. Then imagine a line drawn from the centre of the Earth, <u>through</u> the position of the observer on the surface of the Earth and projected up as far as the celestial sphere, as in **Figure 13**. The point at which this line intersects with the sphere vertically above the observer is called the **zenith (Z)** of the observer.

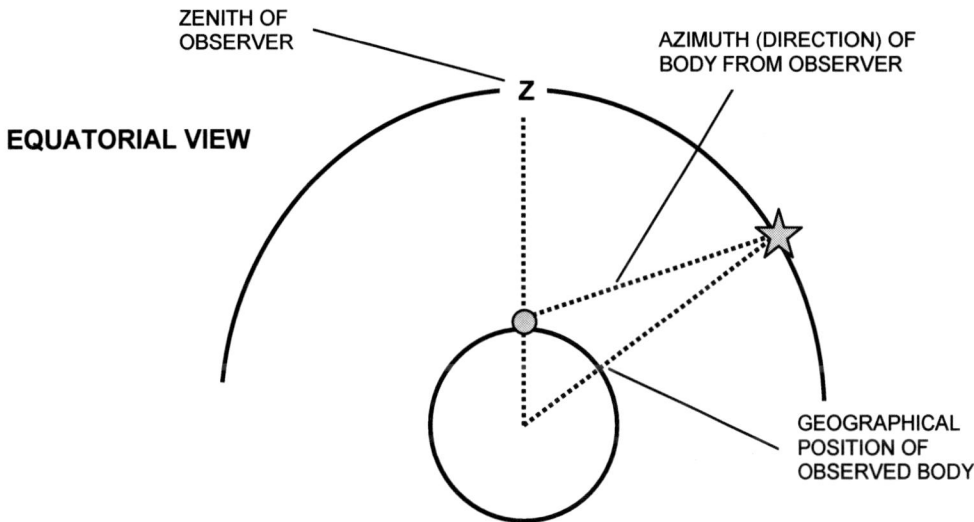

Figure 13. Showing zenith, azimuth and geographical position

Around and above the observer lies a portion of the celestial sphere limited at its edge by the visible horizon, where the imagined sphere seems to intersect with the sea. In the day, in both hemispheres, the Sun moves from the eastern horizon to the western, higher in the sky in summer. At dawn and dusk several stars will be visible and possibly a planet or two, while during the day or at dawn or dusk, the Moon may be seen and all may be useful in astro-navigation.

Astro-navigation from Square One

If watched for a period, all the heavenly bodies will be seen to move in a <u>westerly</u> direction as the celestial sphere appears to us on Earth to rotate. Now focus on just one star. The <u>direction</u> from the observer to the star is termed the **azimuth** of the star (Arabic *as-sumut*: - the direction). **Azimuth** is measured from 0-360° from North, similar to a compass bearing, but measured in degrees true.

Now imagine a line drawn from the star, <u>straight</u> to the centre of the Earth. This line will intersect the Earth's surface at a point directly below the star and this is called the **geographical position (GP)** of the star on the Earth's surface, demonstrated in **Figure 13**.

The distance over the surface of the Earth from the observer to the star's geographical position is called the **azimuth distance**. Its <u>angular</u> length is identical to that between the observer's **zenith (Z)** and the position of the star. The distance between these two latter points is called the **zenith distance**. Azimuth distance and zenith distance are illustrated in **Figure 14**. The equality of these arc-angles is fundamental to astro-navigation and considered in detail in **Section 2**. You should be able to begin to see that knowledge of one provides important information on the other.

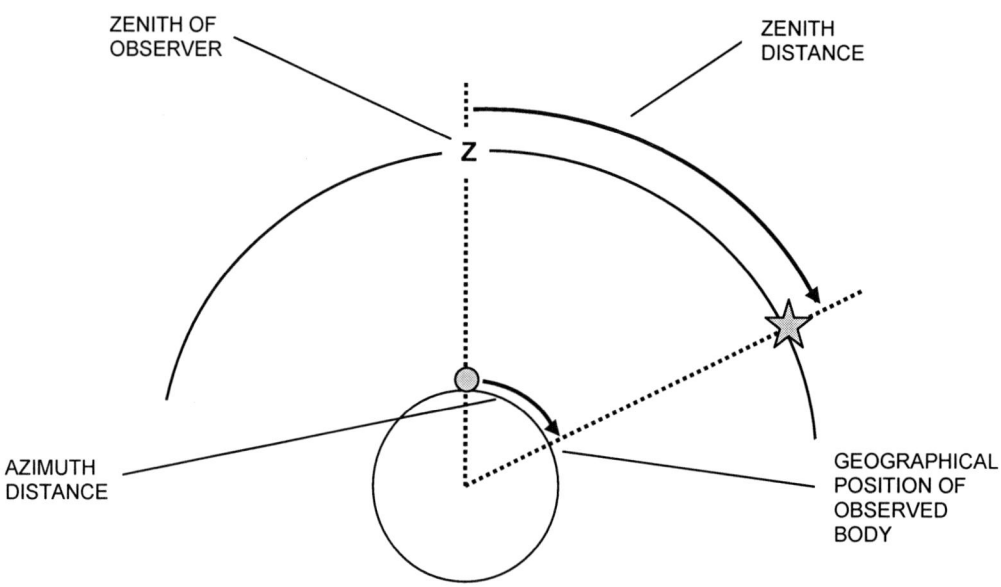

Figure 14. The angular equality of azimuth distance and zenith distance - equatorial view

– the basics

1.11 The strength of the Sun – why are some places warm and others cold?

The previous paragraphs of this section have introduced and defined much of the basic astro-navigational terminology. It is now time to begin the first of several examinations of Earth's relationship to the Sun, the linch-pin of astro-navigation. Copernicus (1473-1543)[1.5] first suggested the rather heretical hypothesis that the Sun and not the Earth lay at the centre of our solar system. The Sun is in reality, a vast sphere of burning gases lying 93 million miles from Earth (150 million km) and without which, life on Earth could not exist.

Our Earth spins on an axis that runs through the North and South Poles, an axis which is tilted at an angle, an understanding of which we will see, is vitally important. The Earth rotates in an easterly direction, once in every 24 hours as we orbit around the Sun in a very slightly eccentric manner, taking in all around 365 days. We receive radiant energy as both light and heat at an average power of 1 kilowatt per square meter. Travelling so far to reach us, the Sun's rays can be considered to be parallel to one another. At the equator, the radiant energy per unit area is highest, resulting in a greater heating effect (**Figure 15**).

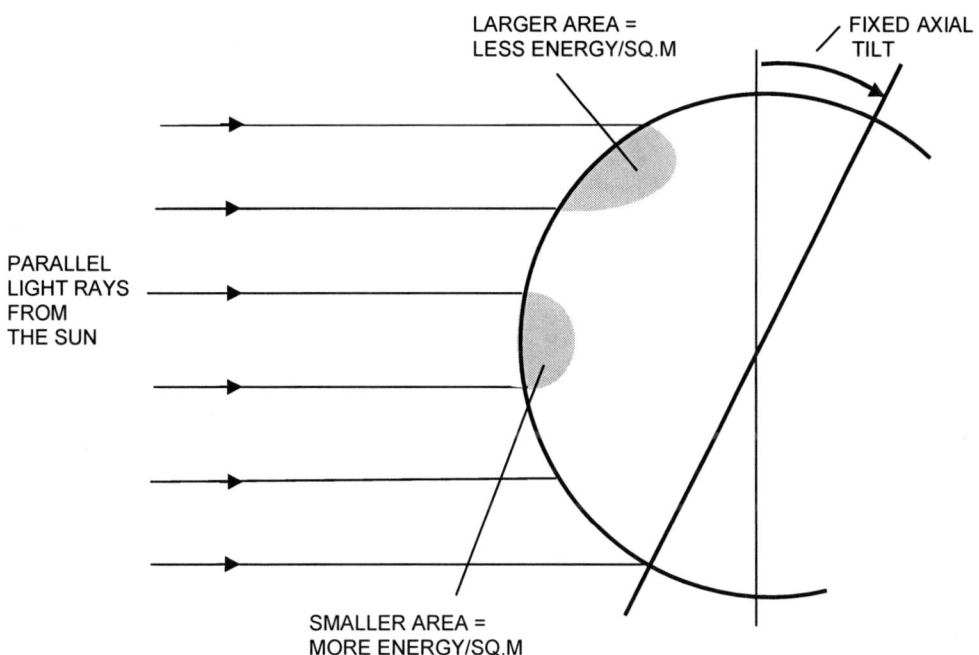

Figure 15. The effect of the Earth's tilt on incident energy from the Sun

Towards the poles, because of the Earth's tilt, the Sun's rays strike the Earth at an increasingly oblique angle and a given amount of energy will be dissipated over a larger surface area, which in turn results in less heating per unit area. So unsurprisingly, the Poles are colder.

1.12 The Earth's tilt and the seasons – how does one produce the other?

Let us now describe a year's passage of the Earth around the Sun, an understanding of which will clarify the change in both the seasons and day-length. Both these phenomena are a consequence of the fact that the axis on which the Earth spins is <u>not vertical</u> in relation to its <u>plane of movement</u> around the Sun, but tilted at an angle of 23° 27′ (**Figure 16**). The Earth's axis maintains this <u>fixed</u> angle <u>throughout</u> its movement around the Sun.

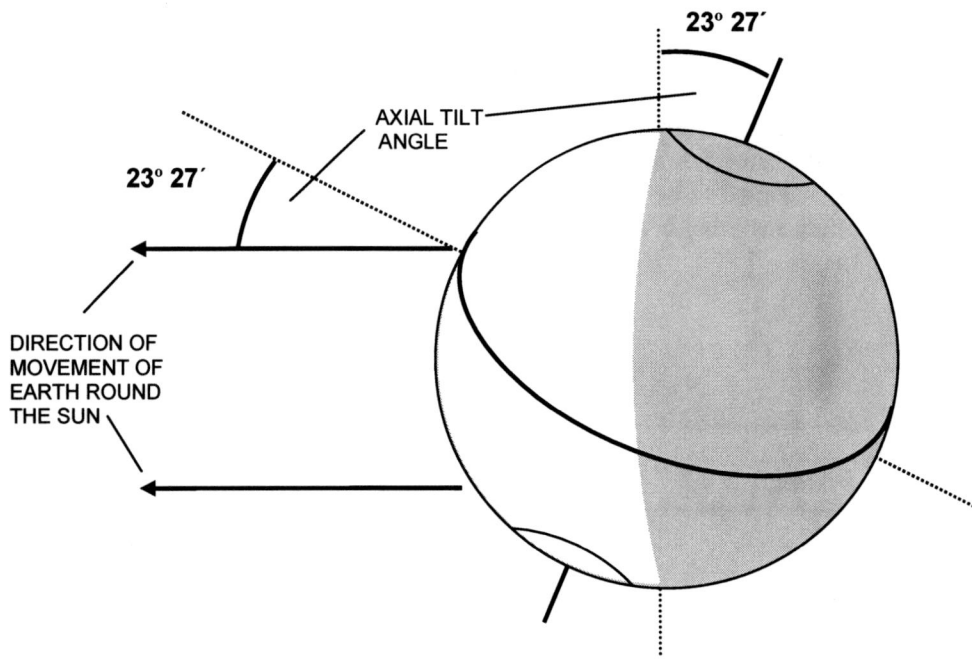

Figure 16. The Earth's Tilt with respect to its rotation round the sun

The fixed axis angle results in an orientation that exposes either the northern or southern hemisphere more directly to the Sun. The aspect of the Earth's surface directed more towards the Sun will experience the seasons of spring and then summer. Concurrently, the other hemisphere, directed away from the Sun receives less energy due to the angulation of the axis, resulting in the seasons of autumn and winter.

Carefully consider **Figure 17**, imagining the globe to be rotating once every 24 hours and moving from position A to B and then from B to C and C to D, each over a period of 3 months, returning to point A at the end of 12 months. Focus on the northern and southern hemispheres in turn, to best understand the seasonal and day-length changes. At point A, the northern hemisphere faces the Sun more directly and as the Earth rotates each 24 hours, at no point is the Arctic in complete darkness, the Sun staying on or above the horizon. Contrastingly, the southern hemisphere faces less directly at the Sun and the whole of the Antarctic faces away from the Sun during each 24 hour period, giving continuous twilight or darkness as the Sun

– *the basics*

stays below the horizon. Point A depicts midsummer in the northern hemisphere and midwinter in the southern hemisphere, on or around June 21st.

Focus next on point C, some 6 months later in the yearly cycle. Here the reverse situation pertains and the Arctic is in twilight or darkness, not facing the Sun at any point during its 24 hour rotation, whereas the Antarctic is in constant light as it faces the Sun relative to the Arctic. Point C depicts midwinter in the northern hemisphere and midsummer in the southern hemisphere, on or around December 21st.

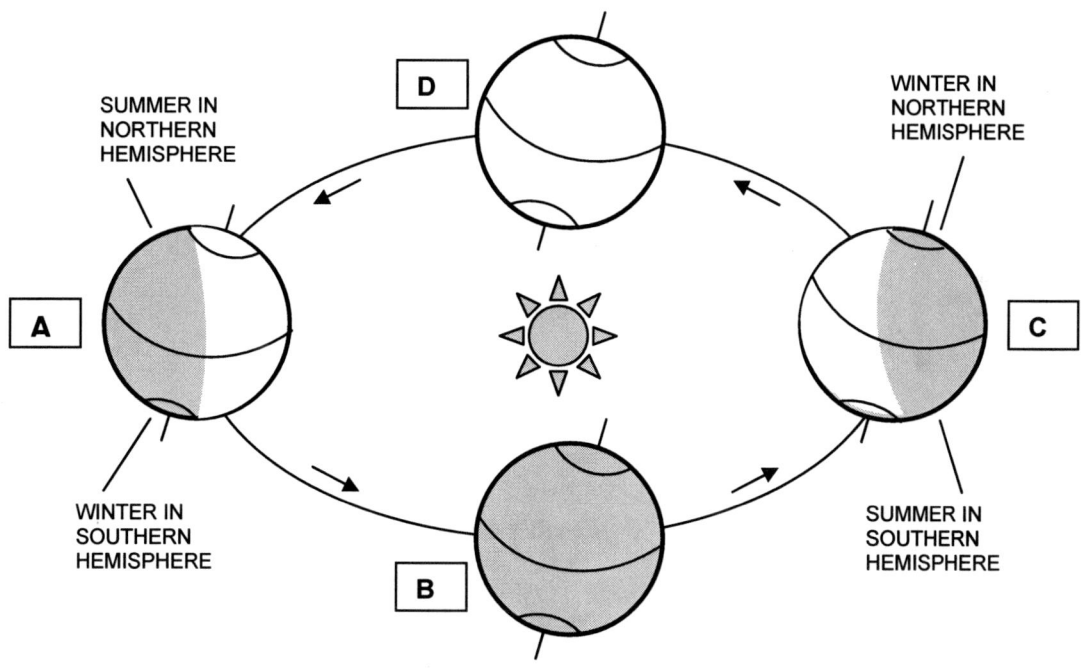

Figure 17. Earth's rotation round the Sun with the resulting seasons

Points B and D can be considered together. At these points on its yearly circuit, both northern and southern hemispheres face the Sun to an equal degree, as do both the Arctic and the Antarctic. Points B and D depict the positions at the **equinoctial points**, the **equinoxes** (Latin: *equinoctial*: equal nights), point B being approximately September 23rd and point D March 21st. At these points, the day and night are of equal length. Between each 3 monthly point, day length and season is gradually changing, the northern hemisphere progressing through spring and summer with increasing day length up to midsummer's day, as the southern hemisphere experiences its autumn and winter and vice versa.

It is worthy of note that the Arctic and Antarctic circles are defined as the latitudes above and below which, in the winter months, the Sun does not rise above their respective horizons for

Astro-navigation from Square One

a period of at least 24 hours. Their parallels of latitude are 66° 33´ North and South. Note the symmetry with the Earth's axial tilt of 23° 27´ and that they together add up to 90°. The <u>fixed</u> tilt is the key.

1.13 Converting to the geocentric view of the Sun – middle Earth?

In the heliocentric (Sun at the centre) solar system that we have begun to describe, the Earth revolves on its axis once each day and whilst doing so, moves around the Sun just once in a year, as depicted in Figure 18. In the "geocentric" (Earth at the centre) view of our navigator's celestial sphere, the Earth lies at the centre and the Sun moves around it once per day.

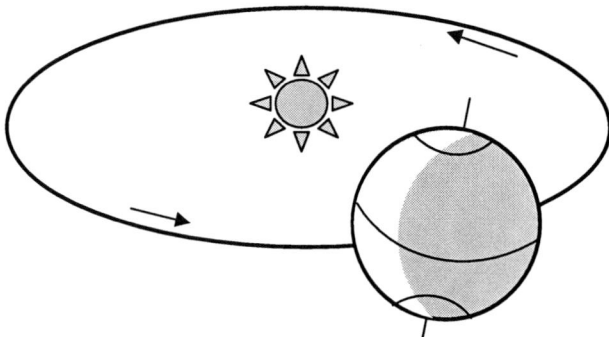

Figure 18. The heliocentric view of the Earth's annual orbit

Rotate the diagram in **Figure 18** until the Earth's axis reaches a vertical position. Then move the Earth to the centre of the picture and visualise the Sun rotating around the Earth over the period of a year as in **Figure 19**.

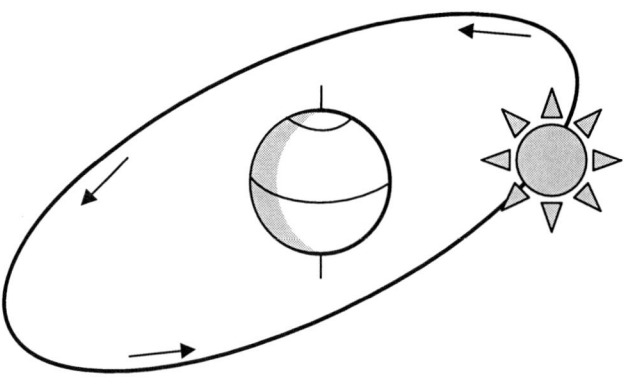

Figure 19. The geocentric view of the Sun on the ecliptic round the Earth

– the basics

The Sun's apparent movement ("apparent" meaning: the movement we see from Earth) when the Earth is geocentric, is complex, the ecliptic being a simplification resulting from the summation of the Sun's position of a daily snap-shot image. It ignores the fact that in the geocentric arrangement, the Sun also rotates about the Earth on a daily basis.

A more useful overview of the Sun's changing path through a year is shown in **Figure 20**. Beginning from a position in the northern hemisphere at the Summer Solstice in June, the Sun's plane of rotation is parallel to the equator for a short time and it is overhead at midday at the northern limit of the tropics. Its daily rotation from thereon, although moving continuously to the West, has a slight southerly leaning or drift, averaging 15′ per day.

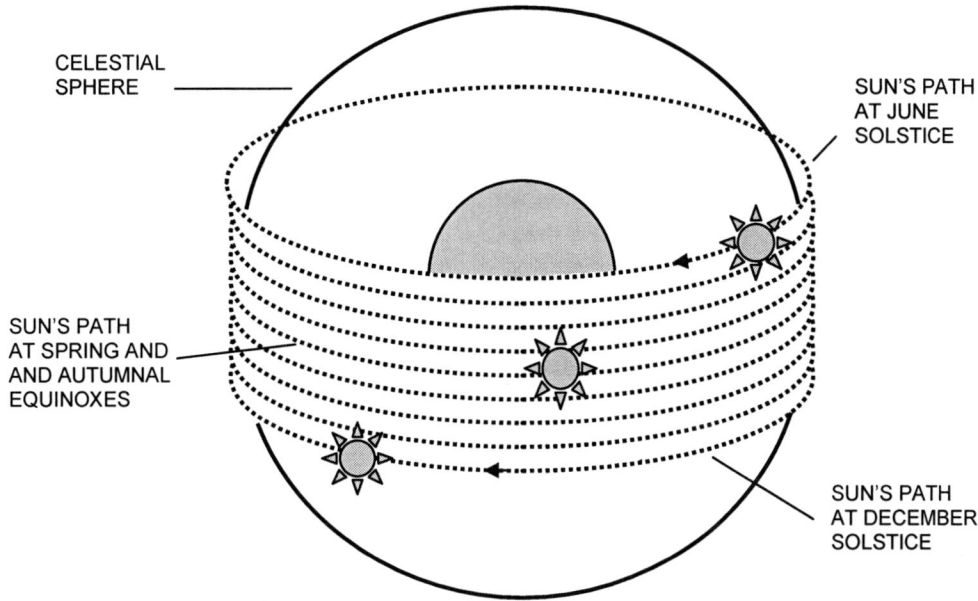

Figure 20. The complex apparent motion of the Sun in the geocentric system

This results in the Sun crossing the equatorial plane in September at the Autumnal equinox and eventually reaching the southerly limit of its course at the Winter solstice in December. Here again its plane of rotation is briefly parallel to the equator and it is overhead at midday at the lower limit of the tropics.

Its course then changes slightly such that its plane of movement has a small northerly leaning averaging 15′ per day, crossing the equatorial plane again at the Spring equinox in March and finally finishing its year's journey in the northern hemisphere at midsummer's day. Viewed from a fixed position on Earth, the astronavigator's view, the Sun would be seen to rise in the East, move across the sky and then set in the West. It would reach its highest point in the sky in mid-summer at its point of maximum northerly declination and its lowest point in mid-winter, at its most southerly, producing respectively, the longest and shortest periods of daylight hours.

On the longest day in the northern hemisphere, the Sun reaches its highest point in the sky for the whole year at which point its declination will be 23° 27′ North. On the shortest day in the northern hemisphere the Sun will be at its maximum southerly declination of 23° 27′ South. Between these two extremes, as each day passes, the Sun's declination changes by the 15′ (0.25°) average. Significantly, the reader should now be able to appreciate that the angle of 23° 27′ as that between the Earth's axis and its annual path round the Sun, in the true "heliocentric" arrangement of our solar system. So in effect and not very surprisingly, the heliocentric and geocentric views of our solar system and beyond are simply the same, but seen and elaborated from differing viewpoints.

1.14. Why does the Sun rise and fall – what makes it seem to move as it does?

Why then when observing on Earth, in the sky overhead, do we see the daily rise and fall of the Sun; a strange question perhaps, but how does this fit in with our geocentric view and its imposed geometry? The answer is that which we appear to see, i.e. what is apparent, is again a function of the observer's position. It is a three dimensional image that is difficult to visualise and one which changes with the fourth dimension of time.

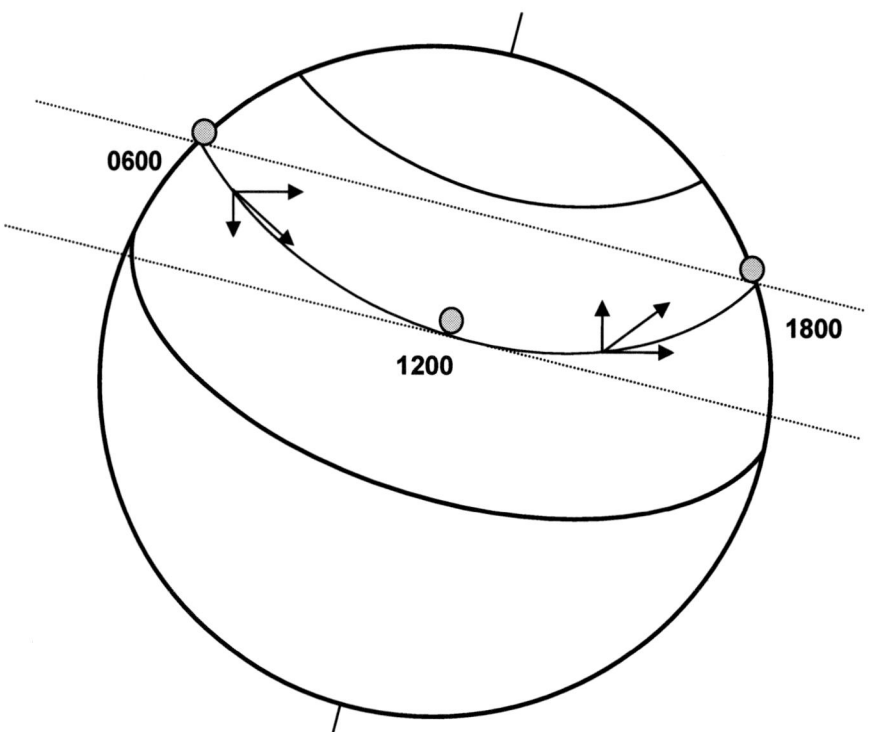

Figure 21. The changing view of the Earth <u>from</u> the Sun over a 12 hour period

– the basics

It is apparent to the observer with a geocentric view that the Sun climbs higher in the sky as the morning period progresses. So far we have looked from the observer's view on Earth and from an elevated view of the Earth-Sun complex from further out in space.

To understand the apparent motion of the Sun from Earth, we have to site ourselves on the Sun. It is the view of the Earth from the Sun that **Figure 21** endeavours to show. The relative orientation shown is that at the equinox in March or September. The observer on Earth at 0600 has just seen the Sun rise over his eastern horizon. The Earth continues to rotate and 6 hours later the observer has reached the position shown at 1200 hours. Note carefully that when seen from the Sun, the observer's movement can be resolved into a horizontal and vertical plane.

It is the vertical element of the movement, an apparent southward movement of the observer on Earth as seen from the Sun that is responsible for the apparent rise in the sky by the Sun when seen by the observer on Earth. It is produced by the rotation of the Earth on its axis. From the Sun, after 1200 hours, the observer on Earth appears to rise again as the Earth rotates towards 1800 hours, coinciding with the observer's apparent view of the Sun falling in the sky and then setting. Once again, what you see depends hugely on your position. The curve of the apparent rise and fall of the Sun is explored further in **Section 6.5** p.146.

Figure 22 shows the relative orientation of the Earth when viewed from the Sun around the solstices in June and December. Remember the angle of the Earth's axis to the plane of its solar orbit remains fixed, but its passage round its orbit causes the

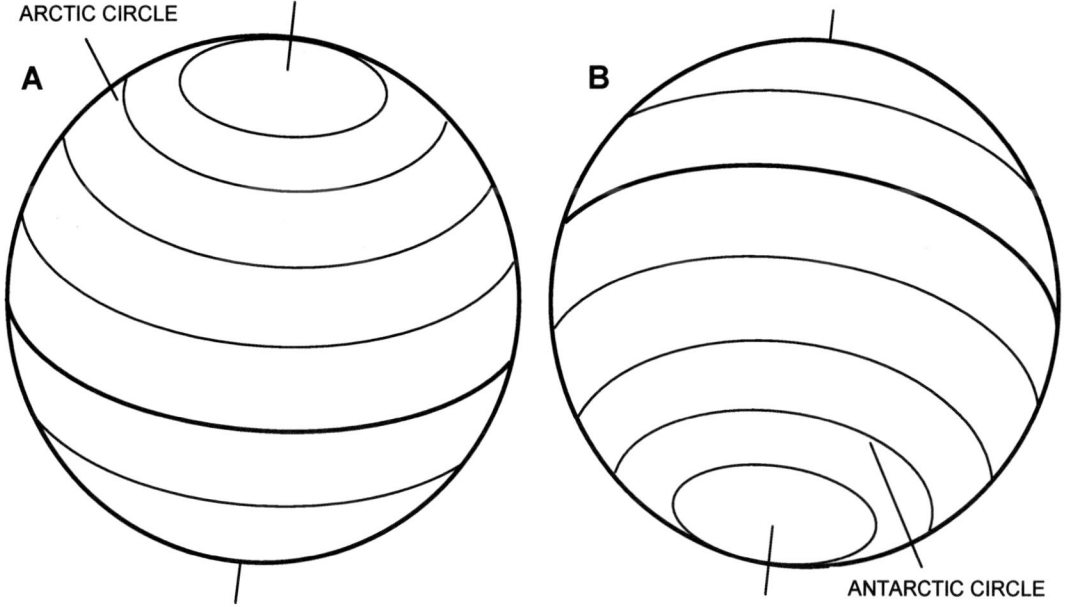

Figure 22. The Earth from the Sun at the solstices in June (A) and December (B)

Earth's northern hemisphere to be tilted toward the Sun in summer months and away from it during the winter period, being at a maximum at the time of the solstices.

1.15 *Local apparent noon* – what is it and how is it defined?

Think again about the movement when changing from the heliocentric to the geocentric view. The Earth's axis moves 23° 27′ to become vertical and correspondingly, the axis of the Sun's movement, when recorded at the same time each day moves from the horizontal to that of a line angled at 23° 27′ to the horizontal: the ecliptic.

At the March and September equinoxes, the Sun is exactly half way along its ecliptic course between the solstices. Here it is over the equator, its declination is 0° and the Sun will come to lie <u>directly</u> over an observer's head, i.e. at the observer's **zenith** in the middle of the day at this latitude of 0°, at all longitudes on the equator. At the summer and winter solstices in the northern hemisphere, the Sun's declination will be respectively 23° 27′ North and 23° 27′ South, marking the upper and lower limits of the tropics. Here at these two specific latitudes at the times of the solstices, the Sun will be overhead in the middle of the day, at all longitudes. Between these extremes, the Sun will be overhead at specific latitudes on each and every day, depending upon the position of the Sun within its yearly cycle.

At latitudes both North and South of the tropics, the Sun never reaches the observers zenith, but its height in the sky varies inversely with latitude; the lower the latitude, the higher the Sun reaches and vice versa. On each and every day and at all latitudes, the Sun will be observed to reach a peak height in the sky, at a time <u>exactly</u> halfway between sunrise and sunset at that particular latitude. This moment is called the **local apparent noon**, when the Sun <u>appears</u> (meaning: it is seen) to be at its highest point in the sky as it crosses the meridian on which stands the observer. It is the most defining point in the time kept by the Sun and is crucial to the navigator in his search for an estimation of latitude. As you will see in subsequent sections, time can become something of a sticky wicket for the navigator.

1.16 *Linking Angle to Time* – what is apparent solar time?

The East-West position of an observer on the Earth's surface is described in meridians of longitude, the **Prime meridian**, from which all others are measured, being at Greenwich. In our navigator's **geocentric** view of the **celestial sphere** and if observed from the theoretical position of the celestial pole, the Sun would be seen to move around the Earth in a period that man has decided to called 24 hours of time. During this period the Sun traverses 360°.

As we have described, the Greenwich meridian is used as the datum line for Earthly longitude and for the measurement of celestial longitude, properly called Greenwich Hour Angle. In the latter case the angular measurement is made in a direction that is always <u>westerly</u>. When the Sun crosses the Greenwich meridian, its **Greenwich Hour Angle** is 00° 00′. Exactly six hours later the Sun will have moved on Westwards to a **Greenwich Hour Angle (GHA)** of

– *the basics*

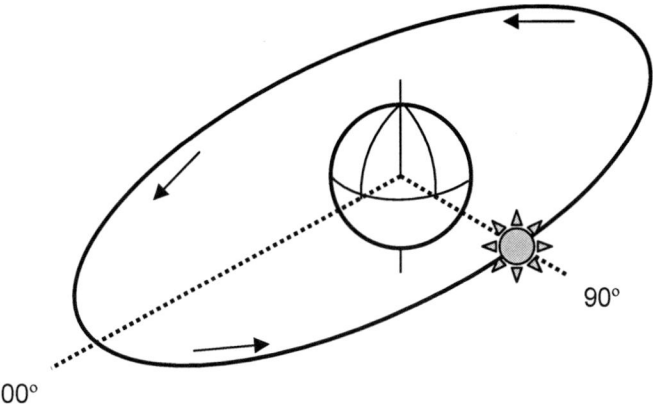

Figure 23. The link between the Sun's movement and time

90° 00′ (6 x 15° = 90), the Sun moving through 15° of Hour Angle per hour, as depicted in **Figure 23**. One further hour would see the Sun at a **GHA** of 105° 00′ (90+15=105) as in **Figure 24**. When 24 hours have passed, our Sun will have moved 24 x 15° = 360° around our geocentric Earth. You should be able to begin to see an inextricable link developing between time and the position of the Sun.

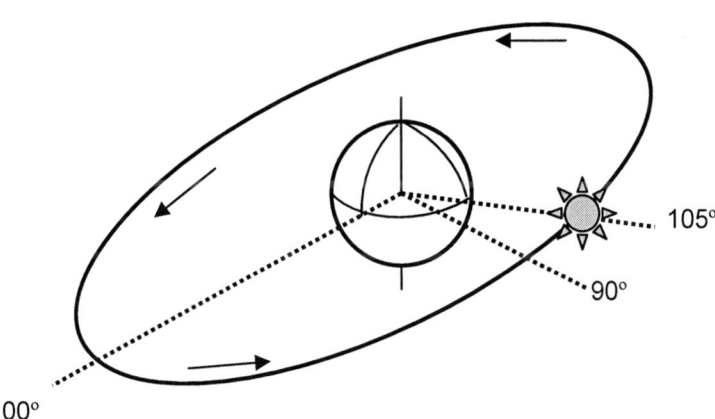

Figure 24. The Sun and its day's circuit of the Earth - one hour (15°) further on

Timekeeping on Earth is a consequence of the position of the **apparent Sun**; the "apparent" meaning "that which we see". Starkly put, the position of the Sun is time. The link between angle and time is summarized in an "Arc to Time Table" and an example is shown in **Table 1**.

CONVERSION OF ARC TO TIME

0°–59°			60°–119°			120°–179°			180°–239°			240°–299°			300°–359°				0ʹ.00		0ʹ.25		0ʹ.50		0ʹ.75	
°	h	m	°	h	m	°	h	m	°	h	m	°	h	m	°	h	m	ʹ	m	s	m	s	m	s	m	s
0	0	00	60	4	00	120	8	00	180	12	00	240	16	00	300	20	00	0	0	00	0	01	0	02	0	03
1	0	04	61	4	04	121	8	04	181	12	04	241	16	04	301	20	04	1	0	04	0	05	0	06	0	07
2	0	08	62	4	08	122	8	08	182	12	08	242	16	08	302	20	08	2	0	08	0	09	0	10	0	11
3	0	12	63	4	12	123	8	12	183	12	12	243	16	12	303	20	12	3	0	12	0	13	0	14	0	15
4	0	16	64	4	16	124	8	16	184	12	16	244	16	16	304	20	16	4	0	16	0	17	0	18	0	19
5	0	20	65	4	20	125	8	20	185	12	20	245	16	20	305	20	20	5	0	20	0	21	0	22	0	23
6	0	24	66	4	24	126	8	24	186	12	24	246	16	24	306	20	24	6	0	24	0	25	0	26	0	27
7	0	28	67	4	28	127	8	28	187	12	28	247	16	28	307	20	28	7	0	28	0	29	0	30	0	31
8	0	32	68	4	32	128	8	32	188	12	32	248	16	32	308	20	32	8	0	32	0	33	0	34	0	35
9	0	36	69	4	36	129	8	36	189	12	36	249	16	36	309	20	36	9	0	36	0	37	0	38	0	39
10	0	40	70	4	40	130	8	40	190	12	40	250	16	40	310	20	40	10	0	40	0	41	0	42	0	43
11	0	44	71	4	44	131	8	44	191	12	44	251	16	44	311	20	44	11	0	44	0	45	0	46	0	47
12	0	48	72	4	48	132	8	48	192	12	48	252	16	48	312	20	48	12	0	48	0	49	0	50	0	51
13	0	52	73	4	52	133	8	52	193	12	52	253	16	52	313	20	52	13	0	52	0	53	0	54	0	55
14	0	56	74	4	56	134	8	56	194	12	56	254	16	56	314	20	56	14	0	56	0	57	0	58	0	59
15	1	00	75	5	00	135	9	00	195	13	00	255	17	00	315	21	00	15	1	00	1	01	1	02	1	03
16	1	04	76	5	04	136	9	04	196	13	04	256	17	04	316	21	04	16	1	04	1	05	1	06	1	07
17	1	08	77	5	08	137	9	08	197	13	08	257	17	08	317	21	08	17	1	08	1	09	1	10	1	11
18	1	12	78	5	12	138	9	12	198	13	12	258	17	12	318	21	12	18	1	12	1	13	1	14	1	15
19	1	16	79	5	16	139	9	16	199	13	16	259	17	16	319	21	16	19	1	16	1	17	1	18	1	19
20	1	20	80	5	20	140	9	20	200	13	20	260	17	20	320	21	20	20	1	20	1	21	1	22	1	23
21	1	24	81	5	24	141	9	24	201	13	24	261	17	24	321	21	24	21	1	24	1	25	1	26	1	27
22	1	28	82	5	28	142	9	28	202	13	28	262	17	28	322	21	28	22	1	28	1	29	1	30	1	31
23	1	32	83	5	32	143	9	32	203	13	32	263	17	32	323	21	32	23	1	32	1	33	1	34	1	35
24	1	36	84	5	36	144	9	36	204	13	36	264	17	36	324	21	36	24	1	36	1	37	1	38	1	39
25	1	40	85	5	40	145	9	40	205	13	40	265	17	40	325	21	40	25	1	40	1	41	1	42	1	43
26	1	44	86	5	44	146	9	44	206	13	44	266	17	44	326	21	44	26	1	44	1	45	1	46	1	47
27	1	48	87	5	48	147	9	48	207	13	48	267	17	48	327	21	48	27	1	48	1	49	1	50	1	51
28	1	52	88	5	52	148	9	52	208	13	52	268	17	52	328	21	52	28	1	52	1	53	1	54	1	55
29	1	56	89	5	56	149	9	56	209	13	56	269	17	56	329	21	56	29	1	56	1	57	1	58	1	59
30	2	00	90	6	00	150	10	00	210	14	00	270	18	00	330	22	00	30	2	00	2	01	2	02	2	03
31	2	04	91	6	04	151	10	04	211	14	04	271	18	04	331	22	04	31	2	04	2	05	2	06	2	07
32	2	08	92	6	08	152	10	08	212	14	08	272	18	08	332	22	08	32	2	08	2	09	2	10	2	11
33	2	12	93	6	12	153	10	12	213	14	12	273	18	12	333	22	12	33	2	12	2	13	2	14	2	15
34	2	16	94	6	16	154	10	16	214	14	16	274	18	16	334	22	16	34	2	16	2	17	2	18	2	19
35	2	20	95	6	20	155	10	20	215	14	20	275	18	20	335	22	20	35	2	20	2	21	2	22	2	23
36	2	24	96	6	24	156	10	24	216	14	24	276	18	24	336	22	24	36	2	24	2	25	2	26	2	27
37	2	28	97	6	28	157	10	28	217	14	28	277	18	28	337	22	28	37	2	28	2	29	2	30	2	31
38	2	32	98	6	32	158	10	32	218	14	32	278	18	32	338	22	32	38	2	32	2	33	2	34	2	35
39	2	36	99	6	36	159	10	36	219	14	36	279	18	36	339	22	36	39	2	36	2	37	2	38	2	39
40	2	40	100	6	40	160	10	40	220	14	40	280	18	40	340	22	40	40	2	40	2	41	2	42	2	43
41	2	44	101	6	44	161	10	44	221	14	44	281	18	44	341	22	44	41	2	44	2	45	2	46	2	47
42	2	48	102	6	48	162	10	48	222	14	48	282	18	48	342	22	48	42	2	48	2	49	2	50	2	51
43	2	52	103	6	52	163	10	52	223	14	52	283	18	52	343	22	52	43	2	52	2	53	2	54	2	55
44	2	56	104	6	56	164	10	56	224	14	56	284	18	56	344	22	56	44	2	56	2	57	2	58	2	59
45	3	00	105	7	00	165	11	00	225	15	00	285	19	00	345	23	00	45	3	00	3	01	3	02	3	03
46	3	04	106	7	04	166	11	04	226	15	04	286	19	04	346	23	04	46	3	04	3	05	3	06	3	07
47	3	08	107	7	08	167	11	08	227	15	08	287	19	08	347	23	08	47	3	08	3	09	3	10	3	11
48	3	12	108	7	12	168	11	12	228	15	12	288	19	12	348	23	12	48	3	12	3	13	3	14	3	15
49	3	16	109	7	16	169	11	16	229	15	16	289	19	16	349	23	16	49	3	16	3	17	3	18	3	19
50	3	20	110	7	20	170	11	20	230	15	20	290	19	20	350	23	20	50	3	20	3	21	3	22	3	23
51	3	24	111	7	24	171	11	24	231	15	24	291	19	24	351	23	24	51	3	24	3	25	3	26	3	27
52	3	28	112	7	28	172	11	28	232	15	28	292	19	28	352	23	28	52	3	28	3	29	3	30	3	31
53	3	32	113	7	32	173	11	32	233	15	32	293	19	32	353	23	32	53	3	32	3	33	3	34	3	35
54	3	36	114	7	36	174	11	36	234	15	36	294	19	36	354	23	36	54	3	36	3	37	3	38	3	39
55	3	40	115	7	40	175	11	40	235	15	40	295	19	40	355	23	40	55	3	40	3	41	3	42	3	43
56	3	44	116	7	44	176	11	44	236	15	44	296	19	44	356	23	44	56	3	44	3	45	3	46	3	47
57	3	48	117	7	48	177	11	48	237	15	48	297	19	48	357	23	48	57	3	48	3	49	3	50	3	51
58	3	52	118	7	52	178	11	52	238	15	52	298	19	52	358	23	52	58	3	52	3	53	3	54	3	55
59	3	56	119	7	56	179	11	56	239	15	56	299	19	56	359	23	56	59	3	56	3	57	3	58	3	59

The above table is for converting expressions in arc to their equivalent in time; its main use in this Almanac is for the conversion of longitude for application to LMT (*added* if *west*, *subtracted* if *east*) to give UT or vice versa, particularly in the case of sunrise, sunset, etc.

Table 1. The standard nautical almanac Arc to Time conversion table

– the basics

The table comprises two parts. The main part consists of six columns each covering a period of 59 degrees of arc (0-59°, 60-119°, 120-179°, 180-239° 240-299°, 300-359°). Alongside each individual degree (°) is given its time equivalent in minutes and seconds. The right hand side of the table comprises 5 columns, giving the relevant time equivalent for 0-59 whole minutes (′) of arc and decimal divisions. Simple addition of the readings from the relevant columns from each side of the table converts arc to time.

An alternative used by the professional navigator, is to divide the arc length by 15°. Both time and arc are "hexadecimal" measures, using base 60 to describe the divisional units and which can be directly handled by many calculators. Examples follow in the sections specific to the Sun, Moon, planets and stars (**Sections 6-9**).

The passage of time is based on the appearance or presence of the Sun at a particular and identifiable point; this is **apparent solar time**. One particular identifiable point on its daily passage being when the Sun crosses the Greenwich Meridian; this brief moment is exactly midday according to the position of the Sun. This is the time of **local apparent noon** at Greenwich, again the word apparent meaning "that which is seen".

To emphasise the point; that for an observer on any **meridian** on the globe, the point at which the Sun crosses that meridian is exactly local apparent noon. This moment is called the Sun's **meridian passage** and is a vital tool in latitude estimation.

In the true "**heliocentric**" solar system, the Earth's passage around the Sun is a little irregular due to certain factors discussed in more detail in **Section 5**. These factors lead to relatively small but significant and measurable differences that cause variation in the time of the Sun's crossing of each meridian on each and every day and in its indication of noon and **apparent solar time**.

The variations in natural time are unhelpful for modern man. The Sun can be early or late in its arrival at each meridian over periods lasting several weeks, depending upon several factors that can work for or against each other. The disparity in regularity of the Sun can reach as much as 16 minutes and is a phenomenon called, somewhat confusingly, the **equation of time**. Man cannot easily organize his world to one in which time seemingly varies, so a scheme was invented to mathematically iron out these variations.

This average time, in which each and every day has an assumed passage of the Sun at Greenwich at exactly 1200 hours and in which midnight occurs exactly 12 hours later is called **Greenwich Mean Time** (**GMT**-now more correctly called Universal Time - **UT**).

Greenwich Mean Time for the trainee navigator might be better termed *Greenwich Mean Solar Time*, to help distance it from **apparent solar time**, based purely upon the true daily movement of the Sun across the sky. This will be further studied in **Section 5**. The Sun's **declination** and **Greenwich hour angle** are tabulated in the astro-nautical almanac for each hour of the day, for each day of the year, together with the necessary increments for minutes and seconds and the day's **equation of time**.

1.17 *Linking GMT/UT to local apparent noon* – the key to longitude?

The feat of acquiring sufficient accuracy in the keeping of time for prolonged navigation at sea was finally achieved in the eighteenth century, due largely to the efforts of John Harrison (1693-1776) in his designs for and construction of a stable, transportable and reproducible chronometer[1.6]. Prior to this and supported strongly by those opposing Harrison, notably Neville Maskelyne (1732-1811) the Astronomer Royal, angular measurements were made between the Moon and certain stars and called **lunars**[1.7]. These were then compared using complex tables, with angular measurement from an unknown position, permitting an estimation of longitude to be made.

In **Section 1.16** attention was drawn to the relationship between the passage of time and the angular arrangement of longitude on the Earth and Hour Angle on the celestial sphere. In reality, in one 24 hour period the Earth rotates 360 degrees. In our view from Earth however, the Sun on the celestial sphere apparently turns through 360° around the Earth in 24 hours, at a rate of 15° per hour. Each individual degree of longitude is equivalent to 4 minutes of time (24 x 60÷360 = 4). This system enables a longitude to be expressed in terms of time e.g. 30 degrees is equivalent to 120 minutes of time (30 x 4 =120m or 2 hrs at 15° per hour).

All this discussion brings us to the supremely important point that: - if you know the **Greenwich Mean Time** at the moment of your **local apparent noon** then you can <u>convert this time period to your longitude</u>. Why? Because your clock has a record of the time since the Sun passed over the prime meridian at Greenwich. You are in effect working out the **Greenwich Hour Angle** of the Sun when it lies exactly on your own particular meridian.

Example 2. If by observation the Sun reaches its greatest height and is therefore crossing your meridian at exactly 16 hours and 30 minutes GMT, what is your longitude?

By converting the hours and minutes of time after 1200 GMT/UT to degrees and minutes of angle you will arrive at the longitude of your observation: 4 hours and 30 minutes have passed since 1200 hours GMT/UT and this is equivalent to a change in the Sun's Greenwich Hour Angle of 67.5° (4 hours x 15° = 60° and 30 minutes of time is 0.5 hours. 0.5 x 15° = 7.5°). In converting time to longitude, either the arc-time table can be consulted or the number of hours multiplied by 15, to give the number of degrees of longitude.

This observation was therefore made at a longitude of 67.5° which is 67° 30´. But is this longitude East or West of Greenwich? Given that the local noon had occurred at 1630 hours GMT/UT, which is <u>after</u> noon GMT/UT then the longitude is 67° 30´ West. Had the **local apparent noon** occurred <u>before</u> noon GMT/UT then the longitude would have been East. Remember, the Earth rotates towards the East, seen on Earth as the Sun moving to the West. I have given a very simple example. In fact longitude estimation needs to be a little more accurate than that obtainable by the simple conversion of the GMT/UT of local noon, but the principle is sound.

Take care to clarify in your mind the confusion that might arise between a "minute of time" i.e. 1/60th of an hour, and a "minute of a degree" i.e.1/60th of a degree. As 360° equates to 24

– the basics

hours, 15° equates to 1 hour (360÷24=15) and 1 degree equates to 4 minutes of time (60m ÷15= 4). One minute (1′) of <u>longitude</u> (1/60th of a degree) is equivalent to 4/60^{ths} of a minute of time, which is 4 seconds of time (4/60 x 60s = 4s). Summarising, 15° of the westerly passage of the Sun = 1 hour of time, 1° = 4 minutes of time and 1′= 4 seconds of time.

We have in the above example noted that meridians of longitude are conventionally measured up to 180 degrees East and West of the Greenwich Meridian, but angular time is measured in 360° in a Westerly direction; but with practice the conversion is not difficult. I have not yet included any correction necessary for the **equation of time**, the explanation and discussion of which is detailed in **Section 5.4**. For the present however, bask in the light of understanding the link between GMT/UT and longitude.

1.18 *The Sun to measure latitude* – how is the meridian passage helpful?

We have learned in **Section 1.2** that the latitude of a point on the surface of the Earth is the angle subtended by a line passing from that point to the centre of the Earth and the plane of the equator. It is measured in degrees North or South of the plane of the equator. **Figure 25** represents a vertical plane through the centre of the Earth and celestial sphere, again using the **equatorial view** of the system.

The observer is at point O in mid-latitudes in the Northern hemisphere and it is summertime, so that the Sun is in the northern celestial hemisphere. We will use the position of the Sun at <u>exactly</u> local midday, at its **meridian passage** of point O.

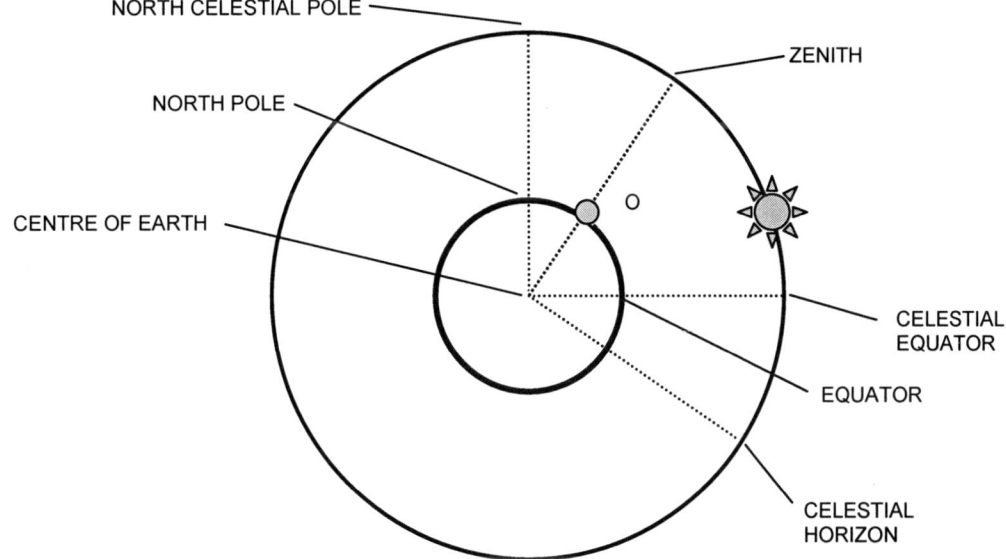

Figure 25. Determining latitude from meridian passage – 1st stage

Astro-navigation from Square One

What can we measure to assist us with finding the latitude of O? Firstly, let us state what we already know, based on a vertical cross-section of the Earth and celestial sphere. We know that a line from the centre of the Earth (CE) up to the North pole (NP) and extended to the North celestial pole (NCP), makes a right angle with a line from the centre of the Earth (CE) to the equator (EQ) and its extension, to the celestial equator (CEQ).

We also know that a line from the centre of the Earth (CE) through point O can be extended to meet the celestial sphere at a point called the **zenith** (Z) of point O. We can draw in a further line from the centre of the Earth (CE) out to the celestial sphere, at a right angle to line CE-Z. Accept for the moment that this line meets the celestial sphere at a point called the **celestial horizon** (CH).

Now consider what angles we know. We have stated that there are two right angles; one made by a line from the North Pole to the Earth's centre then extending to the celestial equator and one from the observer's zenith to the Earth's centre and then to the celestial horizon. We can add another line from the Sun to the centre of the Earth (**Figure 26**). We know the angle between that line and the celestial equator; this angle is the **declination (Dec)** of the Sun (its celestial latitude), found tabulated in the nautical almanac.

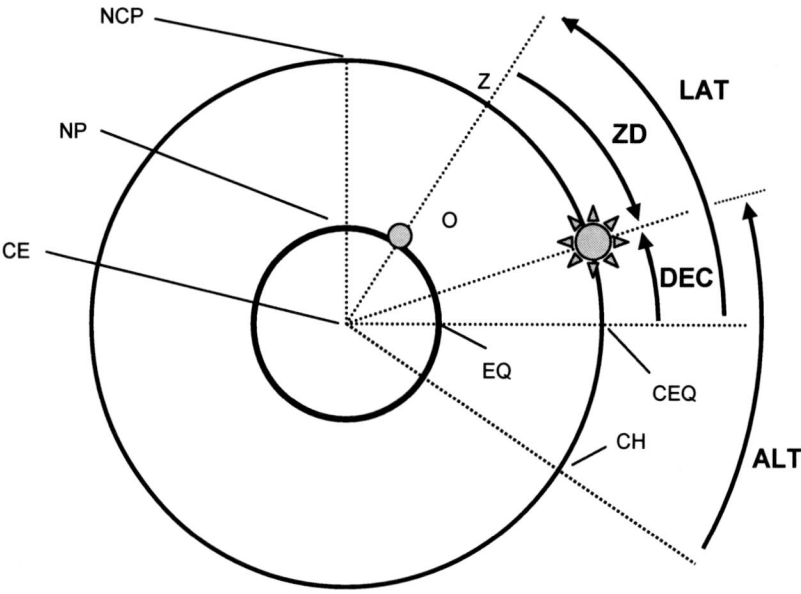

Figure 26. The 2nd stage of latitude assessment by meridian passage

– the basics

What angles do we not know? Plainly we do not know the angle between the observer and the Earth's equator, this being the **latitude (Lat)** of point O – our goal. We also do not know the angle between the observer's zenith and the Sun, called the **zenith distance (ZD)**. This is the angular distance at the centre of the Earth, between the zenith (Z) of the observer (O) and the Sun. We do not know this angle but it can be worked out. Lastly, we do not know the angle between the Sun and the celestial horizon, but we could work it out by measuring the **altitude (Alt)** of the Sun above the visible horizon and adjusting the figure to that which would be arrived at if it were possible to measure altitude from the centre of the Earth. Accept for the moment that the sextant is used to measure the altitude angle, being the essential tool in our navigational armoury.

So in summary up to this point, we have four angles: **latitude** (which we need to find), **declination** (which we can look up), **altitude** (which we measure) and **zenith distance** (which we can work out). Now for the clever part! Look again at **Figure 26** and you should be able to see that added together, the angles **zenith distance** and **altitude** equal a right angle (**90°**). Therefore, using the rules of simple equations:

As **zenith distance + altitude = 90°**

then **zenith distance = 90° – altitude**   (**Equation 1**).

You should also be able to see that:

latitude = zenith distance + declination   (**Equation 2**).

By substituting **90° – altitude**, which we have just found equals **zenith distance**:

Then: **latitude = (90° – altitude) + declination**   (**Equation 3**).

We have looked up **declination**, we have measured **altitude** and taken it away from 90° and then added the result to **declination** to give us our **latitude**. Let us put in some figures that will further clarify things (**Example 3**). Let us assume that the astro-nautical almanac for the day in question states that the **declination** of the Sun at 1200 is exactly 20°N and that the Sun is carefully observed with a sextant from a position in the northern hemisphere and the **observed altitude** of the Sun at exactly midday was found to be 50° 00′.

Example 3. Substituting these simple figures in the **Equation 1**:

zenith distance = 90° – altitude

then: **zenith distance = 90° – 50°**

Substituting for zenith distance in **Equation 2** (latitude = zenith distance + declination):

latitude = (90° – 50°) + 20°

latitude = 40° + 20° = 60° North

We have worked out the latitude of point **O** to be 60° North – Eureka!

I have not yet mentioned any detail of the use of a sextant, nor yet explained our use of the **celestial horizon** and have deliberately placed our position and the Sun in the same celestial hemisphere But a lot of ground has been covered. If you have followed things to this point then there is every reason to think that you can manage what is to come; mere refinements of the methods we have so far used. Everything is based on angles and time; their measurement (altitude, GMT/UT, local time), their computation (Greenwich and local hour angles) and finally, their conversion into position lines to be plotted on to a chart.

1.19 *The importance of the Sun at local noon* – why does it give us latitude?

Why does knowing the altitude of the Sun whose declination you can look up, allow you to work out latitude? It is simply a matter of proportions, using angles you do know to find an angle you don't know. The most important thing is that the **altitude** measurement of the Sun must be made at its peak height, which occurs at exactly local midday. Why must this be the case?
We know that for a very brief moment of local noon, the Sun is at its highest point in the sky and at that moment the Sun lies exactly on the particular meridian of longitude of the observer, on a North-South plane and exactly at right angles to the parallels of latitude, on one of which the observer is sited. For that moment the local hour angle of the Sun is 0° 00´. This is depicted in **Figure 27**.

This technique, using these very specific conditions is called **latitude** by **Sun's meridian altitude**. It only works at local noon as complicated angles are introduced if the Sun is still rising or already falling in the sky. At those times it does not sit on a plane exactly at right angles to the parallels of latitude and more complicated formulae are needed.

A word of warning: the example used has the Sun and observer in the same hemisphere. There are two slightly differing equations for other circumstances, such as when the observer and Sun are in different hemispheres, as would be the case if the latitude of O was worked out in the winter months, when the Sun on its course along the **ecliptic** resides on the southern part of the celestial sphere. A third equation is needed if the Sun passes North of the observer, when the observer's latitude is less than the Sun's declination (see **Section 6.9** p.152).

1.20 Summarising Section 1

This section began by defining the **celestial sphere** in relation to the Earth, with their common centre and axis passing through their poles. **Latitude** and **longitude** were defined as the means of localizing position on the Earth. Their celestial sphere counterparts, **declination** and **Greenwich hour angle** were described. The positions of the heavenly bodies on the celestial sphere were explored, the stars being fixed and the Sun, Moon and planets moving over its surface. The **First point of Aries** and **sidereal hour angle** were explained in fixing a star's position in an East-West direction.

– the basics

The relationships between the Sun, time and longitude were outlined and the concept of **local hour angle** was introduced. The terms **zenith**, **azimuth** and **geographical position** were described. The consequences of the Earth's fixed **axial tilt** were explained, leading on to a description of the differences between the true **heliocentric** and artificial **geocentric** positions of the Sun and Earth and the apparent movement that the latter imparts to the Sun when viewed from Earth. The significance of **local apparent noon** was explored, which led on to an explanation of the importance of the Sun's **meridian passage** and an outline of the method used to find an observer's latitude from a measurement of Sun's altitude at local noon.

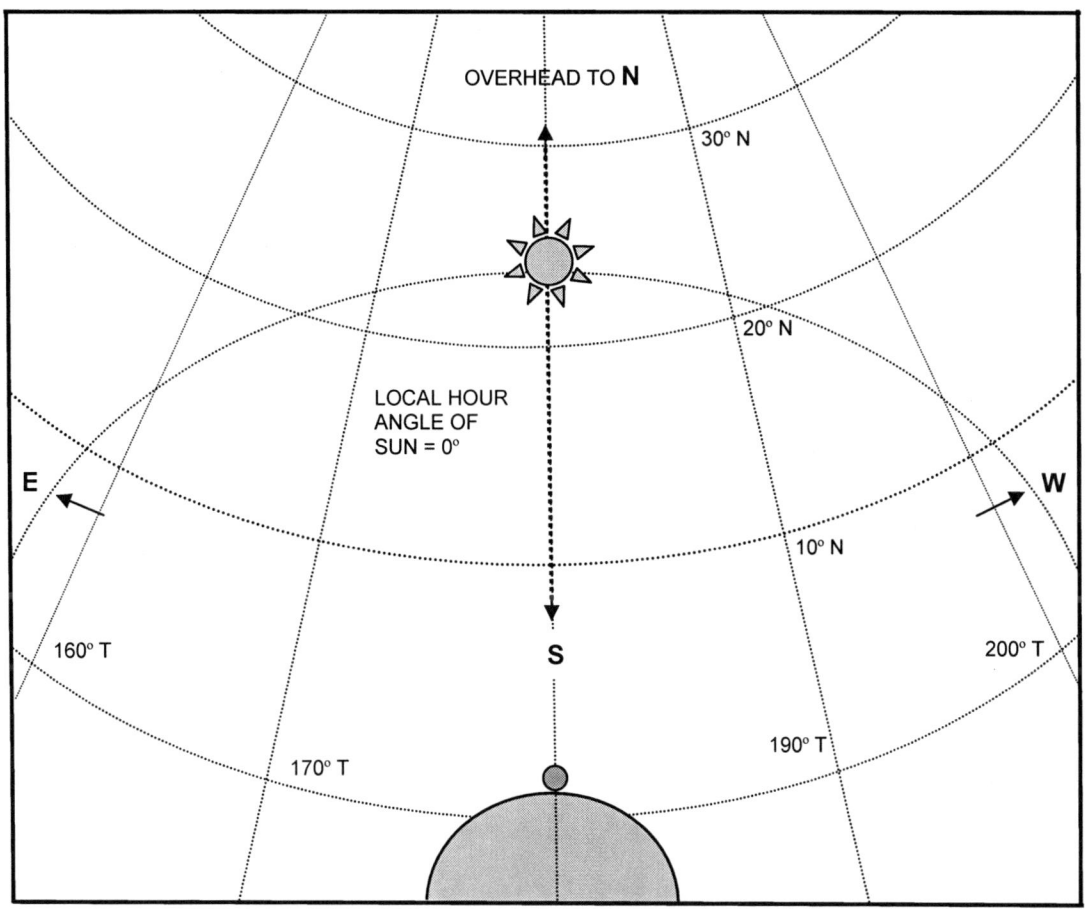

Figure 27. The special features of the Sun's position at local noon

Section 2: Down to the details

Section 2 describes the manipulation of an altitude measurement of a heavenly body so as to arrive at a position line on a chart: beginning with making an angular measurement at the Earth's surface and progressing through the process of sight reduction to achieve an **observed altitude**. The limitations of a simple single altitude measurement are explained. The theory involved in the generation of a **calculated altitude** and **azimuth** follows, including a discussion of the basic theory of **spherical trigonometry** and the calculation of both **local hour angle** and **azimuth angle**. Observed and calculated altitudes are then used to produce an **intercept** on the azimuth of a heavenly body. A discussion on the use of single and multiple **position lines** and the technique of **transferred position lines** ends the section.

By the end of Section 2, the following KEY FACTS should be understood:-

- **A sextant altitude reading from the visible horizon is corrected to give the altitude that would have been found if measured from the Earth's centre.**

- **All sextant altitudes need a correction for refraction and those bodies close to Earth within our solar system, need corrections for parallax and semi-diameter.**

- **An altitude figure gives rise to a circle of position of the Earth, at the centre of which lies the geographical point of the observed body.**

- **The intersection of two or more position circles from simultaneous sights of two or more bodies will fix an observer's position on the Earth's surface.**

- **Arcs on the surface of the celestial sphere, joining the celestial pole (P), observer's zenith (Z) and observed body (X) form the "PZX triangle".**

- **The angle at P is the local hour angle, at the angle formed by lines from Z and X.**

- **The azimuth is the direction of the body (X) from the observer's zenith (Z).**

- **The angle at Z is the azimuth angle, lying between lines from P and X.**

- **The PZX triangle solution is used to calculate the altitude of the observed body from a chosen position near the unknown position of the observation point.**

- **The azimuth angle is used to derive an azimuth, the direction of the body.**

- **The relative size between an observed and calculated altitude for a body enables an intercept; a part of the position circle, to be drawn on a body's azimuth.**

- **Two or more bodies sighted simultaneously, OR one body sighted sequentially, can produce suitable position lines to give a position fix.**

– *down to the details*

2.1 *The Horizon when measuring altitude* – how is it defined?

Section 1.18 introduced the concept of a **celestial horizon**, which was drawn in **Figure 25** p.31, running from the centre of the Earth to meet the **celestial sphere**. At right angles to this a second line ran from the centre of the Earth, through the position of the observer and up to the observer's **zenith** on the celestial sphere as in **Figure 28**.

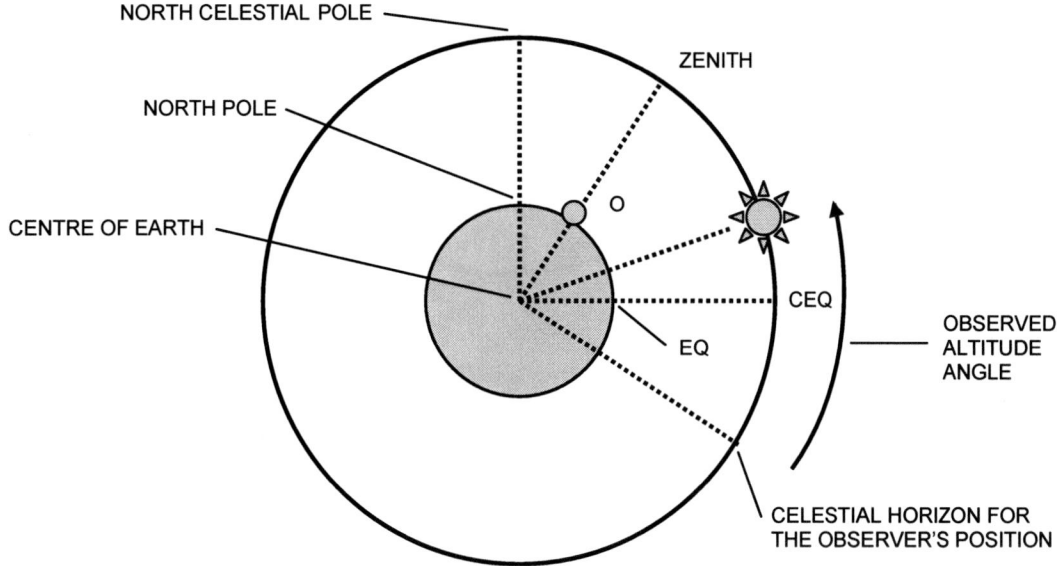

Figure 28. The observed altitude of the Sun: measured from the celestial horizon

Measurement of the angle at the Earth's centre made between the celestial horizon and the observed body, in this case the Sun, gives an altitude called the **observed altitude** (abbreviated to H_o). Plainly it is not practical to measure angles from the centre of the Earth. A way is needed to measure altitude at the Earth's surface and this we do using a sextant measurement of altitude above the horizon. The sextant reading is then adjusted to that which we would have arrived at, had we been able to measure it from the Earth's centre.

From the point of observation on the Earth's surface a line can be drawn to the celestial sphere, parallel to the line from the Earth's centre, as in **Figure 29**. This line, perpendicular to the line to the observer's zenith, is the plane of the so-called **sensible horizon**. It is the horizontal plane formed by the tangent to the Earth's surface at the level of the eyes of the observer and it disappears into oblivion a little above the visible horizon.

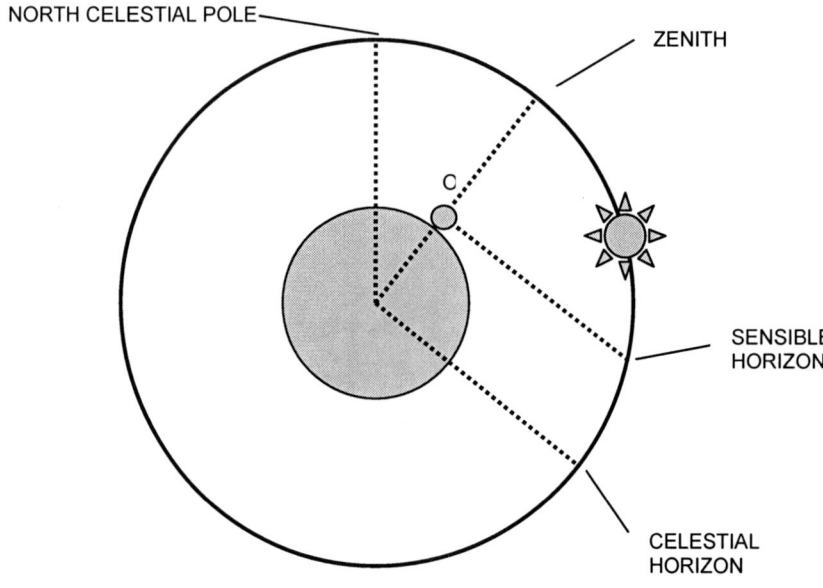

Figure 29. Showing the parallel lines to the "sensible" and celestial horizons

There is <u>nothing</u> to mark the intersection of the sensible horizon with the celestial sphere and taking a reliable sextant sight from this invisible point is not feasible. We can however see the **natural horizon**, which is the view of the sea curving away into the distance to the point at which it appears to meet the celestial sphere. It is a useful "edge" that can be used as a **datum line** from which measurements can be made.

2.2 *The "visible sky" view of the horizons* – changing the orientation to what we see?

By altering the scale and rotating the image of **Figure 29**, we can then redraw it as the more realistic **visible sky** view as in **Figure 30**. The observer (O) surmounted by the observer's zenith (Z) directly above, sits on the apex of the visible world. This is how we perceive our position on the Earth and importantly, it hugely influences how the sky is orientated above us.

In **Figure 30**, 3 angles are apparent: Angle **A** is the called the **sextant altitude** (abbreviated to H_s) and is the obvious angle that you can see when looking at a heavenly body. It is the angle that the observed body makes with the **visible** horizon, after any correction for known sextant error (**Section 3.4 p.78**).

– down to the details

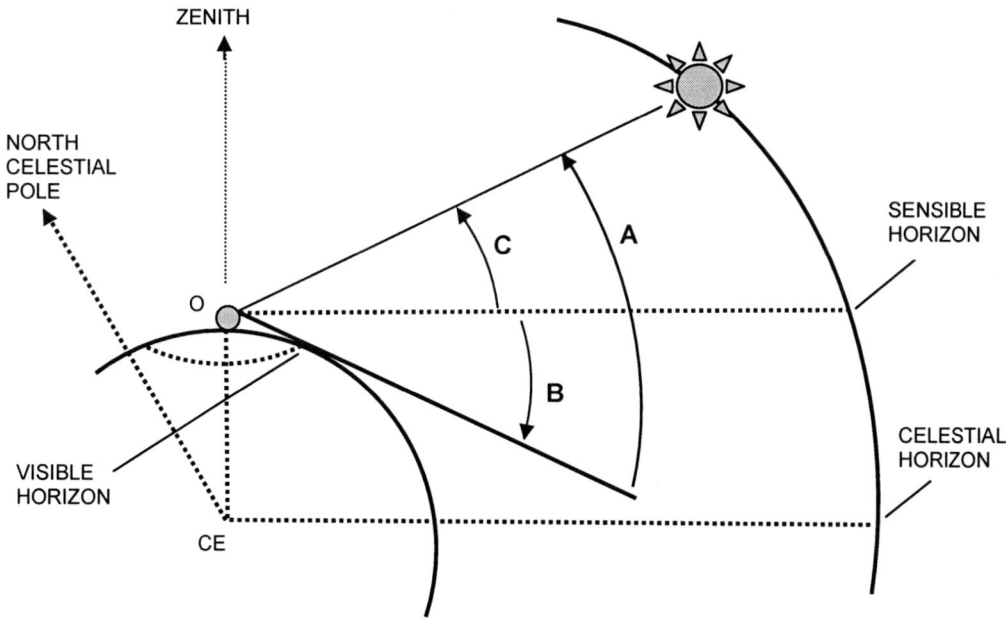

Figure 30. The visible sky view of the sensible and visible horizons

Angle **B** is called the **Angle of Dip** and is the angle between the **sensible horizon**; that horizontal plane at the level of the observer's eye and a line from the eye to the visible horizon. Plainly in **Figure 30** the scale of the observer's height to the Earth's diameter is hugely disproportionate. The actual angle of dip is quite small and is measured in minutes of arc rather than whole degrees, but it is still relevant. The **angle of dip (B)** is zero if the eye is level with the sea, a somewhat impractical place to use a sextant. "Dip" changes in a non-linear fashion over the first two or three metres of increasing eye height, before lessening to a more gradual but continued rise. This is best shown in graphical form as in **Figure 31**.

Angle **C** is arrived at by subtracting the angle of dip (**B**) from the corrected **sextant altitude** (**A**). Angle **C** is the **apparent altitude** (abbreviated to H_a), the angle the observed body makes with the horizontal plane of the sensible horizon.

Most small boat navigators make their sights within a range of eye height of 0-3m above the surface of the sea. The dip correction for this is up to 3´, equivalent to a distance of 3 nautical miles in the computation of the position line, so it is plainly important to know the height of eye to reduce unnecessary error. At a height of 20m on a ship's bridge for example, the dip correction is a significant 8´. Mathematically, the dip angle = 1.76 x square root of eye height in metres.

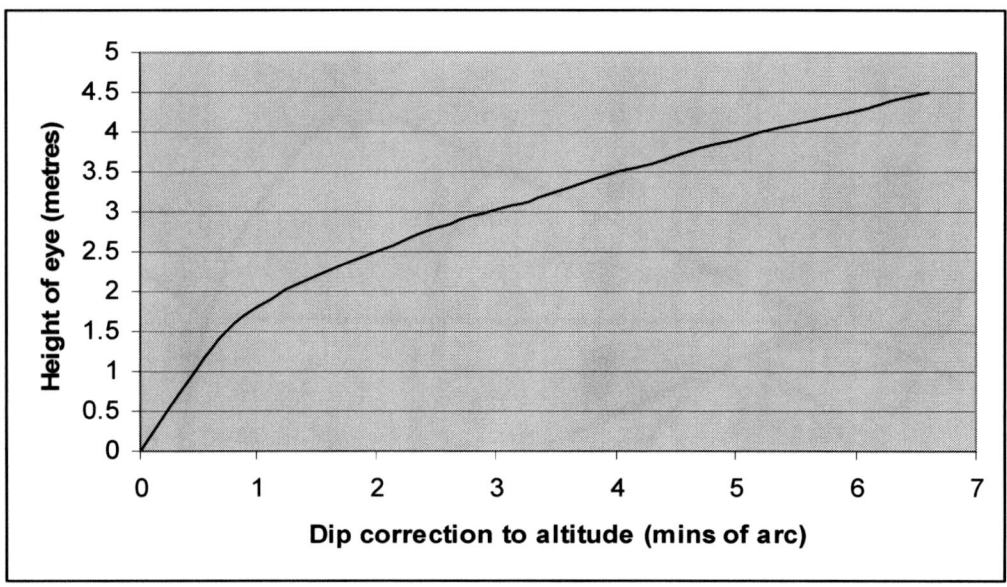

Figure 31. Graph of "dip" correction for height of eye

To summarize so far; the **sextant altitude** (H_s) is first corrected for sextant error. From the resulting figure, the angle of dip is subtracted to give the **apparent altitude** (H_a). The latter needs to be further refined to achieve an **observed altitude** (H_o), (called the **true** altitude by traditionalists) which is the value which would have been found were it possible to take the altitude reading from the centre of the Earth.

The **apparent altitude** may need to be corrected for 3 additional external factors: **parallax**, **semi-diameter** and **refraction**. Their importance varies with the particular heavenly body and the angle at which it is being viewed.

2.3 Parallax – why does what you see vary with the viewing angle?

Parallax is the apparent change in the position of an object caused by a change in the position from which it is being observed. A heavenly body appears to be in different positions in the sky when viewed simultaneously from different points on the Earth. A simple example is seen by holding a finger up to the night sky and alternately closing each of your eyes. The position of a star can be seen to move relative to the finger because the position of the viewing eye has changed.

A larger scale celestial example is the apparent change in the perceived position of the Sun when viewed simultaneously by observers on either side of the Atlantic Ocean. At 6am on the American coast, the Sun would be low in the eastern sky. At the same moment at Greenwich in London, midday would be approaching and the Sun would be high in the southern sky. As each observation is made, the Sun is at exactly the same place in the heavens. The Sun's height

– down to the details

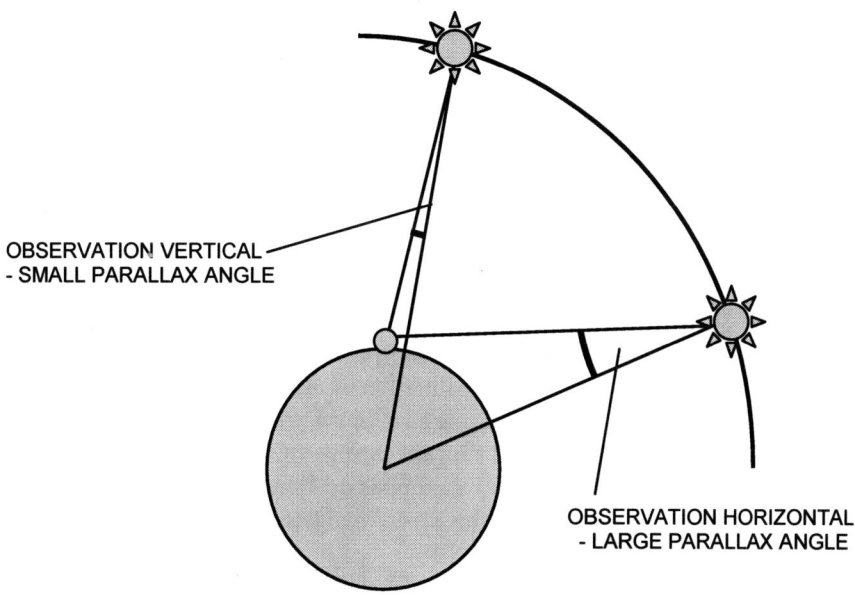

Figure 32. The size of the parallax angle depends on the observation angle

and direction just appears to be different as the observations of the object are made from different viewpoints; from different angles. Similarly and most importantly, the Sun, Moon and planets, being relatively near the Earth, would appear to be in different positions when viewed simultaneously from the surface <u>and</u> the centre of the Earth, if that were possible.

The angular difference at the observed body between straight lines from the position of the observer and one from the centre of the Earth is called the **angle of parallax**. The further away the object, then the smaller is the angle; it is a much greater angle for the Moon than for the Sun and planets and is negligible for the stars.

Parallax is measurable and it can be corrected for when the angle between the points of observation is known. (For the mathematicians amongst us the parallax angle for a body varies as the cosine of its altitude angle: as cosine 90°= 0, cosine 0° =1).

When viewing a body from Earth, the angle of parallax will be at a maximum when looking at an object horizontally and at a minimum, when the object is vertically above the observer. It is simply a matter of the relative size of the angle at the observed body, between a line from the observer and a line from the centre of the earth. In **Figure 32** two observations are being made, one almost vertical in which parallax will be small and one relatively horizontal, in which it will be considerably larger.

In practice, the stars are so very far away that even when observed simultaneously from opposite sides of the Earth, their geocentric parallax (the parallax angle from simultaneous observations from opposite sides of the Earth) is a very tiny fraction of a degree and can be safely ignored. The apparent altitude of the relatively close Sun requires a small parallax correction, whereas the Moon, being very close when compared to other bodies, requires that special Moon parallax correction tables are drawn up. Venus and Mars, also being relatively close, require a tiny correction for parallax.

2.4 *Semi-diameter* – how does the size of a body affect the taking of a sight?

Stars are enormous but so far away that they appear as simple points of light with no measurable diameter. The Sun and Moon when viewed from Earth have appreciable and remarkably similar diameters of 30 arc-minutes (30´) or so. The situation is illustrated in **Figure 33**. The exact centres of either the Sun or Moon are impossible to discern and therefore a reliable measure of their altitude cannot be made by trying to bring the centre of a sextant image to the horizon. Instead, the crisp upper or lower edge, called the upper or lower limb is used and the resulting altitude figure is then corrected to read as if the centre had been used.

The correction is made by taking the value of half the body's diameter, called its **semi-diameter**, measured in arc-minutes and adding it to lower limb observations and subtracting it from upper limb sights. It is usual to use the lower limb for the Sun and Moon, but the appearance of a less than full Moon when viewed from some latitudes, may render the lower limb invisible such that the upper limb may have to be brought to the horizon.

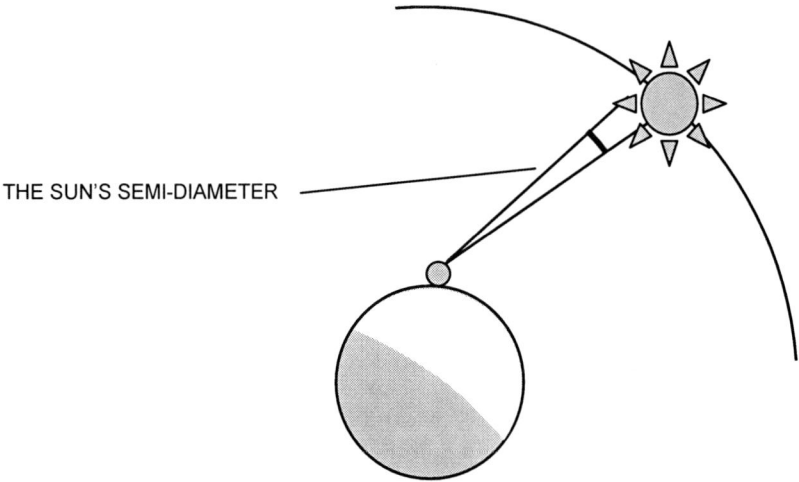

Figure 33. Showing the observed Sun's significant semi-diameter

– down to the details

The Sun's diameter varies slightly and cyclically over a year depending upon the distance between the Sun and Earth, a topic explored in **Section 6**. The Moon's diameter varies not only over a year but also during its monthly lunar cycles, considered later in **Section 7**. Semi-diameters for both Sun and Moon are listed on the daily pages of the astro-nautical almanac.

The planet Venus being relatively near also has a small but definite semi-diameter and Moon-like phases, further detailed in **Section 8**. A very small and varying correction is made depending on the position of Venus in its orbit relative to Earth and the apparent altitude angle.

2.5 *Refraction* – how does it change our perception of position?

Light travels in a perfectly straight line through the vacuum of space. When light passes through our atmosphere a distortion occurs called **refraction**, caused by variations in air density and air turbulence that produce a slight change in the speed of the light ray. This causes a small angular change in the direction of the ray. The reader will be familiar with the Earthly appearance of refraction in a still pond; light appearing to be bent as its speed is decreased on entering and increased on leaving the surface. The image of a stick poked in to the pool is seen to be bent or angled at the entry point. Light rays are similarly bent passing through our atmosphere and this introduces an observational error.

Refraction may make an observed body seem higher than its true position, a most unhelpful complication in navigation (**Figure 34**). At an observer's zenith, refraction is effectively zero and it reaches an angle of approximately 1´ of arc at an apparent altitude of 45°.

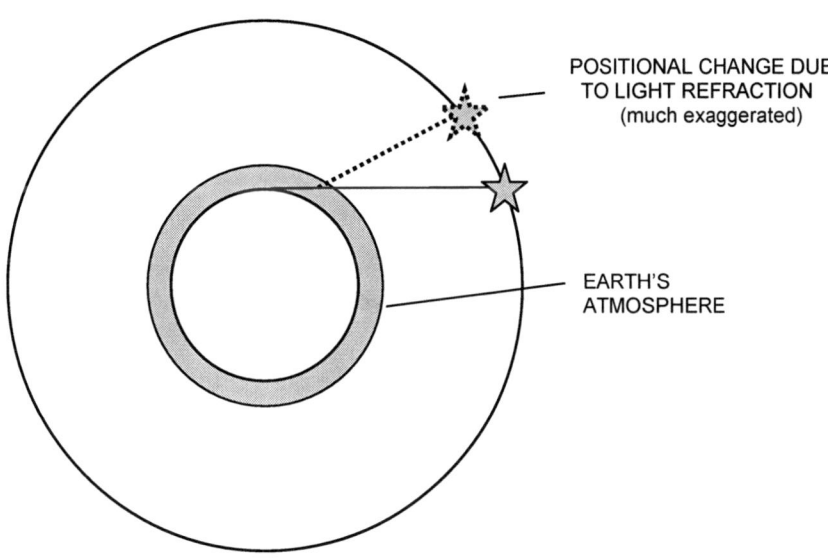

Figure 34. Refractive positional distortion due to Earth's atmosphere

Refractive distortion is greatest when bodies are observed within a few degrees of the horizon; reaching 5´ at 10° and 30´ or more on the horizon. Here light is travelling more tangentially through a greater "thickness" of atmosphere than if coming from a higher angle.

This excessive refraction at low angles makes the Sun appear higher than it really is at sunset and sunrise. The Sun's altitude is increased by half a diameter or so at these times. It is therefore standard practice not to observe any body within 10° of the horizon, unless observational conditions and circumstances make it necessary. The Nautical Almanac does provide a table for adjusting and increasing the reliability of such readings as well as tiny corrections for the effects of temperature and air pressure, but this is of little relevance to the ordinary sea-going navigator.

2.6 *Altitude corrections for parallax, semi-diameter and refraction* – all at once?

Having learned of their presence and variable effect it is comforting that any small adjustments in altitude made necessary in correcting for **parallax**, **semi-diameter** and **refraction** are usually grouped together and accounted for with a single correction for a heavenly body. This is the case for tables relevant to the Sun and the planets.

The Moon, being closest by far to Earth, has complex variations of both parallax and semi-diameter. Its horizontal parallax changes significantly in the space of a few hours and is listed in daily tables. This value when extracted is then used to enter the specific Moon-altitude correction table. There is also a special additional adjustment called **augmentation (Section 7.7 p.187)**, consequent upon the nearness of the Moon to our Earth and which varies with the apparent altitude angle. This too is included within the other corrections.

The stars, being so very far away, have a negligible diameter and no parallax effect, only requiring a correction for refraction. Further detail of these adjustments will be addressed in the relevant heavenly body-specific sections (**Sections 6-9**).

2.7 *Summarising the manipulation of a sextant sight* – sight reduction in a nutshell?

The reader has now reached an important point in that all the various adjustments have now been introduced and an indication has been given as to which adjustments are necessary for a particular heavenly body.

The **sextant altitude** (H_s) is measured from the **visible** horizon. All observations are then adjusted for **sextant error** and **angle of dip** to arrive at the **apparent altitude** (H_a), the altitude of a body above the **sensible horizon**; the level of the observer's eye. Those bodies relatively close to Earth, the Sun, Moon and planets will need additional adjustments for semi-diameter in the case of the Sun, Moon, Venus and parallax adjustments for Sun, Moon, Venus and Mars. Observations of all bodies, including the distant stars, require an adjustment for **refraction**, consequent upon light travelling through Earth's atmosphere. The student should not be unduly concerned however, as multiple adjustments are usually tabulated as a single summed figure.

The resulting adjusted altitude is that which would be achieved if the sight had been taken from the Earth's centre and is called the **observed altitude** (H_o), which by some traditionalist

– down to the details

authors (including the current Royal Yachting Association Yachtmaster Ocean course) is termed the **true altitude**. This text will continue to use the former term: **observed altitude,** which is used throughout the Admiralty-published nautical almanac and sight reduction tables. In short:-

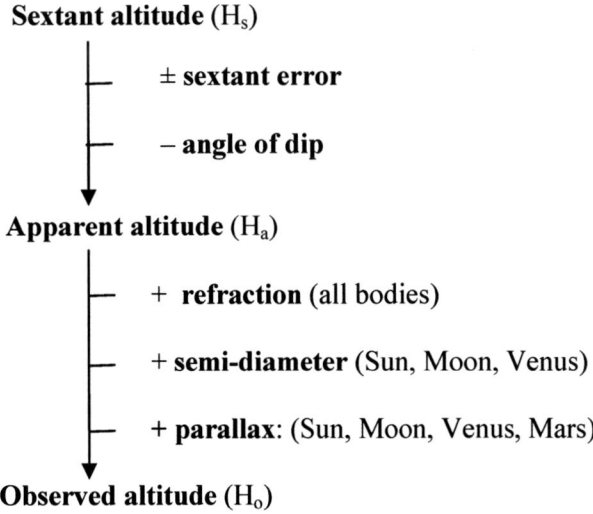

2.8 *What can you deduce from an altitude?* – a lot or a little?

The answer is simple: from an altitude is derived a circle, called a **position circle** and very little more (**Figure 35**). Having made an observation of a heavenly body, correcting it as needed for sextant error, dip, parallax, semi-diameter and refraction, our navigator needs to relate the findings to position. What does an altitude tell you about your position? An altitude can only tell you that you lie on a circle, a rather specific circle on the surface of the Earth, the diameter of which depends upon the altitude of the observed body. The higher the altitude then the smaller is the diameter of the circle.

At the centre of the circle is the **geographical position** of the body on which the observation is being made. The geographical position, as described in **Section 1.10** (p.16) is that point on the Earth's surface through which passes a straight line to the centre of the Earth from the observed body. The observer must lie on some point on the circle around the geographical position of the observed body, because only from a position somewhere on that circle will that particular altitude be found. From any and every point on the circle the altitude of the observed body will be identical.

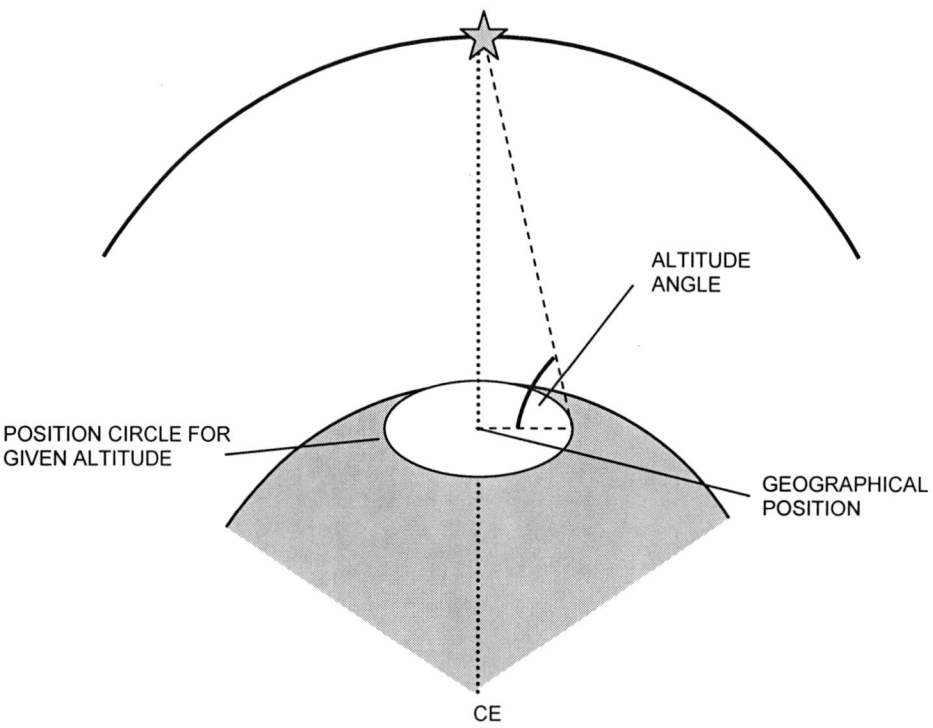

Figure 35. The position circle based on a single observation of altitude

The reader may also be able to see how by using the position circles of 2 or more bodies simultaneously, then the navigator might identify the observer's position as one of the two points at which these circles intersect one another, position 1 and position 2. This is another key point in understanding this section and is illustrated in **Figure 36**.

As one can imagine, the diameter of such position circles could be hundreds or thousands of miles and very difficult to plot on a chart where scale and accuracy are of great importance. The lower the altitude of the observed body, the larger is the circle on which the observer lies and conversely, the higher the altitude the smaller the position circle. Plainly some means of narrowing the field is necessary for a single observation to be of use to the navigator whilst remembering, that if observational conditions are difficult, then one observation must be better than none at all.

– down to the details

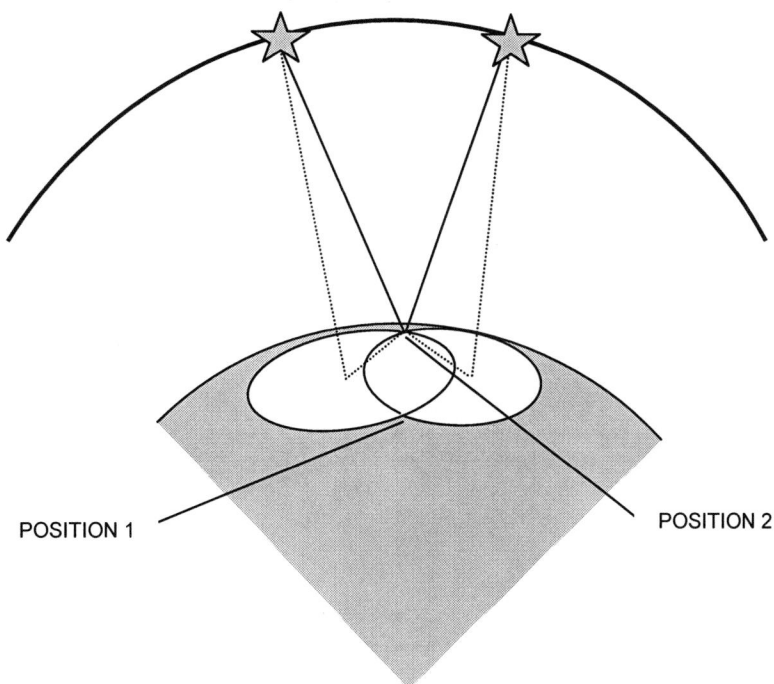

Figure 36. Two star altitude readings localize position to two possible points on position circles

2.9 *Relating an observed altitude to an observer's position* – the PZX triangle revealed

We have found that a single altitude observation places the navigator on a position circle, but more information is necessary for it to be navigationally useful. Some knowledge is necessary of **great circles** together with a little excursion into spherical trigonometry, as it is important for the navigator to determine his position on that circle.

A **great circle** of a sphere has its centre coincident with that of the sphere and was introduced in **Section 1.2** (p.4), the two usual earthly examples being the equator and a circle running through both poles. When three **great circles** intersect, they form on the surface of the sphere, a **spherical triangle**; a triangle with curved sides as in **Figure 37**.

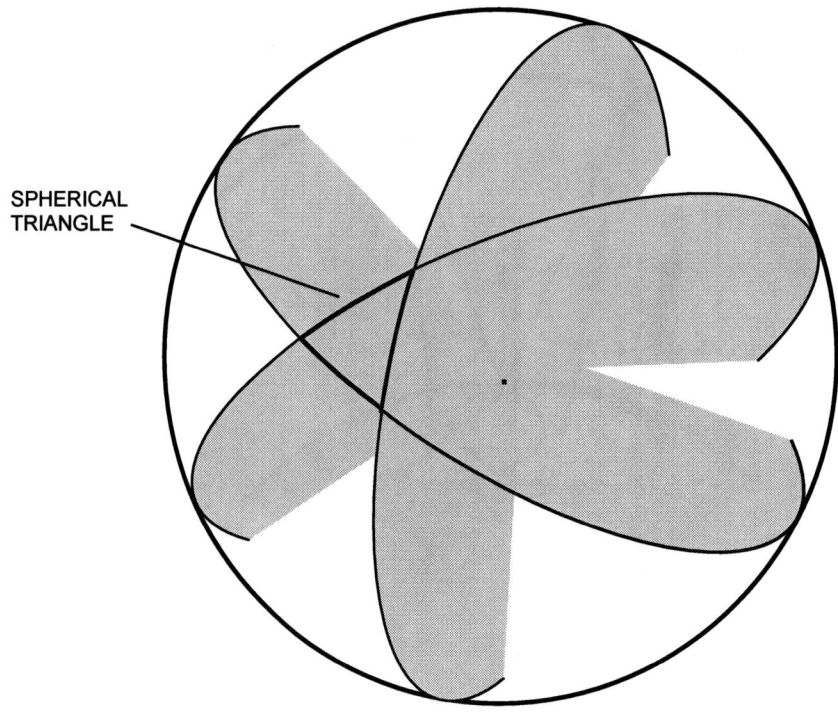

Figure 37. A spherical triangle formed by the intersections of 3 great circles

In **Section 1.18** (p.31) a somewhat simplified means of finding latitude by the meridian passage of the Sun was introduced. For one brief moment, the Sun lies on the meridian of the observer, exactly on the North-South axis. The Sun's geographical position on the Earth and the observer both, for that brief moment, lie on the same great circle, whose circumference also passes through the poles.

Whilst it is possible to take meridian sights of all heavenly bodies, most sights are recorded when the observed body lies on a great circle other than that of the observer's meridian. Fortunately, mathematical means allow one circle to be very usefully related to another.

As well as on the Earth, great circles also exist on the **celestial sphere**, the **celestial equator** being one. An observer's **zenith** (Z) and the North and South celestial poles, all lie on another great circle. An observed heavenly body (X) also lies on one great circle with both celestial poles and a further great circle is formed with the zenith of its observer. These four great circles on the celestial sphere are illustrated in **Figure 38**.

– down to the details

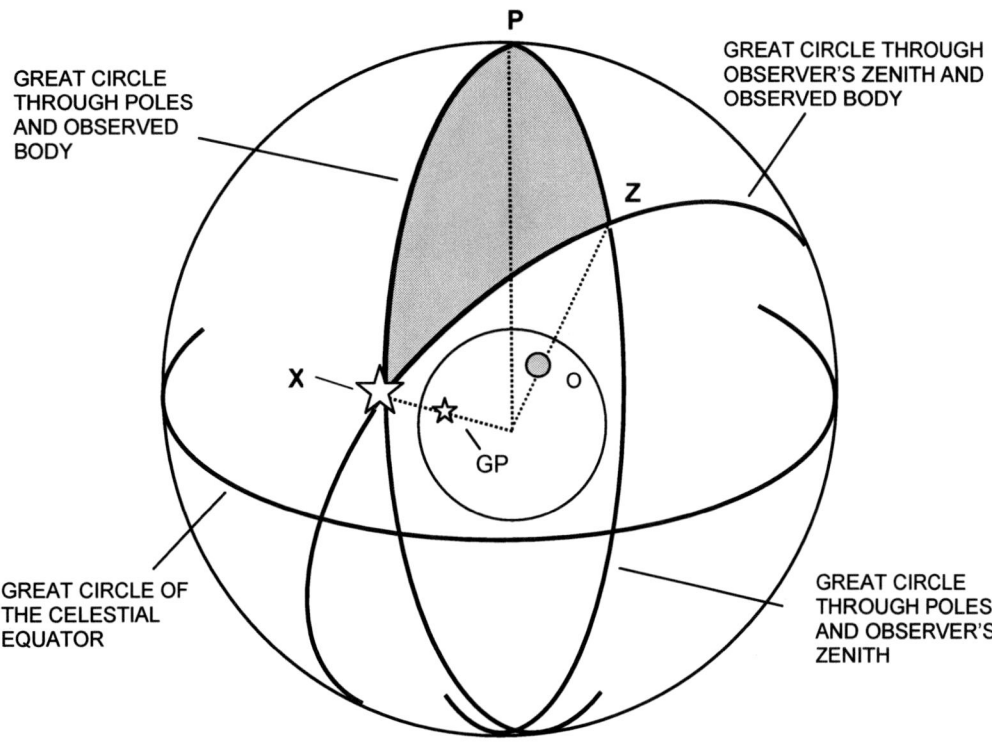

Figure 38. Four great circles and the PZX triangle on the celestial sphere

Given that the Earth and the Celestial Sphere have a common centre, then points on both spheres can have useful relationships. For example, the angle at the Earth's centre between the observer's position (O) and the North Pole of the Earth is the same as that between the observer's zenith (Z) and the North Celestial pole. Similarly, the angle at the Earth's centre between the observed body's geographical position and the North Pole is one and the same as that between the observed body and the North Celestial Pole (P).

A fifth great circle can also be drawn, that of the celestial horizon related to the position of the observer and from which would be derived the **observed altitude** of the heavenly body viewed from the observer's position. The plane of this circle lies at right angles to the line drawn from the Earth's centre through the observer's position and up to his zenith and consequently at a right angle to the great circle passing through the observer's zenith and the observed body, as in **Figure 39**.

49

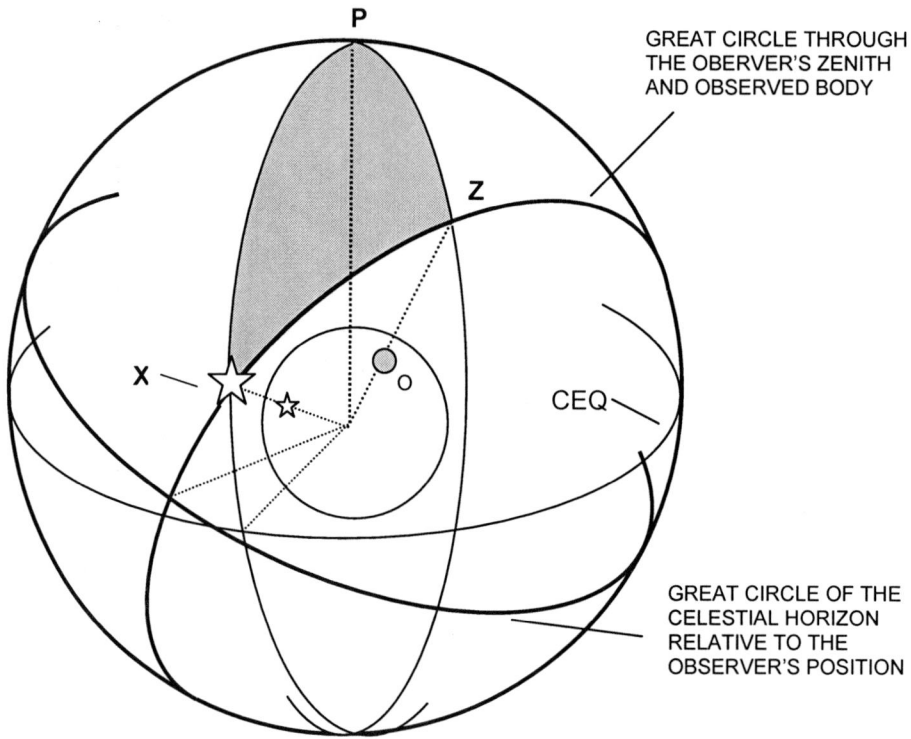

Figure 39. The great circle of the celestial horizon for the observer's position

From this maze of circles one can dissect what has become known as the **PZX triangle**, its three sides formed by the intersection of three main great circles. The first side **PZ** links the observer's zenith to the Pole, the second side **PX** links the observed body to the Pole and the third side **ZX** links the observer's zenith to the observed body. Being a true spherical triangle its vertices (corners) are connected to the Earth's centre by three straight lines of equal length, each being a radius of the celestial sphere.

The PZX triangle is drawn in **Figure 40**, extracted form the maze of circles and where two of the three radii are necessarily foreshortened by perspective.

– *down to the details*

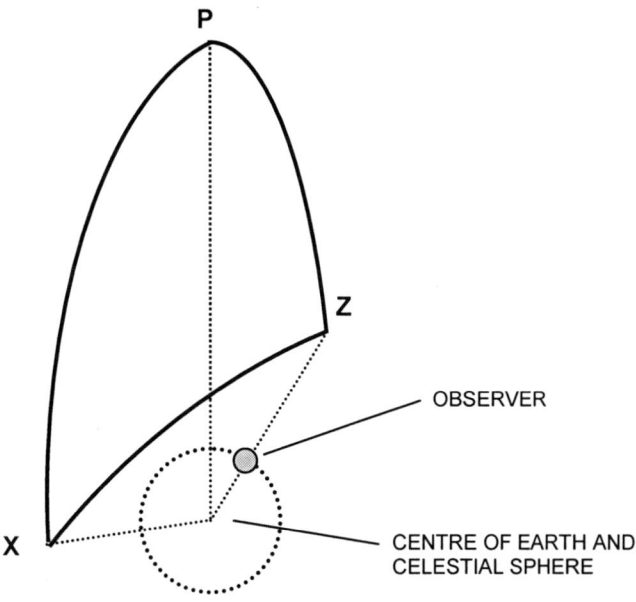

Figure 40. The PZX spherical triangle extracted from Figure 38

Each of the three sides or arcs of the PZX triangle can be extended so that by joining their extremities, right angles are formed at the Earth's centre (**Figure 41**). This is a factor vital in ultimately finding the angular length of each arc.

Arc PZ, called the **co-latitude**, can be extended from Z to the celestial equator so that its arc length is 90°. As the arc length from Z to the celestial equator is equal to the latitude of the observer in the northern hemisphere, then the arc length PZ, the co-latitude, must equal 90° – latitude of an observer in the northern hemisphere (or 90° + latitude South, if the observer is in the southern hemisphere).

In a similar way, the arc PX, called the **polar distance**, from the celestial pole to X the heavenly body, can also be extended to the celestial equator, so its arc length is also 90°. The arc length from X to the celestial equator is the declination of X, so it follows that the arc length PX = 90° – declination of a body with a northern declination (or 90° + declination South, if the observed body lies in the southern hemisphere).

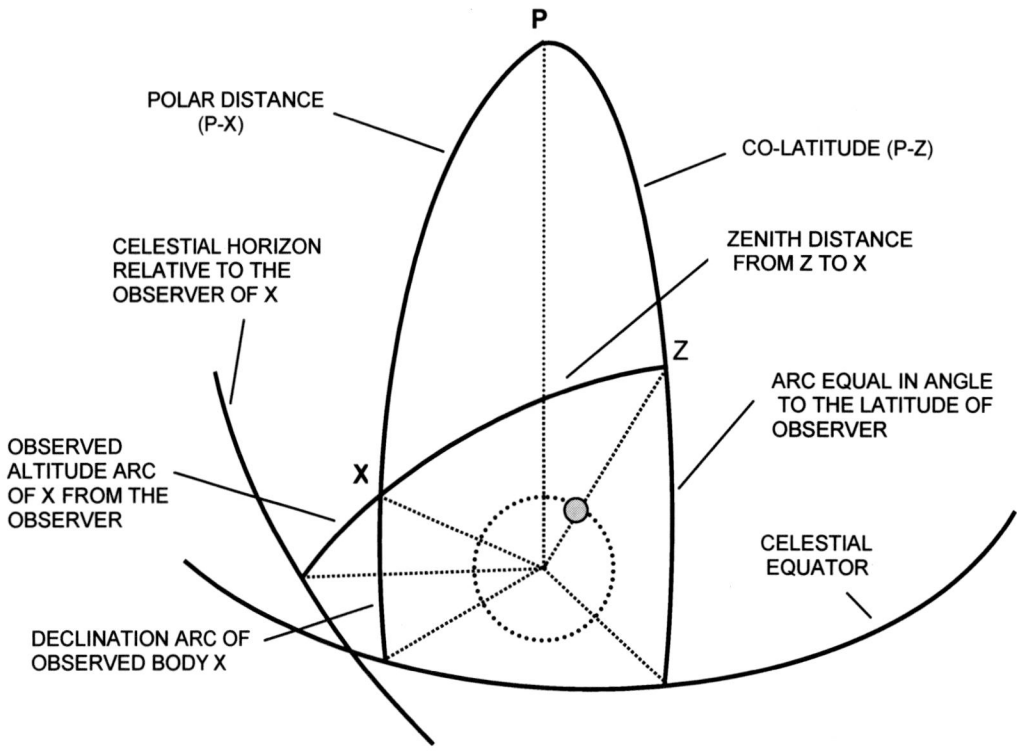

Figure 41. The extended PZX spherical triangle – with the celestial horizon

Lastly, Arc ZX from the observer's zenith (Z) to the observed body X, called the **zenith distance**, can also be extended until it meets the celestial horizon (not the celestial equator) of the observer. The arc length from X to the celestial horizon is the observed altitude of X above the celestial horizon, so the arc length ZX, the zenith distance, must equal 90° − observed altitude.

In this way, the three sides of the PZX triangle have been defined in terms that we recognise. If the observed body X is a star, its declination can be looked up in a star table (as the declination of a star does not significantly change (**Section 9.14** p.252) but if X is the Sun, the Moon or a planet, declination for the time of the observation can be calculated from tabled information in the astro-navigational almanac. Altitude can be measured with a sextant and corrected to give the observed altitude. Latitude can by assessed from the meridian passage of the Sun as introduced in **Section 1.18** (p.31), to be explored in detail in **Section 6.9** (p.152), or estimated using the star *Polaris* (**Section 9.18** p.269).

– down to the details

2.10 *Local Hour Angle and the PZX triangle* - how does one affect the other?

In **Section 1.6** (p.10) the concept of **hour angle** was introduced as the means of measuring **celestial longitude**, which is properly called **Greenwich hour angle** (GHA). It is the angular distance, measured at the North celestial pole, in degrees West of the Greenwich Meridian. GHA is used to position the heavenly bodies that move across the celestial sphere: the Sun, Moon and visible planets. The fixed stars are positioned in relation to "Aries". The GHA for the Sun, Moon, planets and Aries are tabulated for every hour of each day in the astro-nautical almanac with additional tables of adjustment for minutes and seconds.

The PZX triangle has as its uppermost polar angle (P) the **local hour angle** (**LHA**) of the observed body, measured in degrees <u>West</u> of the observer. It is the angle between arcs <u>PZ</u> and <u>PX</u> in **Figure 42**.

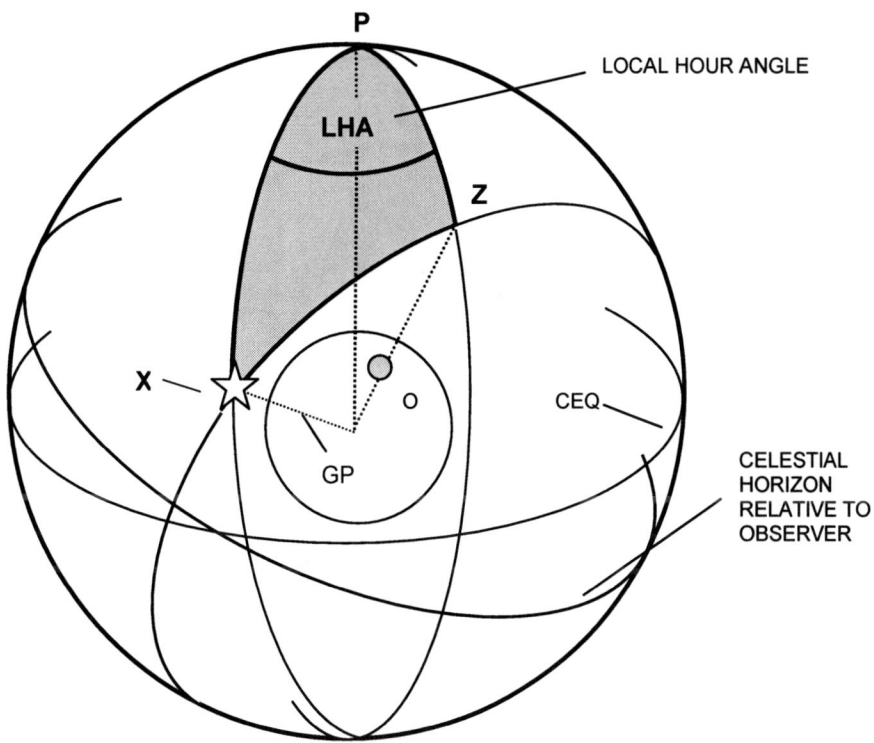

Figure 42. Local Hour Angle is the angle between arcs PZ and PX

Astro-navigation from Square One

The Local Hour Angle must be accurately determined from the observer's longitude, always taking great care, subtracting westerly longitude from GHA and adding easterly longitude to GHA to give the **local hour angle** (LHA). Remember that the value of the LHA ranges from 0-360°, always measured to the West of the observer.

2.11 *Azimuth Angle and the PZX triangle* – how does azimuth angle change?

The second important angle in the PZX triangle is that made at the observer's zenith (Z) on the celestial sphere, between the meridian on which observer and the observer's zenith lie and the great circle on which both the zenith and the observed body lie. In short it is the angle between PZ and ZX.

This angle is called the **azimuth angle**, the word azimuth coming from the Arabic: *as sumut*, meaning "the direction", which is an explanation in itself. The azimuth angle measures the direction of the observed body (X) from the zenith of the observer (Z) as in **Figure 43**. The **azimuth angle** is often given the abbreviation Z.

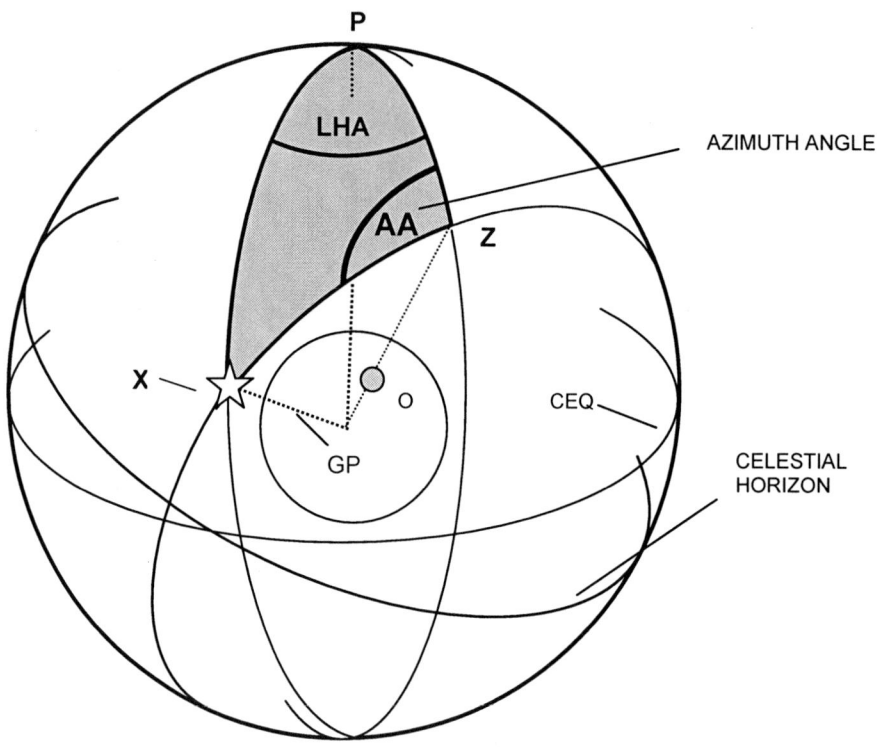

Figure 43. Azimuth Angle is at the zenith, between arcs PZ and ZX

– down to the details

2.12 *Finding the Azimuth from Azimuth Angle* - what is the difference?

On the Earth's surface, directly below the observer's zenith, the <u>bearing</u> of the geographical point of the observed body from the observer can be plotted. This bearing, measured in degrees true <u>from</u> <u>North</u> is called the **azimuth** (often abbreviated to Z_n). In the northern hemisphere, if the LHA is less than 180° then the azimuth = 360° – azimuth angle. If the LHA is more than 180° then the azimuth angle = the azimuth.

If the observer is in the southern hemisphere, then if the LHA is less than 180°, the azimuth = 180° + azimuth angle and if LHA is greater than 180° then the azimuth = 180° – azimuth angle. The two diagrams in **Figure 44** will help to clarify things in the northern hemisphere. Note carefully the relative positions of the observer's zenith (Z) and the observed body (X) as the orientation of the PZX triangle alters when Local Hour Angle exceeds 180°. This potential complication arises because LHA is always measured to the West of the observer from 0-360° and azimuth is measured in degrees true from 0-360°.

Do not despair at these seemingly complicated terms and their confusing arrangements, all will become much more clear in the body-specific sections (**Sections 6-9**), where worked examples are shown.

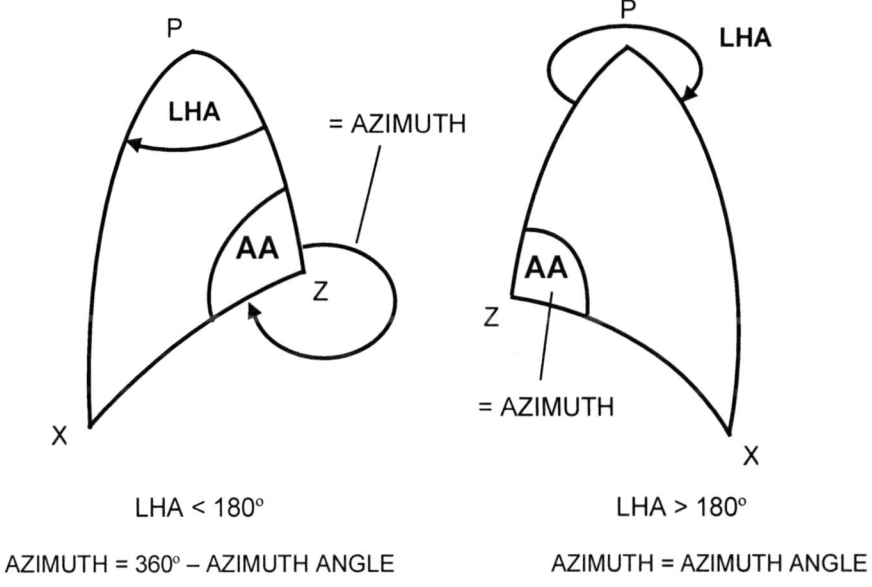

Figure 44. The relationship between azimuth angle, azimuth and local hour angle

2.13 *Summarizing the PZX Triangle* – what information does it give you?

The PZX triangle is a **spherical triangle** drawn of the surface of the celestial sphere. The sides of the triangle are curved lines, parts of great circles on the celestial sphere and all great circles, as their centres are coincident with the centre of both the Earth and celestial sphere, have the same radius. Points P, Z and X are therefore equidistant from the centre of the Earth.

The angular length of the arc PZ is 90° ± **latitude** of the observer (– if the observer lies in the northern hemisphere and + if the observer is in the southern hemisphere). The arc length of the side PX is 90° ± the **declination** of the observed body (– if the body lays the northern hemisphere, a northerly declination and + if it is in the southern hemisphere, having a southerly declination). The arc length of the arc ZX, the **zenith distance**, is 90°– **altitude** (H_o) of the observed body from the observer.

The angle at the Pole P is the **Local Hour Angle** of the observed body and is found from the longitude of the observer and the Greenwich Hour Angle of the observed body. The important angle at Z, the observer's zenith is the **azimuth angle**, which, according to the magnitude of the LHA, gives the direction or **azimuth** (Z_n) in degrees true, of the geographical position of the observed body on the Earth's surface. A great deal is therefore known, or can be calculated, regarding the arcs and angles of this triangle.

2.14 The difficulty with Zenith Distance and Azimuth – how useful are they?

From an observed altitude, a zenith distance arc angle can be worked out (zenith distance = 90° – observed altitude). This arc angle can then be converted to a true distance on the Earth's surface and is the distance between the observer and the geographical position of the observed body. As we have learned from **Section 1.3**, one minute of arc at the Earth's centre equals 1 nautical mile in distance on the surface, so that a definite distance could be worked out.

The azimuth angle, together with the local hour angle of the observed body, permit the azimuth to be calculated, so distance and bearing are known. We therefore have, akin to fixing a position in coastal navigation, a bearing and "distance off" and should be able to accurately pin-point our position on the Earth from these two measurements.

There are considerable problems however, similar to those of the position circle detailed in **Section 2.8**, (p 45), because of the potentially large magnitude of the distance involved and the difficulty in accurately plotting the azimuth. Let us illustrate this by substituting some very simple figures into the equation; zenith distance = 90° – observed altitude.

If the observed altitude (H_o) was found to be 50° then the zenith distance arc = 40° (90 – 50 = 40). By converting this arc angle to a distance on Earth, puts the observer some 2400 nautical miles from the geographical position of the observed body (40° = 40 x 60´ = 2400 nautical miles). Even if you had a chart with a scale such as to allow you to plot this distance, the accuracy in plotting your position would be unacceptably poor and any small angular

– down to the details

discrepancy horribly magnified, rendering the exercise useless.

But do not despair! A solution is at hand for the observer's hard-won measurements and one for which we must thank Captain (later Admiral) Marcq St Hilaire of the French Navy. In 1875 he introduced his method of finding a ship's position, which British navigators subsequently called the **intercept method**[2.1]. This compares a **calculated altitude** with the **observed altitude** and uses the difference between the two, together with a **calculated azimuth** (direction of the observed body) to produce a position line. It is an incredibly simple but quite brilliant concept. Then, using two or more simultaneous observations, position line intersections can be arrived at to produce an accurate positional fix.

2.15 *Calculated altitude* – how and why do you work it out?

We accept that a sextant altitude (H_s) for a body can be corrected to an observed altitude (H_o) at a moment in time, but it is also possible to work out a **calculated altitude** (H_c) for the same body at the same time, from a **chosen** or **assumed position**. Such a position may be arrived at by dead reckoning and called a **dr position**; a position based on time and distance run, or an **estimated position** which includes allowances for tidal streams, currents and leeway.

It might also be chosen purely for the numerical convenience of whole numbers of latitude and longitude or LHA. The actual calculation is complex and involves spherical trigonometry, but it enables the PZX triangle to be solved to give the calculated altitude and azimuth of an observed body from the chosen position.

The navigator can (and should) learn to do the mathematics "longhand" using especially designed tables (**Section 4.7** p.96). A standard scientific calculator or computer program using specific software can help to do the necessary sums. My own preference is that the navigator really ought to learn to do the sums longhand using a set of "log tables", so that on that fateful day when everything else fails, the finding of position is still possible. It only takes a few minutes after a little practice and is surprisingly rewarding. A reasonable compromise however is that the navigator learns to do the mathematics with a calculator, so as to survive that fateful day when the boat electrics fail. The basic concepts of spherical trigonometry are relatively straightforward and a brief explanation here is helpful. A detailed exposition can be seen in the **Appendix** p.277.

2.16 Explaining spherical trigonometry - is it possible to grasp?

The word trigonometry comes from the Greek: *trigona* meaning 3 angles and *metron* meaning measure. Remember that a spherical triangle is formed on the surface of a sphere by the intersection of three great circles and all great circles (such as meridians of longitude) have their centres at the centre of the Earth. Given a spherical triangle with sides **a**, **b** and **c** and with angles **A**, **B** and **C**, as in **Figure 45**, the two fundamental points to understand are these:

1. If the length of two sides of the spherical triangle and the angle between them are known,

then the length of the third side can be calculated.

2. If the lengths of the three sides of the spherical triangle are known, then any of the angles contained within the triangle can be calculated.

Put simply, the angle opposite to any one side must be proportional in some way to that side and as the angle between two sides of the triangle increases, the length of the third side must also increase, again in some proportional way. In addition, as one angle increases in size, the others will correspondingly change.

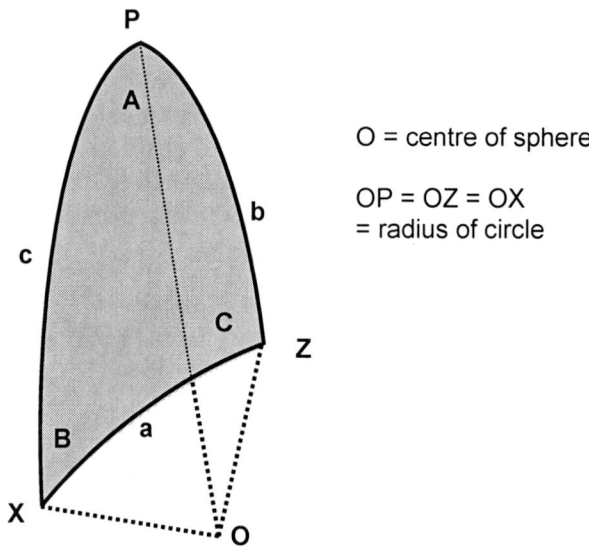

Figure 45. Describing a typical spherical triangle

As with the plane geometry of a flat 2-dimensional right-angled triangle, the proportional relationship of an angle and the opposite side is not a direct one, i.e. if one doubles the other does <u>not</u> necessarily double and the maths governing the relationships are a little complicated. But if you, the reader, is content to accept for the moment, the fundamental points 1 and 2 above as facts, then you will be accepting the following equations which transform those statements into the language of mathematics.

$$\cos a = \sin b \sin c \cos A + \cos b \cos c \quad \text{(where a is the unknown side)}$$

$$\cos A = \frac{\cos a - \cos b \cos c}{\sin b \sin c} \quad \text{(where A is the unknown angle)}$$

The terms *cos* and *sin* are abbreviations of cosine and sine. Both are mathematically derived

— down to the details

ratios used to relate the sides to the angles of triangles. For non-mathematicians (this includes the author), the **Appendix** outlines a step-by-step approach to basic plane and spherical trigonometry in the context of navigation. My own experience finds that astro-navigation texts are prone either to delve deeply into complex and seemingly impenetrable mathematics, or to skate across the surface, blindly accepting an inevitable incomprehension. I have attempted to bridge the gap between these approaches.

In any event, <u>accept for now</u> that a **calculated altitude** from a given position is arrived at by solving the PZX triangle with trigonometry and that this will prove to be immensely useful.

Substituting the astro-navigational terms into **Figure 45**: side a = 90 – calculated altitude, side b = 90° – latitude and side c = 90° – declination. Angle A = local hour angle and angle C = azimuth angle. The spherical triangle then appears as in **Figure 46**.

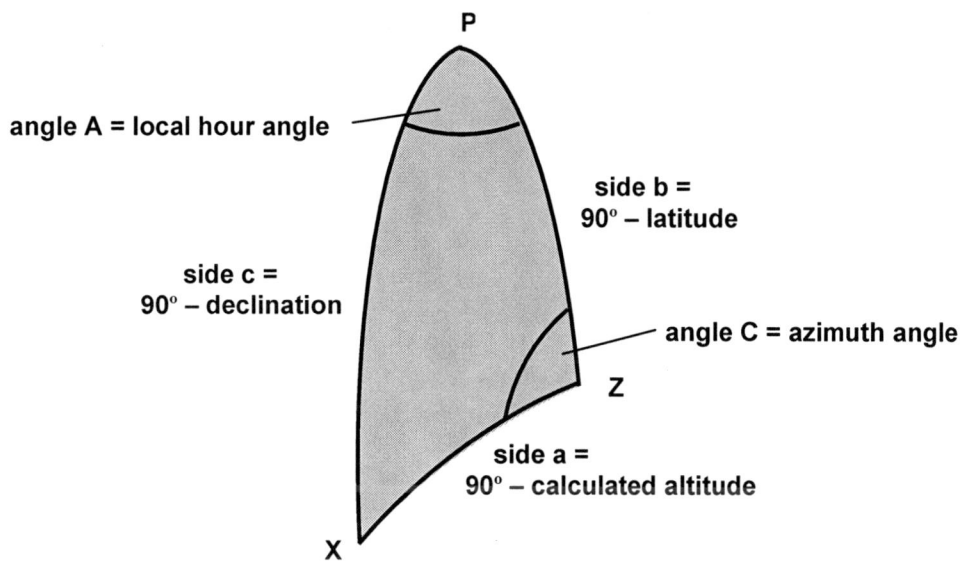

Figure 46 Substituting into the spherical triangle

Finding the value for calculated altitude is the goal. The LHA is known (found from GHA and chosen longitude), as are latitude (estimated from the observer's likely position) and declination of the observed body (found in the astro-nautical almanac). Thus we have details of two sides and the angle between them. According to the rules governing spherical triangles we can then calculate the length of the third side and from it the altitude by subtracting it from 90°. Putting some very simple figures into the navigational equation will help clarify things. Finding side **<u>a</u>** is our goal which we know is equals 90°– calculated altitude. Using the equation outlined above, a straight-forward example will help to clarify the theory.

Example 4. Find the calculated altitude of the Sun at a chosen position: 40° North, 60° East, at 1200 GMT/UT, when the declination of the Sun at the time of the observation is exactly 20° North.

side b = 90° − latitude = 90° − 40° = 50°. *cos* 50° = 0.64   *sin* b = *sin* 50° = 0.76

side c = 90° − declination = 90° − 20° = 70°. *cos* 70° = 0.34   *sin* c = sin 70° = 0.93

local hour angle = GHA + Longitude East = 00° + 60° = 60° and *cos* 60° = 0.5

Substituting these values in: *cos* a = *cos* b *cos* c + *sin* b *sin* c *cos* A:

cos a = [*cos* (90° − lat) *cos* (90° − dec)]+ [*sin* (90° − lat) *sin*(90° − dec) *cos* 60°]

cos a = [*cos* (90° − 40°) *cos* (90° − 20°] + [*sin* (90° − 40°)] *sin* (90° − 20°) *cos* 60°]

cos a =   [*cos* 50°   x   *cos* 70°]   +   [*sin* 50° x *sin* 70° x *cos* 60°]

cos a =   [0.64 x 0.34]   +   [0.76 x 0.93 x 0.5]

cos a =   0.22   +   0.35

cos a =   0.57

angular length of side a = *arccos* 0.57 = 55°

Remember the sides of our spherical triangle are measured in angular lengths, also called arcs or arc-lengths. Since side a = 90° − altitude, then the Sun's **calculated altitude** will be:

$$90 - a = 90 - 55 = \mathbf{35°}$$

We have calculated the <u>altitude</u> for the Sun at 1200 GMT/UT at position 40°N and 60°E.

I emphasize that I have deliberately used simple figures. I do not wish at this stage, for mathematical matters to complicate the explanation of the navigational principles. But using real values is no more difficult than carefully managing a few decimal places.

2.17 *Comparing calculated and observed altitude* – why are they compared?

Having taken a sight of a heavenly body, correcting it to obtain an <u>observed altitude</u> and then, by working out a <u>calculated altitude</u> for the same body from a chosen position near your true (but unknown) position, one arrives at <u>two altitude angles</u>. They will be similar in size but not quite identical, unless the calculated and observed altitudes just happen by some highly unlikely chance to have their origins at exactly the same position.

– down to the details

Usually, the observer's chosen position is a little way from the observer's actual position and the two altitudes will be a small number of arc-minutes in difference. As described in paragraph **2.8**, the altitude angle gives rise to a **position circle**; that theoretical circle on which the observer must lie to have observed the body at the given altitude. Both the observed and calculated altitudes each give rise to a position circle, an observed and a calculated position circle (**Figure 47**).

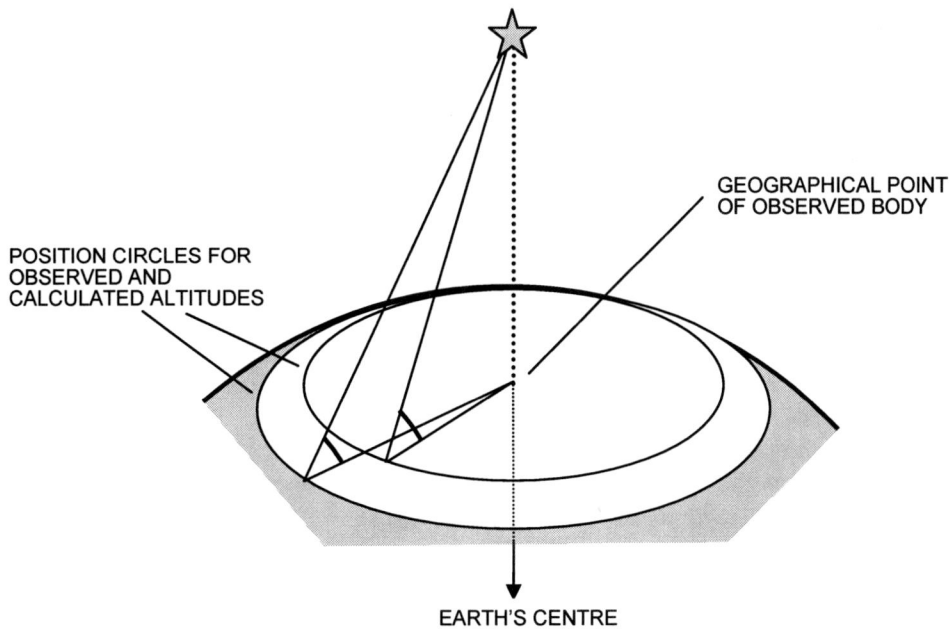

Figure 47. Position circles for a body resulting from observed and calculated altitudes

As both circles relate to the same observed body they will be concentric, their centres lying at the geographical position of the body on the surface of the Earth's sphere. One circle will be slightly larger than the other and the identity of each depends on whether the actual or the assumed observer's position is closer to the observed body. This is a key point.

Put simply, if the sextant-measured and corrected observed altitude is larger than the calculated altitude, then the sextant measured altitude will give rise to the inner of the two circles and vice versa. This is because the angle that is slightly larger must have its circle of origin fractionally nearer to the geographical point of the observed body, which itself is the centre point of the circles.

2.18 *Working out the azimuth* - to give direction to the navigator's work?

As noted in paragraph **2.8**, to have a position simply somewhere on a circle that may be thousands of miles in diameter needs a little refining to be of any navigational use! Fortunately the **azimuth** for the body, its bearing from North in 360° notation, provides the necessary localization on the position circles. An azimuth from any given latitude to a body can be calculated if the body's local hour angle and declination are known.

There are various ways to arrive at a calculated azimuth. A mathematical equation can be used such as the cosine formula as introduced in **Section 2.16**, suitably arranged to solve for the azimuth angle. Alternatively, a formula can be used such as the **ABC formula,** or specifically designed tables, called **ABC tables** are another alternative. The latter can be found in some astro-navigational publications such as *Norie's Nautical Tables* and *Reeds Astro Navigation Tables.*

The tables and formulae require the input of the observer's latitude and both the local hour angle and declination of the observed body. ABC Tables and formulae are explored further in **Sections 4.8** (p.101) and **4.14** (p.110). **Interpolation** may be necessary for LHAs other than whole numbers and for latitude or declination values other than those listed. Because of these limitations, the resulting figures may be less accurate than those derived using a mathematical formula.

The obtained azimuth, when applied to the position circles of **calculated** and **observed altitudes**, will localize the observer to a particular point on the position circles and is therefore of great importance to the navigator (**Figure 48**).

Given that the observed body lays so far away, the infinitesimally small difference in angle that the azimuth would have if drawn from the true rather than the assumed position can be safely ignored and would not be plottable in any event.

2.19 *Converting a quadrantal notation azimuth to 360° notation* – more spin?

The Azimuth figure, that is the angle that the observed body bears from the observer is, slightly awkwardly, not in degrees from true North but in "quadrantal notation" and needs to be converted to a true bearing before it can be plotted on a conventional chart. This is the case whether the azimuth is found by use of the cosine formula, **ABC tables, ABC formulae** or sight reduction tables.

– down to the details

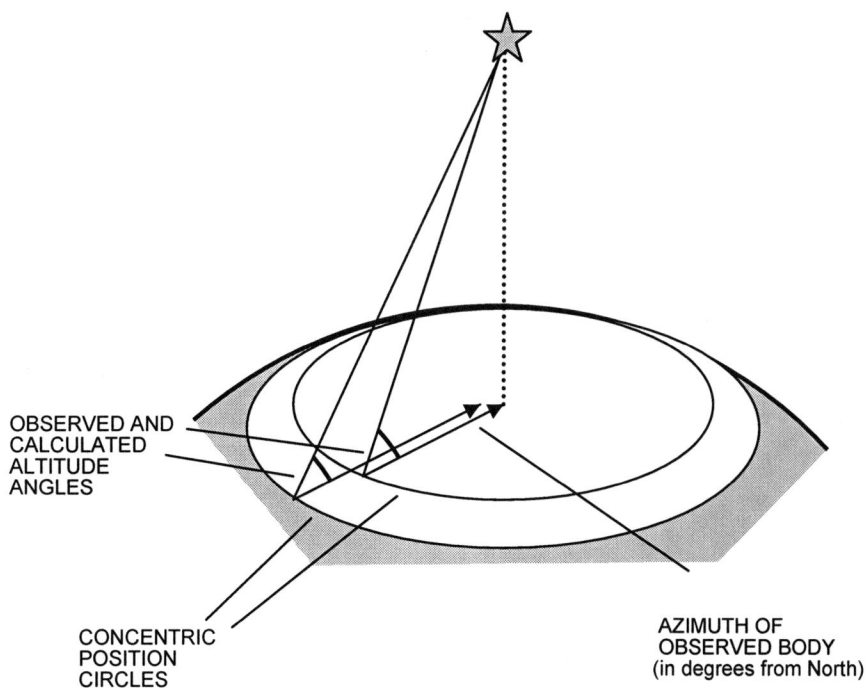

Figure 48. Localising position circle altitudes by the observed body's azimuth

Quadrantal notation is an all but obsolete way of describing the compass card in terms of degrees East or West of North and South e.g. North 20 degrees East is the equivalent of 020° in the more familiar 360° notation. Similarly, South 80° West would be the equivalent of 260° (180° + 80° = 260°). General rules for conversion from quadrantal to 360° notation are:

$$
\begin{aligned}
\text{North Azimuth East} &= 000° + \text{azimuth} \\
\text{North Azimuth West} &= 360° - \text{azimuth} \\
\text{South Azimuth East} &= 180° - \text{azimuth} \\
\text{South Azimuth West} &= 180° + \text{azimuth}
\end{aligned}
$$

Further details are given in **Section 4.12** (p.105) and the body-specific **Sections 6-9**. For the present, do not let the seemingly complex azimuth calculations cloud the simple fact that it merely gives the direction to the observed body from the position at which the calculated altitude was found.

2.20 *The relative sizes of observed and calculated altitudes* – why is this important?

Converting the 3-dimensional view in **Figure 48** to a more 2-dimensional one in **Figure 49** helps to clarify of the relationship between and the identity of the angles of observed and calculated altitude.

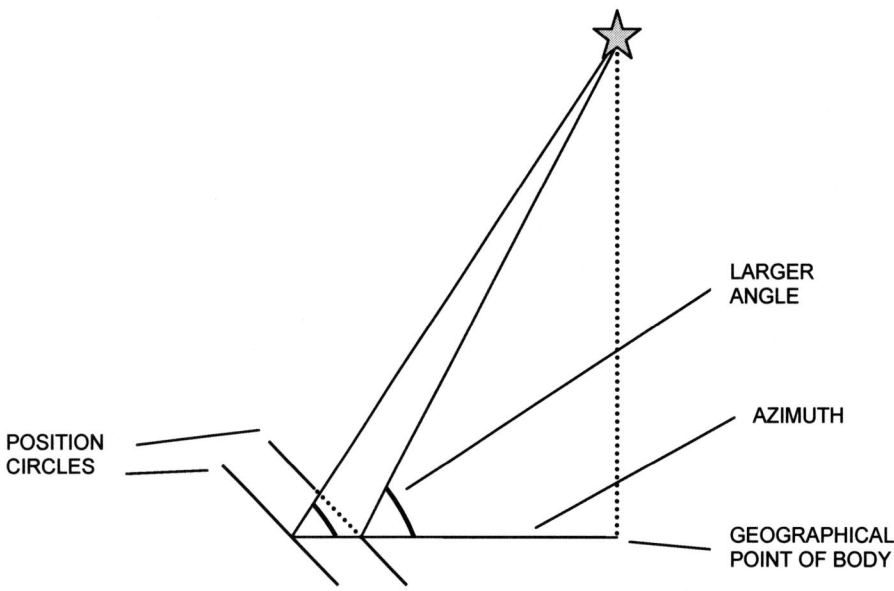

Figure 49. Identifying the observed and calculated position circles

If the calculated altitude angle is larger than that of the observed altitude then the position from which the calculated arises would be sited on the inner of the two position circles, nearer to the geographical position of the observed body. If the situation is reversed and the calculated altitude angle is smaller than that of the observed altitude, then the position for which the calculated altitude was derived would lie on the outer of the two position circles, further away from the geographical position of the observed body.

It is simply a matter of which angle is larger. A larger angle must arise from a point of origin nearer to the geographical position of the observed body.

2.21 The distance between the actual and assumed positions – how is it found?

The angular difference between the observed and calculated altitudes, when converted to arc length (1′ = 1 nautical mile) gives the distance between the true and assumed positions. Why is this so? The proof can be based on simple geometry. In **Figure 50**, a line **(a-b)** is drawn

– down to the details

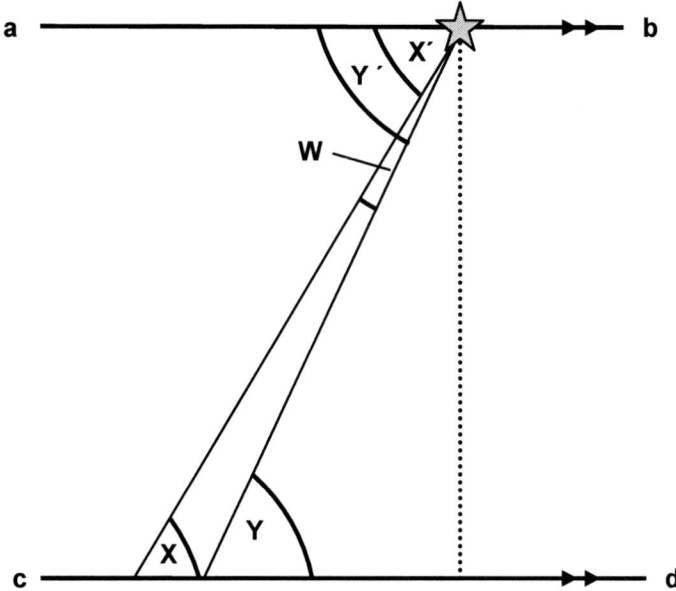

Figure 50. Proving the difference between calculated and observed altitudes

through the observed body parallel to the azimuth (**c-d**). Angles **X** and **Y** have corresponding angles **X′** and **Y′**. Angle **W** represents the angular difference between the altitudes at the observed body, **Y′−X′=W**.

Because lines **a-b** and **c-d** are parallel, then **X′= X** and **Y′= Y** so that **Y − X = W**, proving that the angle subtended at the observed body equals the difference between the calculated and observed altitudes. In actual fact lines **a-b** and **c-d** are curved; they are respectively, small arcs of the concentric circles of the celestial sphere and the Earth's surface, but a similar proof would pertain.

Simply converting angle **W**, to arc length (1 nautical mile = 1′) allows the distance on the Earth's surface to be identified, between the assumed position from which the calculated altitude was computed and that of the unknown point from which the observed altitude must have been measured.

Angle W at the observed body and the length of the base of the triangle are both incredibly small in relation to the actual lengths of the sides of the triangle which are huge in comparison. The triangle is in effect an incredibly tall and very narrow-based isosceles triangle, both long sides being effectively the same length and so the angle between them simply converts from arc to distance.

2.22 *The intercept* - what is it and how is it plotted?

We have a **calculated altitude** (H_c) and **azimuth** (Z_n) for a heavenly body from a chosen or assumed position and an observed altitude (H_o) for that body by taking and correcting a sextant altitude (H_s) from the observer's unknown position. What do we do with this information? The chosen position is first marked on the chart as in **Figure 51**.

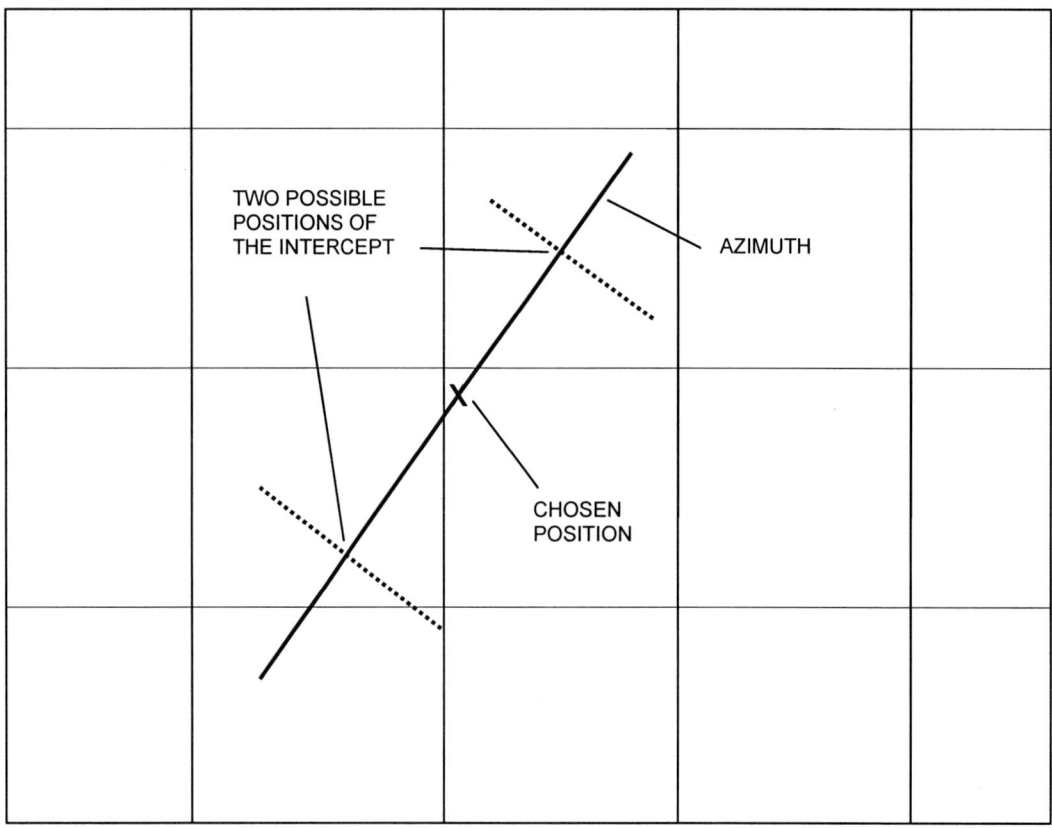

Figure 51. Placing chosen position, azimuth and intercept on the chart

The azimuth (Z_n) of the observed body is marked through the chosen position. The difference between the calculated and observed altitudes for the body in question (angle **W** converted to nautical miles) is measured out with dividers on the latitude scale at the side of the chart. This distance is then laid off along the line of the azimuth of the observed body from the chosen position. This point is the **intercept**.

But which side of the chosen position is the intercept placed I hear you cry? If the observed altitude angle (H_o) is <u>larger</u> than the calculated altitude (H_c), then the position from which the altitude is measured by sextant, must have been sited on the inner position circle and the

– down to the details

intercept therefore lies nearer (towards) the observed body (OLC-T = Observed Larger than Calculated: Towards).

Conversely, if the observed altitude angle is smaller than the calculated altitude then the true position from which it was taken must have been sited on the outer position circle so that the intercept is in the direction away from the observed body (CLO-A = Calculated Larger than Observed: Away).

At the intercept, a line is drawn at right angles to the azimuth. This line is the **position line** and the observer was positioned somewhere on that line when the altitude was measured. It is the point nearest to the actual position of the observation that the techniques used can produce.

The line is in effect, a very small part of the position circle based on the angle of the observed altitude, but so small a part of such a large circle that it is drawn as a straight and not a curved line. Below are worked examples of how the intercept is calculated and drawn for angles of calculated altitude both larger and smaller than the observed altitude.

Example 5:

Calculate the intercept for the following: A Sun sight taken at 1115 GMT/UT which when corrected, gave an observed altitude of 53° 47.5′. A calculated altitude at the dead reckoning position (the assumed or chosen position) of 39° 54.00′ N, 08° 25′ E, for the same time, gave the Sun an altitude of 53° 44.8′. The Sun's azimuth was calculated to be 155°.

The observed altitude is larger than calculated, so that the true position from which the observation was made lies on the inner of the two possible position circles, nearer to the geographical position of the Sun when the sight was taken (remember OLC-T i.e. towards). The difference between the two altitudes is 53° 47.5′ - 53° 44.8′ = 2.7′ (**Figure 52**).

The length of the line to the intercept is calculated simply by taking the difference between the two angles, which equals the angle subtended at the observed body by lines from the assumed and actual positions. This is then converted to nautical miles (remembering that 1′ is equivalent to 1 nautical mile). In this case the angular difference is 2.7′, the equivalent of 2.7 nautical miles along the azimuth of 155°.

Astro-navigation from Square One

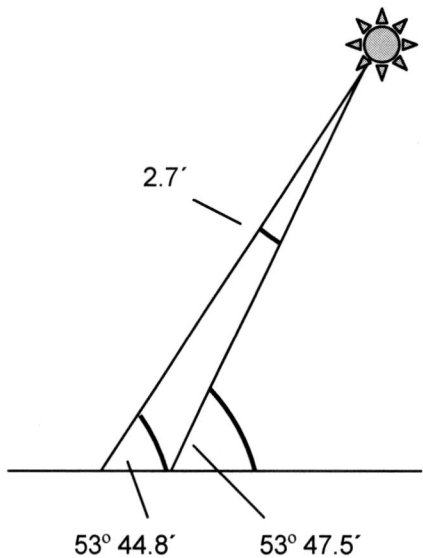

Figure 52. The difference between observed and calculated altitude

The azimuth is drawn through the chosen position from which was produced the calculated altitude. As in this case the observed altitude angle was larger than the angle of the calculated altitude, the intercept is drawn across the azimuth, <u>towards</u> the geographical point of the observed body. The true position of the observer lies on that position line. **Figure 53** illustrates the chart work.

Example 6. Alternatively, if the calculated altitude angle proved to be larger than the observed (measured and corrected) altitude, then the position of the observer will lie on the outer position circle. Let us assume that the difference in angle is again 2.7´, so that in this case the **intercept** on the azimuth of the observed body is drawn across the azimuth at a point 2.7 miles from the chosen position, but <u>away</u> from the direction of the geographical point, as in **Figure 54**. The true position from which the altitude was measured lies on that position line.

2.23 *Summarizing the plotting of a sight* – what do you need for a plot?

In summary, to plot a sight just three things are needed; a **chosen** or **assumed position** (which might be a dead reckoning or estimated position), an **azimuth** and an **intercept**. The chosen or assumed position acts as a datum point. The calculated altitude (H_c) is worked out from this position, either by longhand, using tables and a calculator, or with a programmed calculator or computer program.

– down to the details

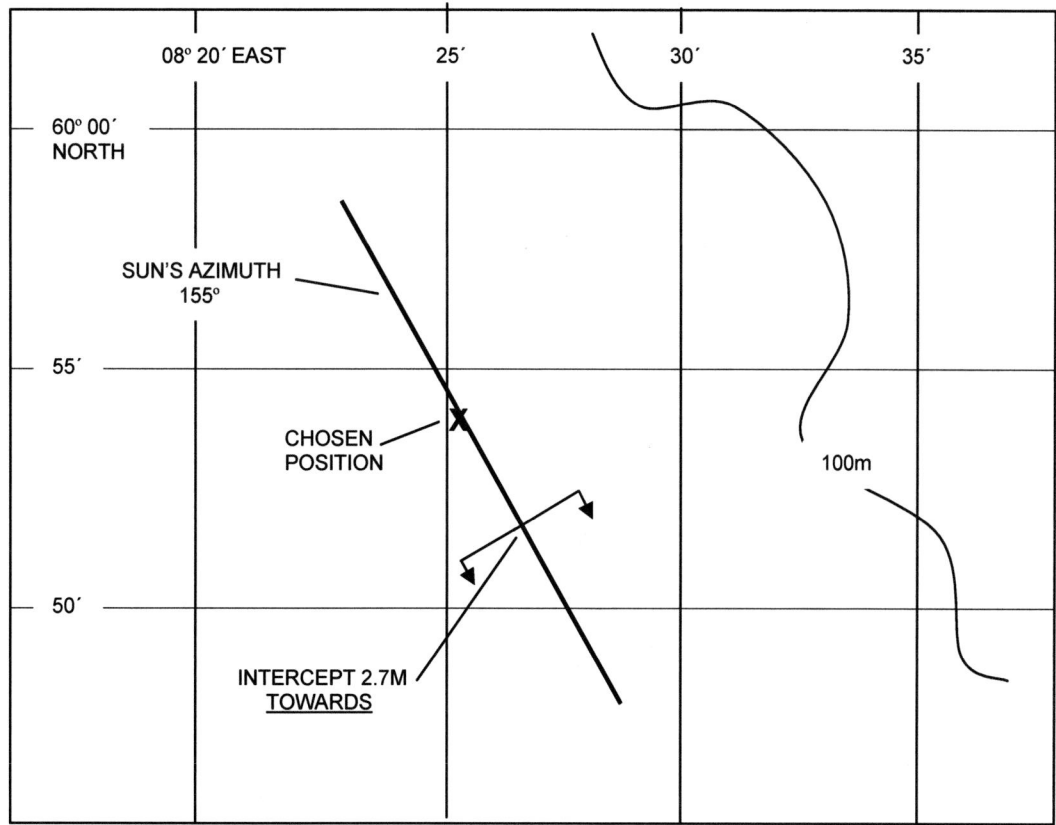

Figure 53. Chartwork for Example 5

The azimuth (Z_n) of the observed body at the assumed position is worked out using either a mathematical formula or an **ABC table** or sight reduction table, converted from quadrantal to 360-notation and is drawn through the assumed position.

The difference between the observed (H_o) and calculated (H_c) altitudes enables the intercept to be defined in terms of both its magnitude and its position on the azimuth. The angular difference between the altitudes, converted to nautical miles is the distance to the intercept from the chosen position. Depending on the relative magnitude of the two altitudes; if the observed altitude (H_o) is larger, the intercept on the azimuth is towards the geographical point of the observed body. If the calculated altitude (H_c) is larger, then the intercept is away from the chosen position.

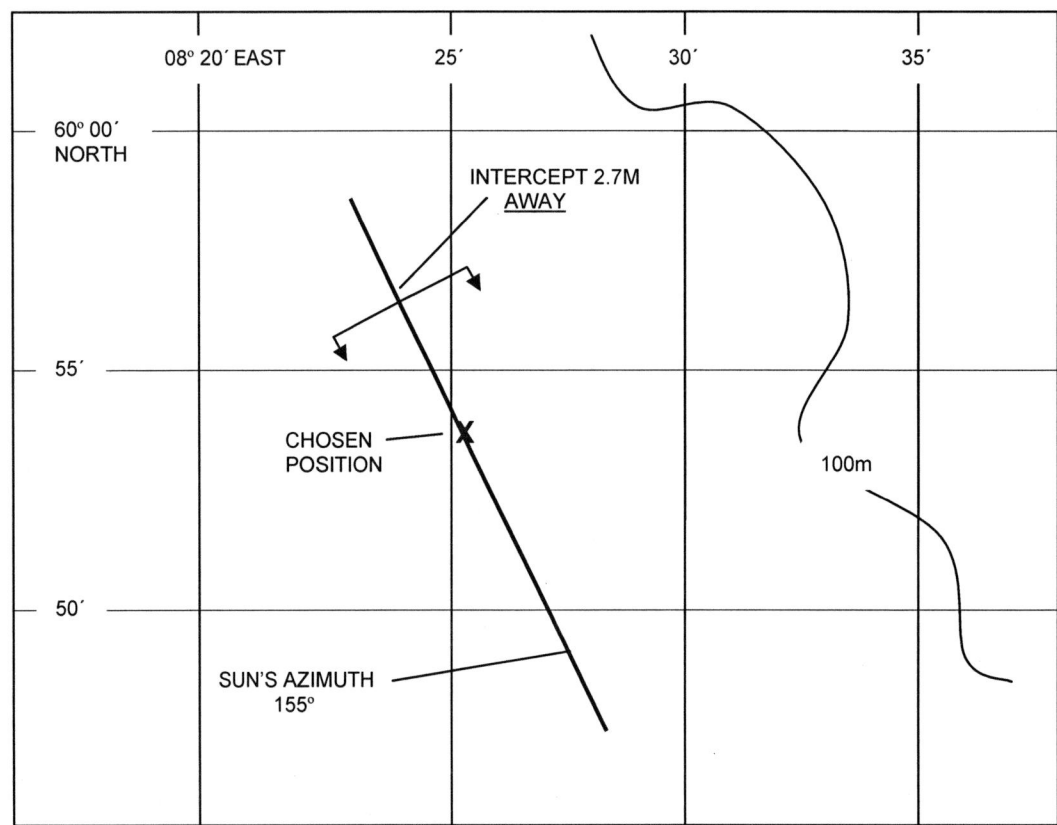

Figure 54. Chartwork for Example 6.

2.24 *Limitations in the plotting of single sight* – the use of multiple sights

In coastal pilotage a single bearing is of limited use and it is standard practice to use bearings of three or more identifiable landmarks, taken closely together in time, then plotted to form a "cocked hat" and then to assume that your position within the cocked hat is on the side closest to the nearest danger. This is much more reliable and seamanlike than using a single bearing and soon shows up an error in the reading or calculation of one observation.

Likewise in astro-navigation, the reliance on a single plotted sight reduction can be unwise. A position line is just that; the sight was taken from a point somewhere on that line. An observational or mathematical error may not be noticed and therefore a single position line should be treated with caution. A good fix needs more than one position line.

It is usual with dawn or dusk star sights that three or more stars or a planet are observed in relatively quick succession, which when plotted will produce several position lines that will with skill, produce a neat intersection; a "cocked hat" of position lines and therefore a more

– down to the details

reliable position fix. Any position line that is wide of the mark can then be relatively safely discarded and assumed to contain an error in the observation or the sight reduction calculation. (see **Section 9.15** p. 259 for an example of a triple star fix).

2.25 *Using transferred position lines* – getting your daily fix?

It may be possible to obtain only a single sight because of observational difficulties or, if the navigator is sure of both a sight and its reduction, it can be used together with earlier or later sights to obtain a reasonably reliable position fix using the technique of transferring position lines. With this technique lines can be moved either forwards or backwards in time to intersect with others so as to produce a "cocked hat"-like fix.

An example of this would be **Sun-run-Sun** observations, where a morning Sun sight position line is transferred up to a noon latitude meridian line or afternoon Sun sight position line. Alternately, the morning position line can be transferred up to and the afternoon sight position line can be transferred backwards in time to produce a position fix with both latitude and longitude at the point of meridian passage, illustrated in **Figure 55**.

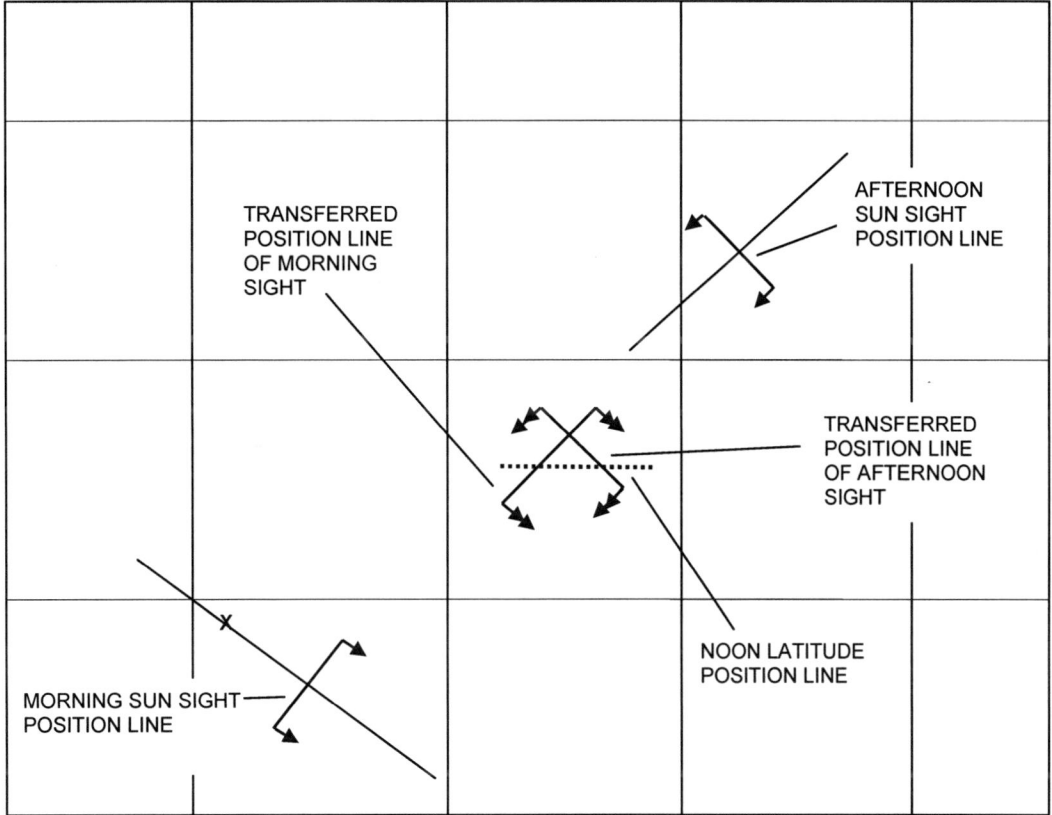

Figure 55. Producing a position fix by transferred position lines

There is a fully worked example of a Sun-run-Sun sight in **Section 6.16** p.172, which of all the techniques used in astro-navigation, is the most commonly used.

Section 2.26 Summarising Section 2.

This section has taken the navigator from an initial sextant sight, through its manipulation to produce an **observed altitude** by making allowances for the **angle of dip** and where necessary for **parallax, semi-diameter** and **refraction**. There followed the description of the process to obtain a **calculated altitude** and **azimuth** from a **chosen position** by solving the **PZX triangle**. The **observed** and **calculated altitudes** were then compared in order to produce an **intercept** on the **azimuth** of the observed body and to construct a **position line**.

This method, the **intercept method** can be distilled to:-

An altitude of a body from a chosen position is compared with an altitude from the observer's nearby but unknown position, the angular difference being plotted as the intercept along the azimuth of the body to produce a position line, giving the best estimate of actual position.

Section 3: The Sextant and its use

This section is relatively brief and does not pretend to be a definitive work on the sextant. For an authoritative text on all aspects of the sextant I recommend *Reed's Sextant Simplified* by Dag Pike, published by Adlard Coles Nautical. This section aims to give a basic description of a typical modern sextant and a working knowledge of its use, commensurate with that required by a relative newcomer to astro-navigation.

3.1 The predecessor of the Sextant – the seaman's quadrant

The sextant has evolved into a precision instrument over more than two centuries from more primitive devices developed primarily by mariners for finding their way across oceans, out of sight of land, judging latitude by *Polaris* and the Sun. The word sextant derives from the Latin: *sextans:* a sixth, its curved arc being one-sixth of the circumference of a circle.

The modern sextant was preceded by the quadrant (Latin: *quadrans*: a fourth part), a quarter circle of wood or brass. It is useful to look at the quadrant's basic design as a key to understanding aspects of the modern instrument. One radial edge was fitted with sighting pins to line up an object in the sky. A plumb-line was hung from the centre-point of the device from which was read the angular elevation of the sighting edge, marked out in degrees on the circumferential arc (**Figure 56**).

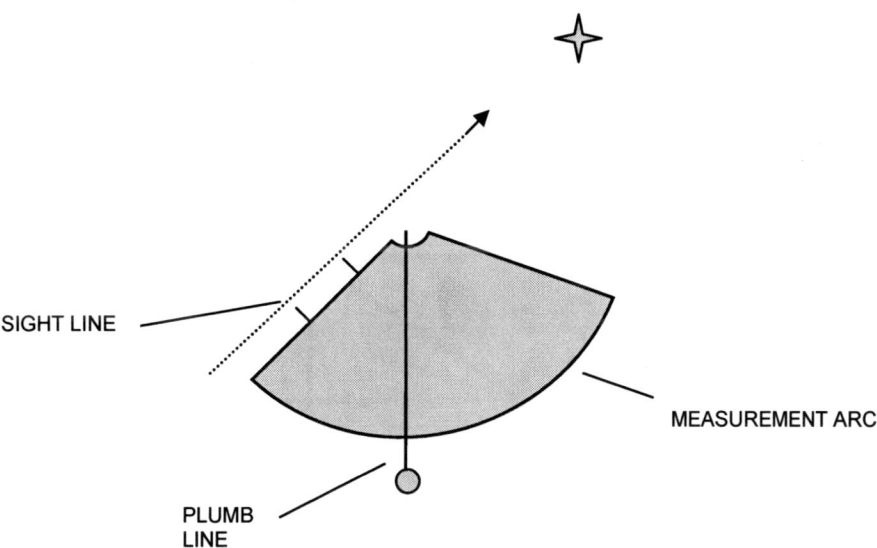

Figure 56. The principle parts of a quadrant

3.2 The development of the modern sextant

The possibility of using a reflected image by using a mirror for the sighting of an object was probably suggested, though not apparently exploited by Isaac Newton (1643-1727). It was later explored by John Hadley (1682-1744), reportedly following a sketch by Newton[3.1].

The modern sextant has developed from technical advances in the three main elements of the quadrant: in the sighting of the object, in the mechanism of determining a defined datum line from which the elevation measurement is made and in the attainment of precision in the measurement of the elevation of the object. The quadrant measured the elevation of an object in relation to itself but the sextant measures an elevation with respect to the horizon, a much more stable datum line, the plumb-line being replaced by reflected light rays, enhanced by the manufacture of plane mirrors of increasing quality.

There is no doubt that the view of a heavenly body and horizon is remarkably stable when viewed through a sextant, even from a remorselessly heaving deck. This is as a consequence of the sighted body and horizon both being still and the relatively advanced state of development of man's hand-eye coordination which combats movement of the observer. The accuracy of measurement of the angle of elevation has been increased by the addition of a micrometer, a means of making an accurate measurement by the movement of a high-precision screw through a thread. The basic components of a modern sextant are illustrated in **Figures 57** and **58**.

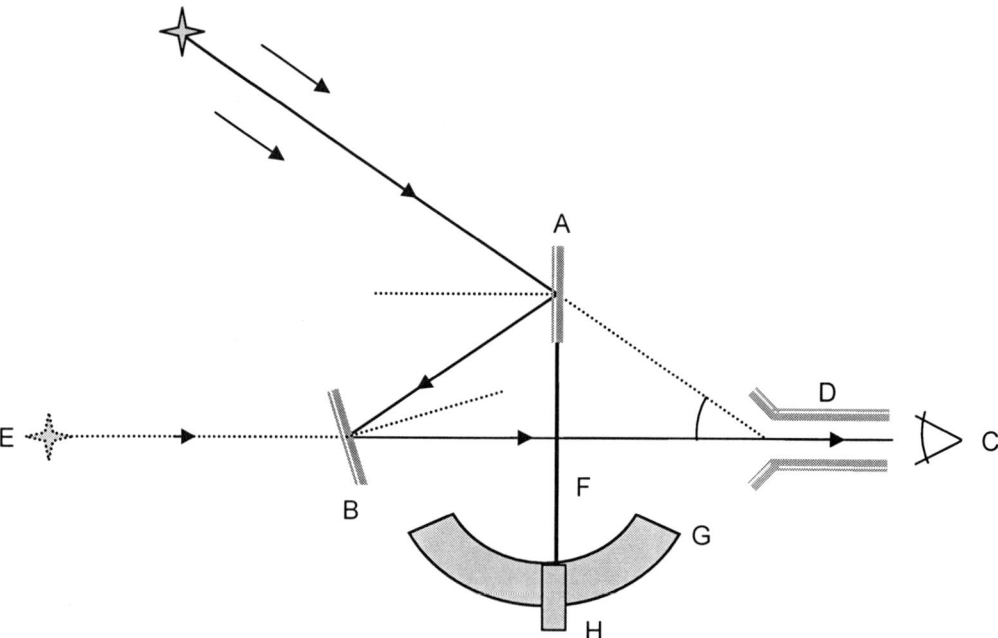

Figure 57. The basic components of the modern sextant

– the sextant and its use

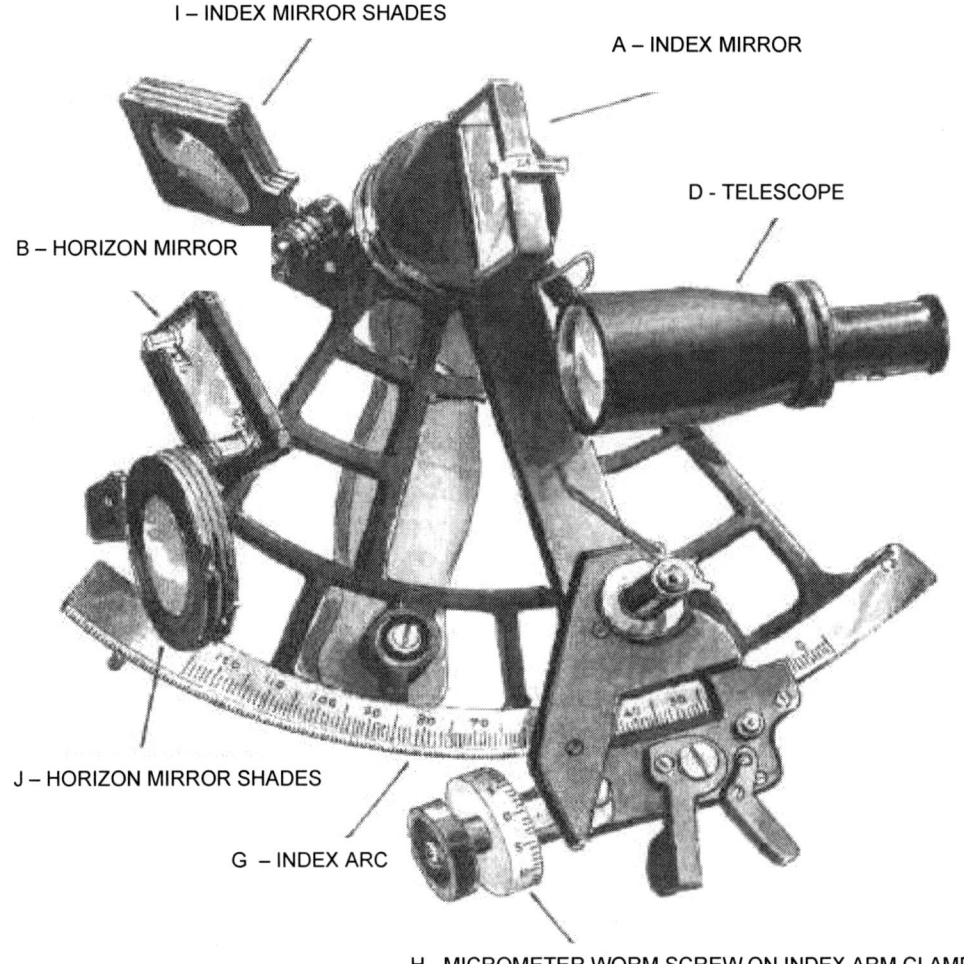

Figure 58 - a typical sextant showing its component parts

3.3 *Basic sextant structure and optics* - how does it really work?

Parallel light rays from the region of the observed body that would otherwise reach the observer's eye, strike the reflecting surface of the **index mirror** (A). They are turned through an angle twice that of the angle between the incident ray and a line perpendicular to the surface of the mirror. The reflected rays strike and are then reflected by the **horizon mirror** (B), towards the eye of the observer (C), either through a plain sighting tube or a low-power telescope (D). The horizon mirror is so designed that an image of the observed body is superimposed upon the image of the distant horizon (E), in line with the observer's eye. This mirror is usually a traditional half-silvered half-clear variety through which is seen half the horizon with the observed body superimposed.

More recently a type of glass has been developed with complex optical coatings, an invention of Angus Macdonald of Davis Instruments and called a "full-field dielectric beam converger" [3.2]. These coatings reflect certain wavelengths of incident light whilst allowing the unaltered transmission of other wavelengths. This produces a full field view of the subject of the sight and a complete horizon (**Figure 59**).

The professional's opinion probably favours the traditional half-silvered mirror over that of the full field view, especially with difficult sights of relatively dim heavenly bodies, where optical coatings might critically limit the amount of light transmitted. The novice might prefer the full field image of horizon and body.

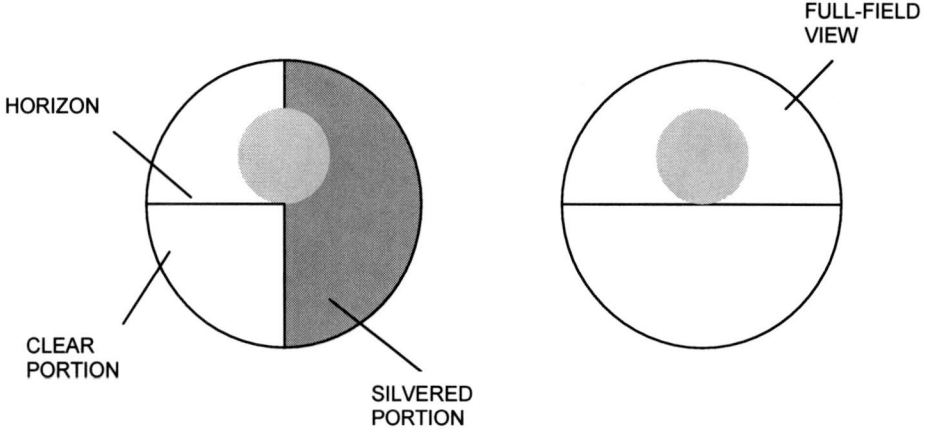

Figure 59. The comparative views of the Sun with conventional (left) and coated mirrors

The index mirror is mounted upon a movable index arm (F) that is arranged to pivot around a plane perpendicular to the index mirror. The lower aspect of the arm is connected to the index arc (G) by a clamp housing a rotating **micrometer** worm-screw (H). The housing may contain a low-power lamp bulb to aid in the reading of the index arc at night. The clamp allows coarse adjustment of the index arm to the nearest degree, altering the angle of the index mirror and permitting the adjustment of the proximity of the reflected image of the observed object in the horizon mirror, superimposed upon the horizon seen through the horizon mirror.

The **micrometer** permits fine adjustments between the two images. The index arc is calibrated and marked in whole <u>degrees</u> of arc and the micrometer is calibrated and marked in whole <u>minutes</u> of arc. The micrometer typically has a further calibrated scale; a vernier scale, introduced by Frenchman Pierre Vernier, in 1631[3.3]. It is so arranged as to increase the precision of the reading on the micrometer scale to 0.2′. This is done by noting the point of alignment of the micrometer divisions above the whole minute reading (**Figure 60**). Only one division on the scale will lie opposite to a decimal division on the vernier and this point indicates the decimal place of the whole minute reading.

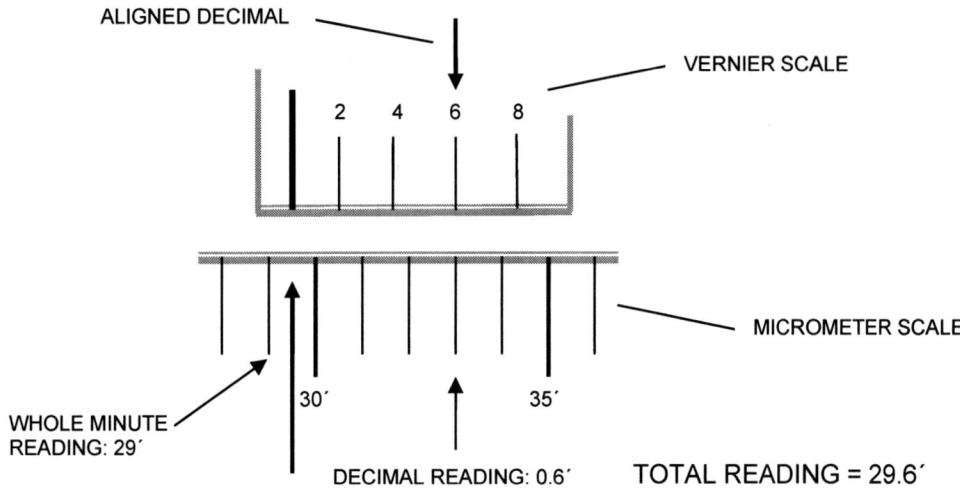

Figure 60. A vernier scale used to indicate decimals of arc-minutes

The essential parts described above are mounted on a supporting frame imparting strength, resilience and congruity and which may be made of a variety of materials, usually lightweight metal alloy or plastics. Both the index and horizon mirrors are secured by supporting springs and have small screws to adjust the plane of the mirrors with respect to one another and the body of the instrument.

The scale on the sextant's arc usually covers more than 120° of angle rather than the 60° actually present in one-sixth of the circumference of a circle (360°/6). This is due to the fact that an incident light ray from an object is reflected through an angle twice its incident value as in **Figure 61**.

This occurrence is based upon the law in physics governing reflection which states: that the angle of incidence of a light ray equals its angle of reflection. This means that when an object is observed at an elevation of 45°, a light ray from the object is reflected through 90°. This increased arc range permits the use of the sextant in addition to standard vertical sights, in a horizontal plane to measure the angular distances between distant landmarks for a useful

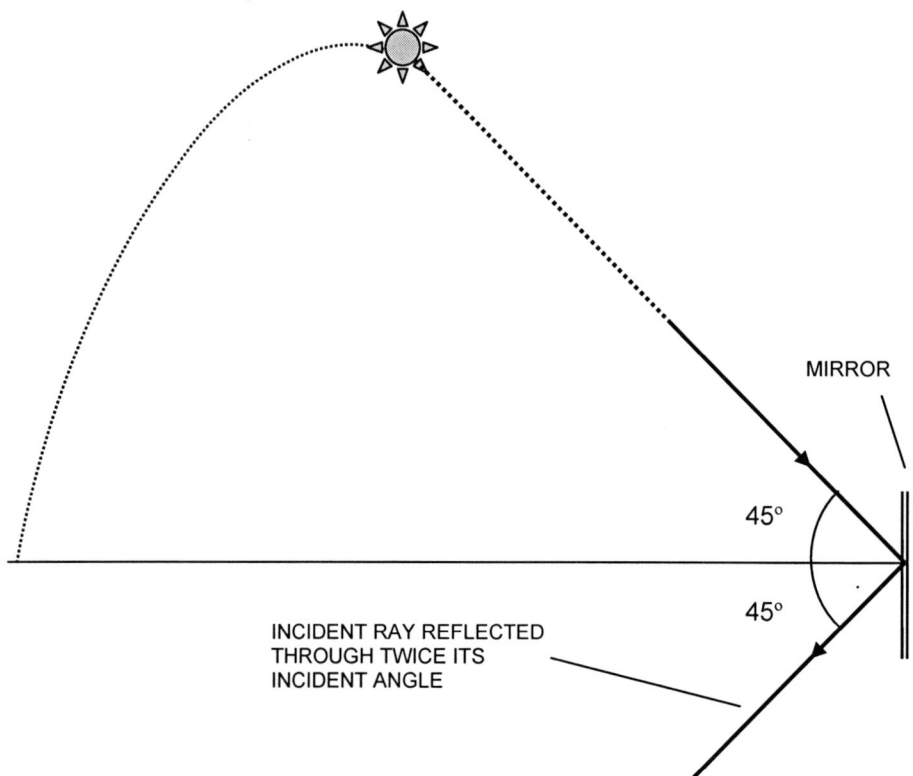

Figure 61. The angle of incidence = angle of reflection

assessment of "distance off" in costal pilotage. Last but by no means least, the standard sextant has two sets of shades of coloured glass or plastic, one set for use between the index and horizon mirrors (I) and a second set lying behind the horizon mirror (J).

The former set MUST be used for Sun sights on <u>every</u> occasion so as to avoid inevitable and probably permanent retinal damage caused by the infra-red (heat) rays in sunlight. Whilst the use of shades is not mandatory for sights other than those of the Sun, they can be useful for changing and adjusting the contrast between other bodies and the horizon.

3.4 *The identification and correction of potential and actual sextant errors*

The sextant is a relatively delicate object and must be treated and stored in its box with great care. The accuracy of a sextant depends upon the precision with which two mirrors reflect light rays, the measuring scale being coupled to the index mirror. If a telescope is used, light rays passing through it must not be abnormally distorted by its lenses.

Three major adjustments are possible in a fixed order, to "tune" a sextant for use, so that unnecessary problems in measurement can be avoided or at least minimized. A fourth type of error relates to the telescope. It is much easier to understand the various errors and their methods of rectification if, while you read this section, you have a sextant to hand.

1. Perpendicularity of the index mirror:- the angle at which the index mirror is held to the plane of the instrument influences the path of the light rays passing through the rest of the sextant. In basic terms, a ray striking the centre line of the mirror needs to be "squarely" reflected such that it is directed to the centre line of the horizon mirror and then on through the telescope to the eye. For true planar reflection, the index mirror must be held as near as possible to an exact right angle (i.e. perpendicular) to the body of the sextant.

The mirror angle can be adjusted with the single screw on the rear of the support of the index mirror. This is done whilst comparing the actual and reflected image of the arc of the instrument, seen at the edge of the index mirror with the instrument held at arms length and the index bar set to 50-60° or so (**Figure 62**). By rotating the screw, the reflected image is adjusted to appear as a smooth curve where it abuts the true image, without the presence of a visible step. When the image has a continuously smooth line, the mirror is perpendicular to the plane of the instrument.

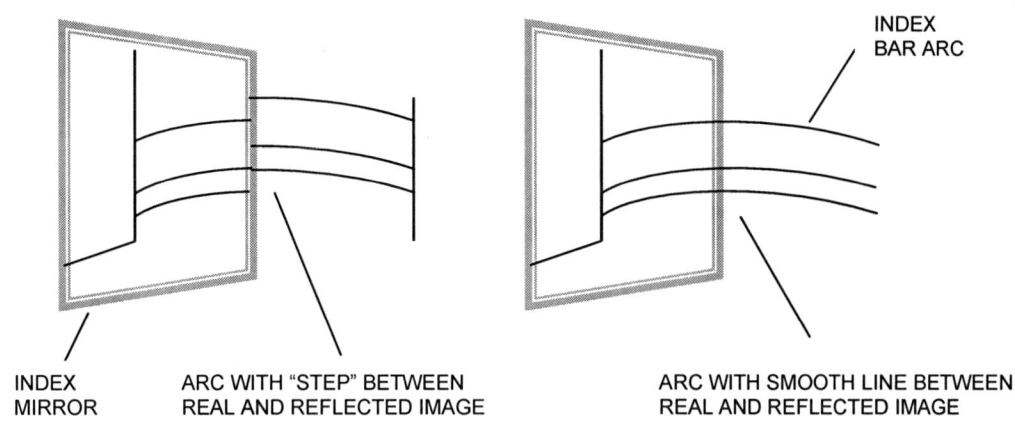

Figure 62. Index mirror perpendicular error and its correction

2. Side error or horizon mirror perpendicularity:- incorrect horizon mirror perpendicularity is generally known as "side error" and like the index mirror the horizon mirror must be perpendicular to the plane of the sextant to correctly reflect a ray received from the centre line of the index mirror to the centre line of the telescope and eye. Perpendicularity is assessed with the index bar set at 0° and the micrometer to 00.0. A star or a distant point light source is then observed through the sextant.

The horizon mirror usually has two rear screws, one of which adjusts the mirror in the <u>horizontal</u> plane (side to side movement), around the shorter axis of the mirror, which corrects any misalignment of perpendicularity (side error) of the mirror to the sextant body. The other screw is for adjustment of the vertical plane (**Figure 63**). This vertical adjusting screw is turned gently back and forth whilst observing the actual and reflected image of a point light source moving vertically past one another.

Any horizontal distance seen to be present between the two images during this manoeuvre is then removed with the horizontal adjusting screw, reducing the side error. This may produce an increasing vertical error called "index error" and this is addressed in the next paragraph.

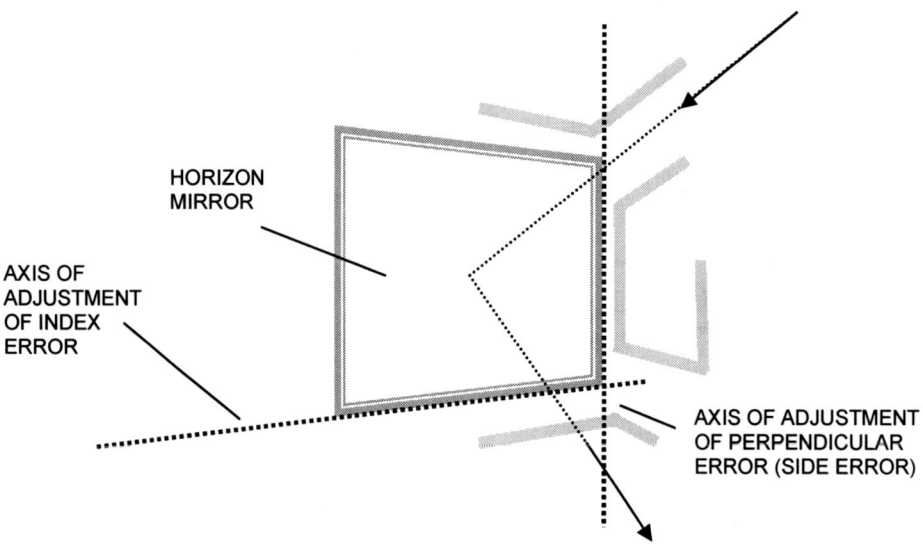

Figure 63. Horizon mirror adjustment – for side error and index error

3. Index error:- this is caused by the vertical misalignment of the <u>horizon</u> mirror (not the index mirror) around its long axis and would perhaps be better called "index arm error". With the index bar at zero the vertical adjusting screw on the horizon mirror is turned whilst looking at a distant light point source, or the horizon. This adjustment may reintroduce some side-error (horizontal misalignment) of the horizon mirror, one being adjusted after the other until with good fortune, both are eliminated. If some vertical misalignment remains then it is adjusted with the micrometer of the index bar.

The resulting reading is called the index error of the sextant. If the value is greater than 0′, e.g. 1′ or 2′, or fractions thereof, then it is described as being an index error "on" the arc. If the reading is less than 0, i.e. 59.8′ or below, it is described as being an error "off" the arc. An index error must then be added (if "off") or subtracted (if "on") from all altitude readings made with the instrument. In short: if it is "off" add it on and if it is "on", take it off.

Example 7. A sight of the Moon gives a sextant altitude of 53° 31.4′. Just before the sight was taken the sextant was adjusted for index perpendicularity, side error and index error, at which point the index bar reading was just less than 0° and the micrometer reading was 57.2′. What was the corrected sextant altitude?

An optically perfect and adjusted sextant without error should read 0° 00.0′. If when adjusted the index bar reads less than 0° and the micrometer reads other than 0.00′ then there is an "off the arc" index error, in this case it is off the arc by 2.8′ (0.00 – 57.2 = –2.8). The instrument will then under-read and 2.8′ should be added to an altitude measured by the sextant to correct for this error. The corrected sextant altitude = 53° 31.4′ + 2.8′ = 53° 34.2′.

From this it follows that if this index error had been "on the arc" then it is subtracted from the sextant altitude to correct for the error. The Moon's corrected sextant altitude would then have been 53° 31.4′ – 2.8′ = 53° 28.6′.

4. Collimation error:- To collimate means to make parallel. In this instance the term relates to the viewing telescope if used and indicates that it is not transmitting the light rays carrying the images in a parallel fashion through its lenses, which leads to the distortion of the image. It is checked for by observing two-point light sources such as stars some 80-90° apart, either vertically or with the instrument held horizontally. By a minimal movement; swinging the sextant slightly, the relationship between the two images should remain the same as they cross the field of view in the telescope.

If their spatial relationship changes during the movement, there is likely to be a collimation error (**Figure 64**), the optical path through the scope being not parallel to the plane of the instrument. This test is easier to do if the telescope has cross hairs. The presence of a probable collimation error indicates the possible need for professional assessment of the optics of the scope or its mountings.

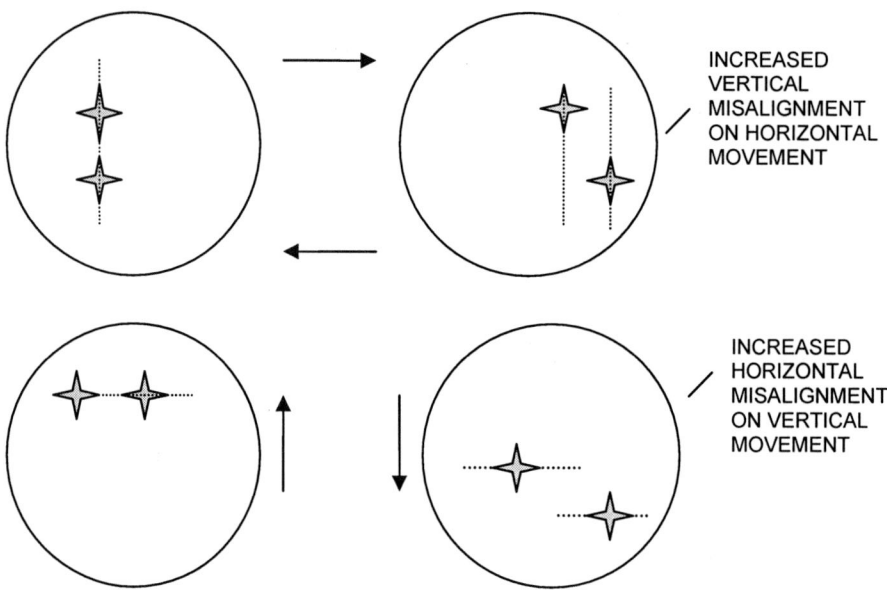

Figure 64. The appearance of a possible collimation error

3.5 *Setting the heavenly body on the horizon* – an art in itself?

There is considerable skill involved in bringing any particular heavenly body down to the horizon and one only developed with practice. It is generally so that the brighter bodies are easier to manipulate with the sextant. A body fairly readily reveals itself by sweeping the horizon to and fro beneath the body while releasing the index arm clamp and increasing the upward tilt of the index mirror. Once caught with the mirror they are retained in view whilst carefully adjusting both clamp and then micrometer. Alternatively, with the index arm at zero, the body is directly observed through the eye-piece, then while maintaining the image, the index arm is slowly manipulated to bring the reflected image to the horizon..

It is vital that the sextant is held in a completely vertical plane or false reading will result. To check this it is routine to gently tilt the sextant to and fro in a small arc whilst viewing the body close to the horizon. The instrument will be vertical when the position of the body just kisses the horizon during its swing, as in **Figure 65**. Further details and examples of the use of the sextant are addressed in **Sections 6-9** on the Sun, Moon, Planets and Stars.

– the sextant and its use

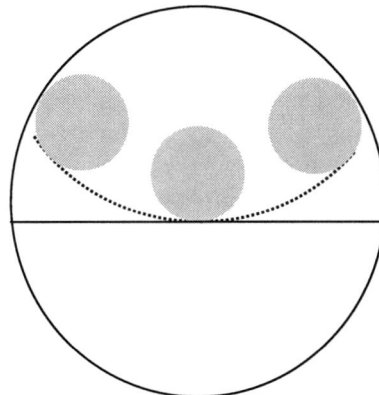

Figure 65. The view whilst swinging the instrument to achieve verticality

3.6 Acquiring a sextant – to see or not to see?

The majority of readers will never have to practice the art of sextant navigation far from land and without the onboard presence of GPS. The purchase of a sextant could therefore be considered as an expensive luxury, except perhaps for those crossing oceans. However, there is absolutely no doubt whatever that the reader will still want to acquire sextant skills. A professional's instrument with a metal-alloy body will cost at least £400-£1000. The best available can cost several thousand pounds.

The purchase of second-hand equipment inevitably risks buying a device with unacceptably large errors due to previous damage producing problems such as a bent frame. A reasonable alternative is the acquisition of a new plastic sextant which entails the lesser expense of two to three hundred pounds or so (e.g. a Davis sextant). These devices are remarkably resilient, made of thermally stable plastics. The "Davis mark 15" has a standard split mirror and the "Davis mark 25" has full-field view optics. Both achieve very acceptable levels of accuracy.

Undoubtedly an understanding of the concepts aired in this book is easier if the reader has a sextant to hand and I would support an exploration of the "pre-owned" internet-based market. Personal experience suggests that the risks of acquiring a useless instrument are acceptably small, but beware of brass reproductions intended for the mantle-piece!

Section 4: Tables, tables, tables.......

Section 1 introduced the basic concepts and terminology of astro-navigation and **Section 2** addressed the details of reducing a sight to produce a position line. **Sections 5-9** will discuss time and then heavenly bodies in detail, showing how each is used and focusing on their individual peculiarities. In these latter sections, frequent reference is made to a multitude of tables of both astro-navigational and mathematical data. Numerous extracts of relevant tables are included in each section. This section aims to throw some light on "Tables" so that when they are referred to in later sections, they seem a little less daunting.

By the end of Section 4, the following KEY FACTS should be understood:-

- **Ephemeris tables contain the GHA and Declinations of the Sun, Moon, planets and Aries, the SHA and declinations of the navigational stars, for every hour of a year.**

- **Increments tables contain the additional adjustments to the hourly ephemeris figures for each minute and second of time.**

- **Correction tables contain the adjustments to GHA and Declination for variations over and above the mean variations used for table construction. Others list the correction values used to convert corrected sextant altitude measurements to observed altitudes.**

- **Calculated altitude tables enable; for a given latitude, declination and LHA, an altitude to be calculated for a given body, for comparison with an observed altitude to arrive at an intercept on a body's azimuth.**

- **Azimuth tables enable; for a given latitude, declination and LHA, an azimuth to be derived for a given body.**

- **Azimuth equations enable, for a given latitude, declination and LHA, an azimuth to be calculated for a given body.**

- **Mathematical tables enable the complex maths of angular measurements to be converted to simpler addition or subtraction, in manual and calculator methods of sight reduction.**

- **Interpolation is the method used to find intermediate values between those given in a particular table.**

- **Generally, figures derived from mathematical equations are more accurate than those derived from tables, where convenience may sacrifice accuracy for speed.**

– tables, tables, tables…………

A summary chart of the tables is shown in **Figure 66**. This is to illustrate the main types of table and to indicate their relevance to each of the heavenly bodies used by the navigator. It is quite complex and not at all difficult to see why the budding astro-navigator might find that tables in general appear as impenetrable jungle, yet the experts wonder just what the fuss is about.

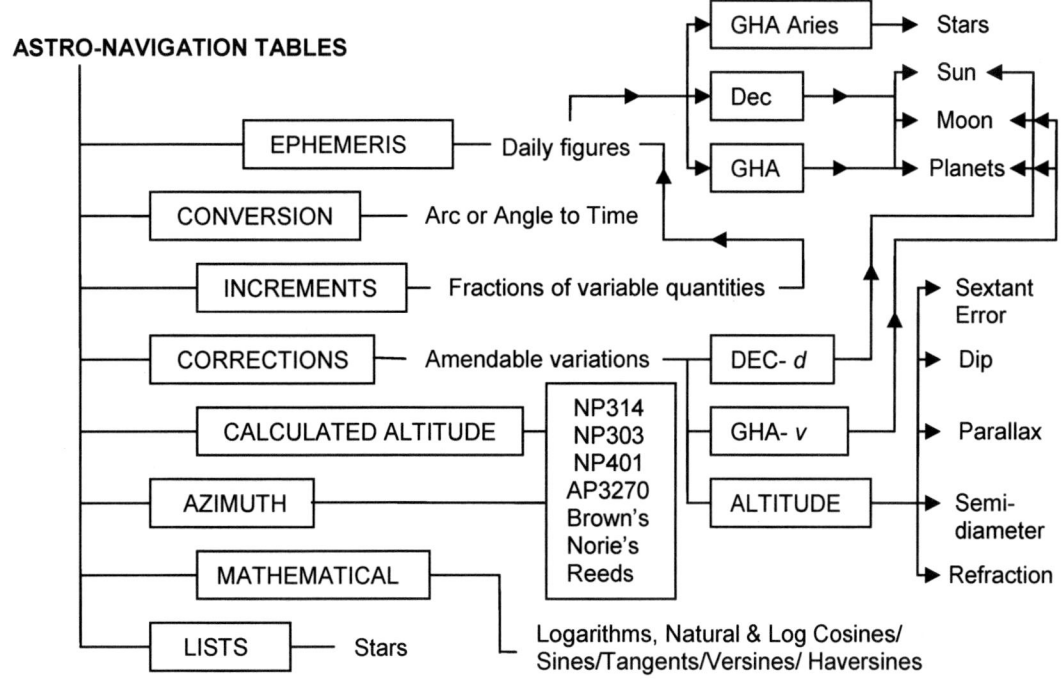

Figure 66. Summarising the variety of astro-navigational tables

4.1 *The Nautical Almanac* -NP 314 – the source of all knowledge.

There is one prime annual publication of tabulated data available to the astro-navigator, *The Nautical almanac* –NP 314[4.1], published by Her Majesty's Nautical Almanac Office of the UK Hydrographic Office. It is the source of astro-navigational data for the professional sailor, contained in some 350 pages but reasonable priced at approximately £37.00.

85

There is a second source, perhaps more suited to the chart table of the amateur yachtsman and published annually by Adlard Coles Nautical: *Reeds Astro Navigation Tables*[4.2]. It too uses H.M. Nautical Almanac Office data, organized with a little less detail, into a compact 70 pages, priced at around £20.00. Less detail of course, may imply less accuracy in the final assessment of position. For the novice, all tables whatever their source are quite daunting but the first key to the successful understanding of this jungle is to define and understand the various <u>types</u> of table. The second is to understand the limitations of certain tables, some applying to all bodies and others being body-specific.

There are 6 main table types: **ephemeris**, **conversion**, **increments**, **correction**, **calculated altitude** <u>with</u> **azimuth**, and **mathematical**.

Once the type of table and its intended use is grasped then its contents will, with practice, fall into place. Within each table type there may be a variety of different "sub-types", as for example, calculations for GHA and Declination require increment tables for minutes and seconds of time and Altitude readings require correction tables particularly tailored for each of the useful heavenly bodies.

Table complexity is commonplace however, as correction and other data for more than one type of body may be presented within one large table. Considerable skill needs to be developed to extract that which is relevant to the navigator's particular concern.

4.2 *Ephemeris tables*: - the daily bread of the astro-navigator?

The ephemeris tables give the daily positions of the navigationally useful heavenly bodies: the Sun, Moon, planets and stars (based on Aries). The word ephemeris comes from the Greek *ephemeros*, meaning living a day and from which we derive the word "ephemeral" used in everyday language and meaning very short-lived. The actual arrangement of the data varies according to the source, one form or another of an astro-navigational **nautical almanac**.

In the professional's annually published **NP 314**[4.1]**,** paired pages give all the necessary data on all the useful heavenly bodies for three consecutive days. The first of each pair (**Table 2**) gives the whole hour values for the GHA Aries, the hourly GHA and Declination of the navigational planets, the SHA and Declination for 57 brighter stars, listed alphabetically, the planet SHA and times of meridian passage. In all there are approximately 1000 items of data per page to an accuracy of one-tenth of a degree.

The second of each pair of daily pages (**Table 3**) gives hourly details of the GHA and Declination of the Sun, the GHA and Declination of the Moon with its hourly rates of variation and an hourly figure for horizontal parallax. Details of sunrise and sunset, moonrise and moonset are also given, together with the timings of twilight, all with adjustments for latitude and the details of the Greenwich meridian passage for both Sun and Moon. This amounts to a further 900 or so data items.

2011 JANUARY 1, 2, 3 (SAT., SUN., MON.)

UT	ARIES	VENUS −4.6		MARS +1.2		JUPITER −2.3		SATURN +0.8		STARS		
	GHA	GHA	Dec	GHA	Dec	GHA	Dec	GHA	Dec	Name	SHA	Dec
d h	° ′	° ′	° ′	° ′	° ′	° ′	° ′	° ′	° ′		° ′	° ′
1 00	100 18.2	228 13.1	S16.5	170 15.5	S23 09.6	103 ...	S 2 32.0	264 01.4	S 4 19.4	Acamar	315 19.4	S40 15.8
01	115 20.7	243 13.2		185 15.9	09.3	...	31.8	279 03.8	19.4	Achernar	335 27.8	S57 11.1
02	130 23.1	258 13.2	17...	200 16.3	09.0	...	...	... 06.1	19.4	Acrux	173 11.3	S63 09.4
03	145 25.6	273 13.3	.. 18.2	... 16.6	.. 08.8	...	...	... 08.5	.. 19.5	Adhara	255 13.5	S28 59.3
04	160 28.1	288 13.3	18.8	...	08.5	...	...	... 10.9	19.5	Aldebaran	290 51.1	N16 31.9
05	175 30.5	303 13.4	19.3	24...	08.3	...	...	... 13.2	19.5			
06	190 33.0	318 13.4	S15 19.9	260 1...	08.0	193 10.8	S 2 31.1	354 15.6	S 4 19.6	Alioth	166 22.1	N55 53.6
07	205 35.5	333 13.4	20.5	275 ...	31.0	208 13.0		9 17.9	19.6	Alkaid	153 00.3	N49 15.1
08	220 37.9	348 13.5	21.0	290 ...	...	...	30.8	24 20.3	19.6	Al Na'ir	27 46.2	S46 54.5
09	235 40.4	3 13.5	.. 21.6	305 ...	...	...	30.7	39 22.7	.. 19.7	Alnilam	275 47.8	S 1 11.8
10	250 42.8	18 13.6	22.2	320 ...	...	...	30.6	54 25.0	19.7	Alphard	217 57.6	S 8 42.5
11	265 ...	33 13.6	22.7	335 ...	...	...	30.4	69 27.4	19.7			
12	280 47...	48 13.6	S15 23.3	350 20.1	S23 06.5	283 23.8	S 2 30.3	84 29.7	S ...	Alphecca	126 12.7	N26 40.5
13	295 50.2	... 13.7	23.8	5 20.4	06.2	298 25.9		...	...	Alpheratz	357 45.4	N29 09.3
14	310 52.7	...	24.4	20 20.8	06.0	313 28.1		...	...	Altair	62 10.3	N 8 53.9
15	325 55.2	93 ...	25.0	35 21.2	.. 05.7	328 30.2	.. ...	...	.. ...	Ankaa	353 17.4	S42 14.9
16	340 57.6	108 ...	25.5	50 21.6	05.4	343 32.4	...	...	...	Antares	112 28.7	S26 27.3
17	356 00.1	12...	...	...	05.2	358 34.6	...	...	...			
18	11 02.6	13...	S15 ...	...	S22 04.9	13 36.7	S 2 29...	...	S ...	Arcturus	145 57.4	N19 07.3
19	26 05.0	153 ...	...	...	04.7	28 38.9	29.3	189 46.2	20.0	Atria	107 32.7	S69 02.7
20	41 07.5	168 ...	...	...	04.4	43 41.0	29.1	204 48.6	20.0	Avior	234 18.2	S59 32.7
21	56 10.0	183 ...	.. ...	...	04.1	58 43.2	.. 29.0	219 51.0	.. 20.0	Bellatrix	278 33.6	N 6 21.5
22	71 12.4	198 ...	...	...	03.9	73 45.3	28.8	234 53.3	20.1	Betelgeuse	271 02.9	N 7 24.5
23	86 14.9	213 14.0	29.5	155 24.2	03.6	88 47.5	28.7	249 55.7	20.1			
2 00	101 17.3	228 14.0	S15 30.0	170 24.6	S23 03.3	103 49.6	S 2 28.6	264 58.1	S 4 20.1	Canopus	263 56.4	S52 42.2
01	116 19.8	243 14.0	30.6	185 25.0	03.1	118 51.8	28.4	280 00.4	20.2	Capella	280 36.6	N46 00.6
02	131 22.3	258 14.0	31.2	200 25.4	02.8	133 54.0	28.3	295 02.8	20.2	Deneb	49 33.1	N45 19.4
03	146 24.7	273 14.1	.. 31.7	215 25.7	.. 02.6	148 56.1	.. 28.1	310 05.1	.. 20.2	Denebola	182 35.3	N14 30.4
04	161 27.2	288 14.1	32.3	230 26.1	02.3	163 58.3	28.0	325 07.5	20.2	Diphda	348 57.6	S17 55.6
05	176 29.7	303 14.1	32.9	245 26.5	02.0	179 00.4	27.8	340 09.9	20.3			
06	191 32.1	318 14.1	S15 33.4	260 26.9	S23 01.8	194 02.6	S 2 27.7	355 12.2	S 4 20.3	Dubhe	193 53.4	N61 41.1
07	206 34.6	333 14.1	34.0	275 27.3	01.5	209 04.7	27.5	10 14.6	20.3	Elnath	278 14.5	N28 37.0
08	221 37.1	348 14.2	34.5	290 27.6	01.2	224 06.9	27.4	25 16.9	20.4	Eltanin	90 47.5	N51 29.2
09	236 39.5	3 14.2	.. 35.1	305 28.0	.. 01.0	239 09.0	.. 27.2	40 19.3	.. 20.4	Enif	33 49.1	N 9 55.7
10	251 42.0	18 14.2	35.7	320 28.4	00.7	254 11.2	27.1	55 21.7	20.4	Fomalhaut	15 26.0	S29 33.9
11	266 44.4	33 14.2	36.2	335 28.8	00.4	269 13.3	27.0	70 24.0	20.5			
12	281 46.9	48 14.2	S15 36.8	350 29.2	S23 00.1	284 15.5	S 2 26.8	85 26.4	S 4 20.5	Gacrux	172 02.9	S57 10.3
13	296 49.4	63 14.2	37.3	5 29.5	22 59.9	299 17.6	26.7	100 28.8	20.5	Gienah	175 54.1	S17 36.2
14	311 51.8	78 14.2	37.9	20 29.9	59.6	314 19.8	26.5	115 31.1	20.5	Hadar	148 50.7	S60 25.4
15	326 54.3	93 14.2	.. 38.5	35 30.3	.. 59.3	329 21.9	.. 26.4	130 33.5	.. 20.6	Hamal	328 02.6	N23 31.0
16	341 56.8	108 14.2	39.0	50 30.7	59.1	344 24.1	26.2	145 35.9	20.6	Kaus Aust.	83 46.5	S34 22.7
17	356 59.2	123 14.2	39.6	65 31.1	58.8	359 26.2	26.1	160 38.2	20.6			
18	12 01.7	138 14.3	S15 40.2	80 31.5	S22 58.5	14 28.4	S 2 25.9	175 40.6	S 4 20.7	Kochab	137 20.4	N74 06.3
19	27 04.2	153 14.3	40.7	95 31.8	58.3	29 30.5	25.8	190 42.9	20.7	Markab	13 40.2	N15 16.0
20	42 06.6	168 14.3	41.3	110 32.2	58.0	44 32.7	25.6	205 45.3	20.7	Menkar	314 16.7	N 4 08.0
21	57 09.1	183 14.3	.. 41.8	125 32.6	.. 57.7	59 34.8	.. 25.5	220 47.7	.. 20.7	Menkent	148 09.8	S36 25.3
22	72 11.6	198 14.3	42.4	140 33.0	57.4	74 37.0	25.4	235 50.0	20.8	Miaplacidus	221 39.4	S69 45.7
23	87 14.0	213 14.3	43.0	155 33.4	57.2	89 39.1	25.2	250 52.4	20.8			
3 00	102 16.5	228 14.3	S15 43.5	170 33.7	S22 56.9	104 41.3	S 2 25.1	265 54.8	S 4 20.8	Mirfak	308 42.6	N49 54.3
01	117 18.9	243 14.3	44.1	185 34.1	56.6	119 43.4	24.9	280 57.1	20.9	Nunki	76 00.9	S26 16.9
02	132 21.4	258 14.3	44.7	200 34.5	56.3	134 45.6	24.8	295 59.5	20.9	Peacock	53 22.5	S56 42.0
03	147 23.9	273 14.3	.. 45.2	215 34.9	.. 56.1	149 47.7	.. 24.6	311 01.9	.. 20.9	Pollux	243 29.5	N27 59.8
04	162 26.3	288 14.2	45.8	230 35.3	55.8	164 49.9	24.5	326 04.2	20.9	Procyon	245 01.2	N 5 11.7
05	177 28.8	303 14.2	46.3	245 35.6	55.5	179 52.0	24.3	341 06.6	21.0			
06	192 31.3	318 14.2	S15 46.9	260 36.0	S22 55.2	194 54.2	S 2 24.2	356 09.0	S 4 21.0	Rasalhague	96 08.4	N12 33.1
07	207 33.7	333 14.2	47.5	275 36.4	54.9	209 56.3	24.0	11 11.3	21.0	Regulus	207 45.1	N11 54.6
08	222 36.2	348 14.2	48.0	290 36.8	54.7	224 58.5	23.9	26 13.7	21.1	Rigel	281 13.4	S 8 11.4
09	237 38.7	3 14.2	.. 48.6	305 37.2	.. 54.4	240 00.6	.. 23.7	41 16.1	.. 21.1	Rigil Kent.	139 54.5	S60 52.6
10	252 41.1	18 14.2	49.1	320 37.6	54.1	255 02.8	23.6	56 18.4	21.1	Sabik	102 14.9	S15 44.3
11	267 43.6	33 14.2	49.7	335 37.9	53.8	270 04.9	23.4	71 20.8	21.1			
12	282 46.1	48 14.2	S15 50.3	350 38.3	S22 53.6	285 07.1	S 2 23.3	86 23.2	S 4 21.2	Schedar	349 42.6	N56 36.2
13	297 48.5	63 14.2	50.8	5 38.7	53.3	300 09.2	23.1	101 25.5	21.2	Shaula	96 24.7	S37 06.6
14	312 51.0	78 14.1	51.4	20 39.1	53.0	315 11.3	23.0	116 27.9	21.2	Sirius	258 34.9	S16 44.0
15	327 53.4	93 14.1	.. 51.9	35 39.5	.. 52.7	330 13.5	.. 22.9	131 30.3	.. 21.3	Spica	158 33.2	S11 13.2
16	342 55.9	108 ...	...	...	52.4	345 15.6	22.7	...	21.3	Suhail	222 53.4	S43 28.6
17	357 58.4	12...	...	...	52.1	0 17.8	22.6	...	21.3			
18	13 00.8	13...	...	...	S22 51.9	15 19.9	S 2 22.4	...	S ... 21.3	Vega	80 40.6	N38 47.7
19	28 03.3	15...	...	...	51.6	30 22.1	22.3	...	...	Zuben'ubi	137 07.6	S16 05.2
20	43 05.8	...	...	...	51.3	45 24.2	22.1	...	...		SHA	Mer. Pass.
21	58 08.2	18...	...	...	51.0	60 26.4	.. 22.0	221 44.5	.. 21.4		° ′	h m
22	73 10.7	198 14.0	55.9	... 22.2	50.7	75 28.5	21.8	236 46.8	21.4	Venus	126 56.7	8 47
23	88 13.2	213 14.0	56.4	155 ...	50.4	90 30.6	21.7	251 49.2	21.5	Mars	69 07.3	12 38
	h m									Jupiter	2 32.3	17 02
Mer. Pass.	17 12.0	v 0.0	d 0.6	v 0.4	d 0.3	v 2.2	d 0.1	v 2.4	d 0.0	Saturn	163 40.7	6 19

Table 2. NP 314-11 Daily page Jan 1st-3rd – Aries, Planet and Star data

Reeds Astro Navigation Tables take a different approach to data presentation[4.2]. Just three pages of tables are devoted to each month of the year in question. The first of each triplet lists the GHA and Declination for the Sun and the GHA of Aries, for every second hour of each day of the month.

The second page is devoted to the Moon and planets, giving the GHA and Declination of the Moon at 6 hourly intervals, together with their mean variation per hour. For each of the four navigational planets, Venus, Mars, Jupiter and Saturn; this page also lists for each day, the time of Greenwich meridian passage together with GHA and Declination with their mean variations per hour.

The third page for each month is initially devoted to the stars, giving their magnitude, time of Greenwich transit, Declination and Sidereal Hour Angle for the first day of the relevant month, of 60 stars deemed useful for navigation. This page also gives further details of our own star, the Sun; for each day is given the Greenwich transit time and semi-diameter, together with the times of twilight, sunrise and sunset at Greenwich and a correction factor for these times for observations at other latitudes.

Lastly for the Moon, for each day is listed the Moon's age, the time of the Moon's Greenwich transit, the extra correction for its orbit around the Earth, its varying semi-diameter and a value for horizontal parallax. The time at Greenwich latitude for moonrise and moonset and details of the Moon's phases are also given.

Polaris, our special North-pointing star that is useful in latitude determination in the northern hemisphere, has its own special place with a table in the ephemerides section in both NP 314 and *"Reeds"*. This is because its celestial position slightly varies on a daily basis, linked as with all stars, to the GHA Aries, as it very closely orbits the North Celestial Pole (**Section 9.18** p.269).

The key to understanding the ephemeral table data is to get access to a set and practice using the tables. The novice might find a set of *Reeds*, less initially daunting but it is less comprehensive. The seriously minded navigator, professional or amateur, soon progresses to a useful handling of the data in NP 314 – *the* Nautical Almanac.

– tables, tables, tables…………

2011 JANUARY 1, 2, 3 (SAT., SUN., MON.)

Table 3. NP 314-11 Daily page Jan 1st-3rd Sun and Moon & rising and setting data

The daily page contains columns for:
- **UT** (date and hour, d h)
- **SUN**: GHA and Dec
- **MOON**: GHA, v, Dec, d, HP
- **Lat.**
- **Twilight**: Naut., Civil
- **Sunrise**
- **Moonrise** for days 1, 2, 3, 4
- **Sunset**
- **Twilight**: Civil, Naut.
- **Moonset** for days 1, 2, 3, 4
- **SUN**: Eqn. of Time at 00h and 12h, Mer. Pass.
- **MOON**: Mer. Pass. Upper and Lower, Age, Phase

Callouts on the page indicate:
- GHA AND DEC OF MOON
- GHA AND DEC OF SUN
- MORNING TWILIGHT AND SUNRISE
- EVENING TWILIGHT AND SUNSET
- DATE
- DAY

Astro-navigation from Square One

4.3 *Tables of Conversion* – time's distance explored?

Arc to Time - this important conversion table is that of arc angle at the surface of the Earth, measured in degrees and minutes, to time, measured in hours, minutes and seconds, **Section 1.16, Table 1**, p.26.

The cornerstone of the arc to Time table being that the Earth turns through 360° of arc in 24 hours. From this follows that 1 hour of time is equivalent to 15° of arc (15 x 24 = 360). As suggested in **Example 2** (p.30), be aware of a possible confusion as "minutes" and "seconds" can refer to both arc and time. It is best to think in terms of "arc minute" referring to angular measure and "minute" referring specifically to time. Likewise, it seems safest to use the terms "arc second" for angular measure and a "second" for time, although arc-seconds are rarely used, being replaced by decimals of an arc-minute. One arc minute (1´) is equal to 4 seconds (4s) of time and fifteen arc minutes (15´) to 1 minute (1m) of time.

The standard arc to time conversion table usually lists whole degrees from 1-360° and in the adjacent column, the equivalent in numbers of hours and minutes. A further column follows for converting arc minutes from 1-60´ to minutes and seconds of time.

Example 8. Using the **Table 1** (p.28) to convert 241° 14´ to time:

```
241°    =    16h 04m
14´     =    00m 56s +
             16h 04m 56s
```

Alternatively the navigator can use a technique involving direct division taught to professional navigators and called "longitude in time", easily handled on a calculator. Angular distance is converted to time by dividing by 15, based on 1 hour of time's equivalence to 15° of the Sun's movement westwards. Modern scientific calculators usually have a specific button for entering degrees, minutes and seconds of arc, so avoiding the necessity of converting minutes of arc to decimals of a degree. A typical calculator might require the input:

241 [° ´ ˝] 14 [° ´ ˝]

Both arc and time are in hexadecimal notation (where 60 is used for the division of both degrees and hours) so that dividing arc by 15 to give time is simple. Some calculators describe divisions of 1´ as seconds of arc (0-60˝) and as decimal divisions of a minute are read from a sextant, then the latter will have to be converted to seconds of arc for calculator input etc. (0.2´ = 12˝, 0.4´ = 24˝, 0.6´ = 36˝, 0.8´ = 48˝). Some automatically convert hexadecimal input to decimals of a degree and others allow the input of decimals of a minute. All need practice.

It is worth investing in a calculator handling the direct input of arc but if an older type is used, the complexity of using decimals of a degree and hours might at first seem cumbersome but it is an easy habit to acquire. Francios Meyrier devotes much of his "astro" book to it [4.3].

– tables, tables, tables…………

4.4 Tables of Increments and Corrections - precautionary words?

First a word or two of caution and explanation: Some almanacs may slightly confuse "correction tables" and "increments tables". An increments table is one that lists fractional changes in a predictably variable quantity and is usually an <u>additional</u> table used subsequent to another table and gives details of fractions of time or angle.

I prefer that the title of <u>correction</u> table indicates that a particular type of reading is <u>amended</u> because of a known variation or potential error much like a compass reading is corrected for variation. A table in which a reading such as an observed altitude is amended, because of the inclusion of a potential error due to the height of observing eye above the surface of the sea is rightly, in my view, a "correction table".

4.5 *Tables of Increments* – how little bits add up to a lot?

An increment is by definition, a <u>finite increase</u> in a <u>variable quantity</u> and in astro-navigation it usually refers to the necessary additions to a whole hour figure so as to adjust it for both minutes and seconds of time. A typical example would be the increments for changes occurring each minute and second in Greenwich Hour Angle of an observed body, that need to be added to the whole hour figure listed in the daily table. Accuracy is vitally important in fixing position and this is usually based on an exact timing to the second of an altitude observation, which in turn relies on an exact to the second comparison of the calculated altitude of the body on the celestial sphere. It is therefore necessary to adjust a whole-hour figure incrementally for the exact moment of observation.

NP 314 has one increments table running to 29 pages, each of which gives adjustment details for each and every second for two specific minutes per page. As an example, part of the increments table for 32 and 33 minutes is given in **Table 4.** Each NP 314 increments table page also and initially rather confusingly, includes two important <u>corrections</u>; *v* and *d* (see next paragraph). These figures are placed here for convenience and ease of use and most importantly, the figure extracted <u>varies</u> with the particular minute used.

32ᵐ INCREMENTS AND CORRECTIONS 33ᵐ

s	32 SUN PLANETS	ARIES	MOON	v or d Corrⁿ	v or d Corrⁿ	v or d Corrⁿ	s	33 SUN PLANETS	ARIES	MOON	v or d Corrⁿ	v or d Corrⁿ	v or d Corrⁿ
00	8 00·0	8 01·3	7 38·1	0·0 0·0	6·0 3·3	12·0 6·5	00	8 15·0	8 16·4	7 52·5	0·0 0·0	6·0 3·4	12·0 6·7
01	8 00·3	8 01·6	7 38·4	0·1 0·1	6·1 3·3	12·1 6·6	01	8 15·3	8 16·6	7 52·7	0·1 0·1	6·1 3·4	12·1 6·8
02	8 00·5	8 01·8	7 38·6	0·2 0·1	6·2 3·4	12·2 6·6	02	8 15·5	8 16·9	7 52·9	0·2 0·1	6·2 3·5	12·2 6·8
03	8 00·8	8 02·1	7 38·8	0·3 0·2	6·3 3·4	12·3 6·7	03	8 15·8	8 17·1	7 53·2	0·3 0·2	6·3 3·5	12·3 6·9
04	8 01·0	8 02·3	7 39·1	0·4 0·2	6·4 3·5	12·4 6·7	04	8 16·0	8 17·4	7 53·4	0·4 0·2	6·4 3·6	12·4 6·9
05	8 01·3	8 02·6	7 39·3	0·5 0·3	6·5 3·5	12·5 6·8	05	8 16·3	8 17·6	7 53·6	0·5 0·3	6·5 3·6	12·5 7·0
06	8 01·5	8 02·8	7 39·6	0·6 0·3	6·6 3·6	12·6 6·8	06	8 16·5	8 17·9	7 53·9	0·6 0·3	6·6 3·7	12·6 7·0
07	8 01·8	8 03·1	7 39·8	0·7 0·4	6·7 3·6	12·7 6·9	07	8 16·8	8 18·1	7 54·1	0·7 0·4	6·7 3·7	12·7 7·1
08	8 02·0	8 03·3	7 40·0	0·8 0·4	6·8 3·7	12·8 6·9	08	8 17·0	8 18·4	7 54·4	0·8 0·4	6·8 3·8	12·8 7·1
09	8 02·3	8 03·6	7 40·3	0·9 0·5	6·9 3·7	12·9 7·0	09	8 17·3	8 18·6	7 54·6	0·9 0·5	6·9 3·9	12·9 7·2
10	8 02·5	8 03·8	7 40·5	1·0 0·5	7·0 3·8	13·0 7·0	10	8 17·5	8 18·9	7 54·8	1·0 0·6	7·0 3·9	13·0 7·3
11	8 02·8	8 04·1	7 40·8	1·1 0·6	7·1 3·8	13·1 7·1	11	8 17·8	8 19·1	7 55·1	1·1 0·6	7·1 4·0	13·1 7·3
12	8 03·0	8 04·3	7 41·0	1·2 0·7	7·2 3·9	13·2 7·2	12	8 18·0	8 19·4	7 55·3	1·2 0·7	7·2 4·0	13·2 7·4
13	8 03·3	8 04·6	7 41·2	1·3 0·7	7·3 4·0	13·3 7·2	13	8 18·3	8 19·6	7 55·6	1·3 0·7	7·3 4·1	13·3 7·4
14	8 03·5	8 04·8	7 41·5	1·4 0·8	7·4 4·0	13·4 7·3	14	8 18·5	8 19·9	7 55·8	1·4 0·8	7·4 4·1	13·4 7·5
15	8 03·8	8 05·1	7 41·7	1·5 0·8	7·5 4·1	13·5 7·3	15	8 18·8	8 20·1	7 56·0	1·5 0·8	7·5 4·2	13·5 7·5
16	8 04·0	8 05·3	7 42·0	1·6 0·9	7·6 4·1	13·6 7·4	16	8 19·0	8 20·4	7 56·3	1·6 0·9	7·6 4·2	13·6 7·6
17	8 04·3	8 05·6	7 42·2	1·7 0·9	7·7 4·2	13·7 7·4	17	8 19·3	8 20·6	7 56·5	1·7 0·9	7·7 4·3	13·7 7·6
18	8 04·5	8 05·8	7 42·4	1·8 1·0	7·8 4·2	13·8 7·5	18	8 19·5	8 20·9	7 56·7	1·8 1·0	7·8 4·4	13·8 7·7
19	8 04·8	8 06·1	7 42·7	1·9 1·0	7·9 4·3	13·9 7·5	19	8 19·8	8 21·1	7 57·0	1·9 1·1	7·9 4·4	13·9 7·8
20	8 05·0	8 06·3	7 42·9	2·0 1·1	8·0 4·3	14·0 7·6	20	8 20·0	8 21·4	7 57·2	2·0 1·1	8·0 4·5	14·0 7·8
21	8 05·3	8 06·6	7 43·1	2·1 1·1	8·1 4·4	14·1 7·6	21	8 20·3	8 21·6	7 57·5	2·1 1·2	8·1 4·5	14·1 7·9
22	8 05·5	8 06·8	7 43·4	2·2 1·2	8·2 4·4	14·2 7·7	22	8 20·5	8 21·9	7 57·7	2·2 1·2	8·2 4·6	14·2 7·9
23	8 05·8	8 07·1	7 43·6	2·3 1·2	8·3 4·5	14·3 7·7	23	8 20·8	8 22·1	7 57·9	2·3 1·3	8·3 4·6	14·3 8·0
24	8 06·0	8 07·3	7 43·9	2·4 1·3	8·4 4·6	14·4 7·8	24	8 21·0	8 22·4	7 58·2	2·4 1·3	8·4 4·7	14·4 8·0
25	8 06·3	8 07·6	7 44·1	2·5 1·4	8·5 4·6	14·5 7·9	25	8 21·3	8 22·6	7 58·4	2·5 1·4	8·5 4·7	14·5 8·1
26	8 06·5	8 07·8	7 44·3	2·6 1·4	8·6 4·7	14·6 7·9	26	8 21·5	8 22·9	7 58·7	2·6 1·5	8·6 4·8	14·6 8·2
27	8 06·8	8 08·1	7 44·6	2·7 1·5	8·7 4·7	14·7 8·0	27	8 21·8	8 23·1	7 58·9	2·7 1·5	8·7 4·9	14·7 8·2
28	8 07·0	8 08·3	7 44·8	2·8 1·5	8·8 4·8	14·8 8·0	28	8 22·0	8 23·4	7 59·1	2·8 1·6	8·8 4·9	14·8 8·3
29	8 07·3	8 08·6	7 45·1	2·9 1·6	8·9 4·8	14·9 8·1	29	8 22·3	8 23·6	7 59·4	2·9 1·6	8·9 5·0	14·9 8·3
30	8 07·5	8 08·8	7 45·3	3·0 1·6	9·0 4·9	15·0 8·1	30	8 22·5	8 23·9	7 59·6	3·0 1·7	9·0 5·0	15·0 8·4
31	8 07·8	8 09·1	7 45·5	3·1 1·7	9·1 4·9	15·1 8·2	31	8 22·8	8 24·1	7 59·8	3·1 1·7	9·1 5·1	15·1 8·4
32	8 08·0	8 09·3	7 45·8	3·2 1·7	9·2 5·0	15·2 8·2	32	8 23·0	8 24·4	8 00·1	3·2 1·8	9·2 5·1	15·2 8·5
33	8 08·3	8 09·6	7 46·0	3·3 1·8	9·3 5·0	15·3 8·3	33	8 23·3	8 24·6	8 00·3	3·3 1·8	9·3 5·2	15·3 8·5
34	8 08·5	8 09·8	7 46·2	3·4 1·8	9·4 5·1	15·4 8·3	34	8 23·5	8 24·9	8 00·6	3·4 1·9	9·4 5·2	15·4 8·6
35	8 08·8	8 10·1	7 46·5	3·5 1·9	9·5 5·1	15·5 8·4	35	8 23·8	8 25·1	8 00·8	3·5 2·0	9·5 5·3	15·5 8·7
36	8 09·0	8 10·3	7 46·7	3·6 2·0	9·6 5·2	15·6 8·5	36	8 24·0	8 25·4	8 01·0	3·6 2·0	9·6 5·4	15·6 8·7
37	8 09·3	8 10·6	7 47·0	3·7 2·0	9·7 5·3	15·7 8·5	37	8 24·3	8 25·6	8 01·3	3·7 2·1	9·7 5·4	15·7 8·8
38	8 09·5	8 10·8	7 47·2	3·8 2·1	9·8 5·3	15·8 8·6	38	8 24·5	8 25·9	8 01·5	3·8 2·1	9·8 5·5	15·8 8·8
39	8 09·8	8 11·1	7 47·4	3·9 2·1	9·9 5·4	15·9 8·6	39	8 24·8	8 26·1	8 01·8	3·9 2·2	9·9 5·5	15·9 8·9
40	8 10·0	8 11·3	7 47·7	4·0 2·2	10·0 5·4	16·0 8·7	40	8 25·0	8 26·4	8 02·0	4·0 2·2	10·0 5·6	16·0 8·9
41	8 10·3	8 11·6	7 47·9	4·1 2·2	10·1 5·5	16·1 8·7	41	8 25·3	8 26·6	8 02·2	4·1 2·3	10·1 5·6	16·1 9·0
42	8 10·5	8 11·8	7 48·2	4·2 2·3	10·2 5·5	16·2 8·8	42	8 25·5	8 26·9	8 02·5	4·2 2·3	10·2 5·7	16·2 9·0
43	8 10·8	8 12·1	7 48·4	4·3 2·3	10·3 5·6	16·3 8·8	43	8 25·8	8 27·1	8 02·7	4·3 2·4	10·3 5·8	16·3 9·1
44	8 11·0	8 12·3	7 48·6	4·4 2·4	10·4 5·6	16·4 8·9	44	8 26·0	8 27·4	8 02·9	4·4 2·5	10·4 5·8	16·4 9·2
45	8 11·3	8 12·6	7 48·9	4·5 2·4	10·5 5·7	16·5 8·9	45	8 26·3	8 27·6	8 03·2	4·5 2·5	10·5 5·9	16·5 9·2
46	8 11·5	8 12·8	7 49·1	4·6 2·5	10·6 5·7	16·6 9·0	46	8 26·5	8 27·9	8 03·4	4·6 2·6	10·6 5·9	16·6 9·3
47	8 11·8	8 13·1	7 49·3	4·7 2·5	10·7 5·8	16·7 9·0	47	8 26·8	8 28·1	8 03·7	4·7 2·6	10·7 6·0	16·7 9·3
48	8 12·0	8 13·3	7 49·6	4·8 2·6	10·8 5·9	16·8 9·1	48	8 27·0	8 28·4	8 03·9	4·8 2·7	10·8 6·0	16·8 9·4
49	8 12·3	8 13·6	7 49·8	4·9 2·7	10·9 5·9	16·9 9·2	49	8 27·3	8 28·6	8 04·1	4·9 2·7	10·9 6·1	16·9 9·4
50	8 12·5	8 13·8	7 50·1	5·0 2·7	11·0 6·0	17·0 9·2	50	8 27·5	8 28·9	8 04·4	5·0 2·8	11·0 6·1	17·0 9·5
51	8 12·8	8 14·1	7 50·3	5·1 2·8	11·1 6·0	17·1 9·3	51	8 27·8	8 29·1	8 04·6	5·1 2·8	11·1 6·2	17·1 9·5
52	8 13·0	8 14·3	7 50·5	5·2 2·8	11·2 6·1	17·2 9·3	52	8 28·0	8 29·4	8 04·9	5·2 2·9	11·2 6·3	17·2 9·6
53	8 13·3	8 14·6	7 50·8	5·3 2·9	11·3 6·1	17·3 9·4	53	8 28·3	8 29·6	8 05·1	5·3 3·0	11·3 6·3	17·3 9·7
54	8 13·5	8 14·9	7 51·0	5·4 2·9	11·4 6·2	17·4 9·4	54	8 28·5	8 29·9	8 05·3	5·4 3·0	11·4 6·4	17·4 9·7
55	8 13·8	8 15·1	7 51·3	5·5 3·0	11·5 6·2	17·5 9·5	55	8 28·8	8 30·1	8 05·6	5·5 3·1	11·5 6·4	17·5 9·8
56	8 14·0	8 15·4	7 51·5	5·6 3·0	11·6 6·3	17·6 9·5	56	8 29·0	8 30·4	8 05·8	5·6 3·1	11·6 6·5	17·6 9·8
57	8 14·3	8 15·6	7 51·7	5·7 3·1	11·7 6·3	17·7 9·6	57	8 29·3	8 30·6	8 06·1	5·7 3·2	11·7 6·5	17·7 9·9
58	8 14·5	8 15·9	7 52·0	5·8 3·1	11·8 6·4	17·8 9·6	58	8 29·5	8 30·9	8 06·3	5·8 3·2	11·8 6·6	17·8 9·9
59	8 14·8	8 16·1	7 52·2	5·9 3·2	11·9 6·4	17·9 9·7	59	8 29·8	8 31·1	8 06·5	5·9 3·3	11·9 6·6	17·9 10·0
60	8 15·0	8 16·4	7 52·5	6·0 3·3	12·0 6·5	18·0 9·8	60	8 30·0	8 31·4	8 06·8	6·0 3·4	12·0 6·7	18·0 10·1

Table 4. NP 314-11 Minute increments and corrections page for 32 & 33m

– tables, tables, tables…………

4.6 *Tables of Corrections* – putting right potential errors?

A variety of astro-navigational values need to be corrected because of amendable variations in a body's movement or due to errors introduced into their measurement by the instrument used or the position from which the observation is taken. This type of table may relate to several measurements and calculations in the computation of Greenwich hour angle, Declination and Altitude. Examples of the *v* and *d* corrections for GHA and Dec respectively are shown in the right hand columns of the minutes tables in **Table 4**.

1. Dip correction

An introduced and correctable error in altitude assessment and one introduced at every observation is that due to the height of eye above sea level. This affects the distance of the apparent horizon; the angle between the sensible horizon level with the eye and the observed horizon being called the "dip" angle, introduced in **Section 2.2** (p.38). The dip correction needs to be subtracted from the corrected sextant altitude to give the apparent altitude figure (H_a).

NP 314 has correction tables combined on one page, for both dip and altitude of the Sun, stars and planets. This page is duplicated in the bookmark to NP 314 (**Table 5**). Both dip and altitude correction tables are designed as "critical tables"; a useful way of presenting correction data, when a defined range of values need to be corrected by a given fixed amount.

For example in **Table 5,** in the top right hand corner, the height of eye correction figures are given for 1, 1.5, 2, 2.5 and 3m eye heights, each value having its own unique correction figure. This is an ordinary arrangement of corrections. If the adjacent column is then examined (it starts with the value: 2.4m), each correction figure is appropriate for more than one height value. Eye heights of both 2.5m and 2.6m will be corrected by – 2.8´. The correction for 2.7m is corrected by the next correction value, 2.9´, as will 2.8m. This is the essence of a critical table. At certain "critical" values, the correction figure used changes to the next higher value.

Now look at the foot of the same column, the correction figures span a larger range of height values. Eye heights between 20.5 and 20.9, will be corrected by a figure of – 8.0. A height of 21.0m will be corrected with the next value: – 8.1´. A careful look at the Sun altitude correction table on the left side of the page also shows a "critical" arrangement. For example, in October to March inclusive, all lower limb apparent Sun altitudes between 9° 34´ and 9° 45´ inclusive are corrected by the +10.8´ figure. A value of 9° 46´ is corrected by the next figure.

2. Apparent Altitude correction

Apparent altitude (H_a= sextant altitude corrected for sextant index error and dip) is corrected to give observed altitude (H_o), the figure that would have been found if it had been possible to view a body from the Earth's centre. It may include one to three elements; corrections for parallax, semi-diameter and refraction.

As outlined in **Sections 2.3-2.5** (p.40-43), stars are so distant that they merely need a correction for refraction, which varies with the height of the body. Planets except for Venus are also essentially point light sources and require a correction for refraction and, as they are relatively near, both Venus and Mars require a very small parallax correction.

Astro-navigation from Square One

ALTITUDE CORRECTION TABLES 10°–90°—SUN, STARS, PLANETS

OCT.–MAR.	SUN	APR.–SEPT.		STARS AND PLANETS		DIP			
App. Alt.	Lower Limb	Upper Limb	App. Alt.	Lower Limb	Upper Limb	App Alt.	Corrⁿ	App. Alt.	Additional Corrⁿ	Ht. of Eye	Corrⁿ	Ht. of Eye	Corrⁿ	Ht. of Eye	Corrⁿ

° ′	′	′	° ′	′	′	° ′	′			m	′	ft.	′	m	′
9 33	+10·8	−21·5	9 39	+10·6	−21·2	9 55	−5·3		2011	2·4	−2·8	8·0		1·0	−1·8
9 45		−21·4	9 50	+10·7	−21·1	10 07	−5·2		VENUS	2·6	−2·9	8·6		1·5	−2·2
9 56	+11·0		10 02	+10·8	−21·0	10 20	−5·1		Jan. 1–Feb. 18	2·8	−3·0	9·2		2·0	−2·5
10 08	+11·1		10 14	+10·9	−20·9	10 32	−5·0		° ′	3·0	−3·1	9·8		2·5	−2·8
10 20	+11·2	−21·1		+11·0	−20·8	10 46	−4·9		41 +0·2	3·2	−3·2	10·5		3·0	−3·0
10 33	+11·3	−21·0				10 59	−4·8		76 +0·1	3·4	−3·3	See table			
10 46	+11·4	−20·9				14	−4·7		Feb. 19–Dec. 31	3·6		11			
11 00	+11·5	−20·8				29	−4·6		° ′	3·8	−3·4				
11 15	+11·6	−20·7				44	−4·5		0 +0·1	4·0	−3·5				
11 30	+11·7	−20·6				00	−4·4		60	4·3	−3·6				
11 45	+11·8	−20·5	11 53	+11·6	−20·2	12 17	−4·3		MARS	4·5	−3·7			24	−8·6
12 01	+11·9	−20·4	12 10	+11·7	−20·1	12 35	−4·2		Jan. 1–Dec. 31	4·7	−3·8	15·7		26	−9·0
12 18	+12·0	−20·3	12 27	+11·8	−20·0	12 53	−4·1		° ′	5·0	−3·9	16·5		28	−9·3
12 36	+12·1	−20·2	12 45	+11·9	−19·9	13 12	−4·0		0 +0·1	5·2	−4·0	17·4			
12 54	+12·2	−20·1	13 04	+12·0	−19·8	13 32	−3·9		60	5·5	−4·1	18·3		30	−9·6
13 14	+12·3	−20·0	13 24	+12·1	−19·7	13 53	−3·8			5·8	−4·2	19·1		32	−10·0
13 34	+12·4	−19·9	13 44	+12·2	−19·6	14 16	−3·7			6·1	−4·3	20·1		34	−10·3
13 55	+12·5	−19·8	14 06							6·3	−4·4	21·0		36	−10·6
14 17	+12·6	−19·7	14							6·6	−4·5	22·0		38	−10·8
14 41	+12·7	−19·6	14 5							6·9	−4·6	22·9			
15 05	+12·8	−19·5	15							7·2	−4·7	23·9		40	−11·1
15 31	+12·9	−19·4	15							7·5	−4·8	24·9		42	−11·4
15 59	+13·0	−19·3	16							7·9	−4·9	26·0			
16 27	+13·1	−19·2	16 43			17 27	−3·1			8·2	−5·0	27·1		44	−11·7
16 58	+13·2	−19·1	17 14			18 01	−3·0			8·5	−5·1	28·1		46	−11·9
17 30	+13·3	−19·0	17 47		−18·8	18 37	−2·9			8·8	−5·2	29·2		48	−12·2
18 05	+13·4	−18·9	18 2		−18·7	19 16	−2·8			9·2	−5·3	30·4		ft.	′
18 41	+13·5	−18·8	19	+13·2	−18·6	19 56	−2·7			9·5	−5·4	31·5		2	−1·4
19 20	+13·6	−18·7	19	+13·3	−18·5	20 40	−2·6			9·9	−5·5	32·7		4	−1·9
20 02	+13·7	−18·6	20 24	+13·4	−18·4	21 27	−2·5			10·3	−5·6	33·9		6	−2·4
20 46	+13·8	−18·5	21 10	+13·5	−18·3	22 17	−2·4			10·6	−5·7	35·1		8	−2·7
21 34	+13·9	−18·4	21 59	+13·6	−18·2	23 11	−2·3			11·0	−5·8	36·3		10	−3·1
22 25	+14·0	−18·3	22 52	+13·7	−18·1	24 09	−2·2			11·4	−5·9	37·6		See table	
23 20	+14·1	−18·2	23 49	+13·8	−18·0	25 12	−2·1			11·8	−6·0	38·9			
24 20	+14·2	−18·1	24 51	+13·9	−17·9						−6·1	40·1		ft.	′
25 24	+14·3	−18·0	25 58	+14·0	−17·8						−6·2	41·5		70	−8·1
26 34	+14·4	−17·9	27 11	+14·1	−17·7	28 5					−6·3	42·8		75	−8·4
27 50	+14·5	−17·8	28 31	+14·2	−17·6	30 2					−6·4	44·2		80	−8·7
29 13	+14·6	−17·7	29 58	+14·3	−17·5	31 5					−6·5	45·5		85	−8·9
30 44	+14·7	−17·6	31 33	+14·4	−17·4	33 43					−6·6	46·9		90	−9·2
32 24	+14·8	−17·5	33 18	+14·5	−17·3	35 28	−1·4			14·2	−6·7	48·4		95	−9·5
34 15	+14·9	−17·4	35 15	+14·6	−17·2					14·7	−6·8	49·8			
36 17	+15·0	−17·3	37 24	+14·7	−17·1					5·1	−6·9	51·3		100	−9·7
38 34	+15·1	−17·2	39 48	+14·8	−17·0					5·5	−7·0	52·8		105	−9·9
41 06	+15·2	−17·1	42 28	+14·9						6·0	−7·1	54·3		110	−10·2
43 56	+15·3	−17·0	45 29	+15·0	−16·8					6·5	−7·2	55·8		115	−10·4
47 07	+15·4	−16·9	48 52	+15·1	−16·7	52 16				6·9	−7·3	57·4		120	−10·6
50 43	+15·5	−16·8	52 41	+15·2	−16·6	56 09	−0·7			17·4	−7·4	58·9		125	−10·8
54 46	+15·6	−16·7	56 59	+15·3	−16·5					17·9	−7·5	60·5			
59 21	+15·7	−16·6	61 50	+15·4	−16·4					18·4	−7·6	62·1		130	−11·1
64 28	+15·8	−16·5	67 15	+15·5	−16·3					18·8	−7·7	63·8		135	−11·3
70 10	+15·9	−16·4	73 14	+15·6	−16·2					19·3	−7·8	65·4		140	−11·5
76 24	+16·0	−16·3	79 42	+15·7						19·8	−7·9	67·1		145	−11·7
83 05	+16·1	−16·2	86 31	+15·8	−16·0	87 03				20·4	−8·0	68·8		150	−11·9
90 00			90 00	+15·9	−15·9	90 00	0·0			20·9	−8·1	70·5		155	−12·1
										21·4					

App. Alt. = Apparent altitude = Sextant altitude corrected for index error and dip.

Annotations on table:
- CRITICAL SUN ALTITUDE CORRECTIONS
- "CRITICAL" DIP CORRECTIONS
- CRITICAL APPARENT ALTITUDE VALUE CORRECTIONS FOR 20° 03′ 20° 46′ p.151
- EXAMPLE 22 p.151 APPARENT ALTITUDE 23° 21.3′ > MAIN CORRECTION 13.8′
- EXAMPLE 26 p.167 APPARENT ALTITUDE 41° 03.3′ > MAIN CORRECTION 14.9′
- EXAMPLE 25 p.159 APPARENT ALTITUDE 82° 07.2′ > MAIN CORRECTION 15.8′

Table 5. NP 314-11 Apparent altitude and Dip correction tables: Sun, Stars, planets

– tables, tables, tables…………

The central column of **Table 5** gives the correction values to be applied to star and planet apparent altitudes (H_a), to achieve an observed altitude (H_o).

The Sun and Moon, having significant and varying diameters according to their apparent orbits, require a semi-diameter correction as well as those for both refraction and parallax. The altitude correction tables for the Sun, planets and stars give a combined figure for the necessary adjustments for each of the bodies, in the left-hand column of **Table 5**.

The Moon's proximity and varied orbit entails hourly adjustments for parallax, requiring a Moon-specific parallax adjustment table which NP 314 combines in column format with those for refraction, the use of upper and lower limb sights and augmentation. More detail will be given in the section specific to this body (**Section 7** p.179) where Moon augmentation will also be explained.

3. Greenwich Hour Angle minutes and seconds increments and the *v* correction

Greenwich hour angle, the equivalent of celestial longitude for a heavenly body would be an exact and unvarying 15° per hour for the Sun, if the Earth's orbit was exactly circular and the Earth's axis was vertical. The effects of orbit and axis cause cyclical variations in the apparent motion of bodies observed from Earth and give us seasons. They are explored in **Section 5**. The daily listed GHA figures for the Sun, Moon, planets and stars in NP 314 are carefully calculated to take into account these small cyclical variations that might otherwise cause errors. These variations are one of the prime reasons that an annual publication of values is necessary.

The increment tables, for additional minutes and seconds over and above the hourly figure are based on fixed rates of change values; 15° per hour for the Sun and planets, 15° 02.46′ for Aries and 14° 19.0′ for the Moon. Each minute of the hour has its own table in NP 314, listing the additional GHA value necessary for each minute period and from 1-60 seconds (**Table 4** p.92).

Additional values termed *v* values, give correction figures for GHA rate changes at variance with the base rates used to produce the tables. These are necessary for the planets, where a 3 day average *v* value is given and the Moon, where adjustments are listed for every hour, its motion being more complex. These values, found at the foot of the daily page for planets and adjacent the hourly GHA reading for the Moon are used to enter the *v* correction columns on the relevant minute page and are generally added (see **Table 2** p.87). Some values for Venus are subtracted, where indicated at the daily page foot.

The *v* value for Aries is always zero and for the Sun is negligible and so none are given. Examples are used in the body specific **Sections 6-9**.

4. Declination correction: *d*

Declination varies slowly for the Sun and planets and so hourly values averaged to give a single figure for the 3 days on each page is given at the foot of the daily page. This *d* value is then used on the relevant minute's increment and corrections page corresponding to the time of the observation. A correction figure is extracted and used to adjust the whole hour value. If declination is increasing (found by inspecting the daily page) it is added, if decreasing it is subtracted. Detail and examples are given in the Sun and planets sections (**Sections 6 and 8**).

Be aware that <u>no specific</u> minute and second increment is needed for the sluggishly changing declination of Sun and planets; it is catered for by the 3 day *d* value alone. This is initially confusing, as its value is taken from the relevant minute's increment page.

For the Moon, declination changes relatively quickly so an <u>hourly</u> rate of change *d* value is listed and this value is then used to enter the *d* correction column on the relevant minute's increment page according to the time of the observation. The value found is then used to adjust the hour figure, again added for increasing declination and subtracted if decreasing. More detail and examples are given in the Moon specific **Section 7** p.179.

4.7 *Tables for Calculated Altitude* – for the conception of the intercept?

For a given set of input data on a heavenly body from a point on the Earth's surface, at a known date and time, the calculated altitude can be worked out in a variety of ways:

1. longhand using mathematical tables

2. using a calculator and a mathematical equation

3. with specific sight reduction tables for calculated altitude and azimuth

4. with a programmed calculator or computer.

All four methods solve the spherical triangle given chosen latitude together with the local hour angle and declination for the body in question, as detailed in **Section 2.15** p.57. The relative size of the observed altitude found from a fully corrected sextant sight taken at the unknown position and the calculated altitude from the chosen position, give the magnitude and positioning of the intercept. The calculator and calculated altitude table methods are used in examples in **Sections 6-9**.

Several Calculated Altitude tables are available, so designed as not to need complex mathematical manipulations. Essentially the work is done for you. They vary in complexity, ease of use, the data required for entry into the tables and importantly, the <u>accuracy</u> of the end result.

Generally, simplicity and speed are traded off against a slight loss of accuracy in position fixing, which is rightly frowned upon by professional navigators and not generally taught in nautical colleges. Some calculated altitude tables require careful <u>interpolation</u> (see **Section 4.11**); the process of manipulating pairs of data to produce a required intermediate figure. The following outlines the major publications containing data from which Calculated Altitude can be derived:

***Sight Reduction Tables for Marine Navigation* NP 401** – Vols. 1-6. Published by the United Kingdom Hydrographic Office 1987.

Each volume covers 15° of latitude (0-15°, 15-30°, 30-45°, 45-60° etc) and all declinations (0-90° North and South), requiring LHA and declination to give altitude to an accuracy of 0.1 ´ and azimuth angle to one tenth of a degree, for all the heavenly bodies.

– tables, tables, tables…………

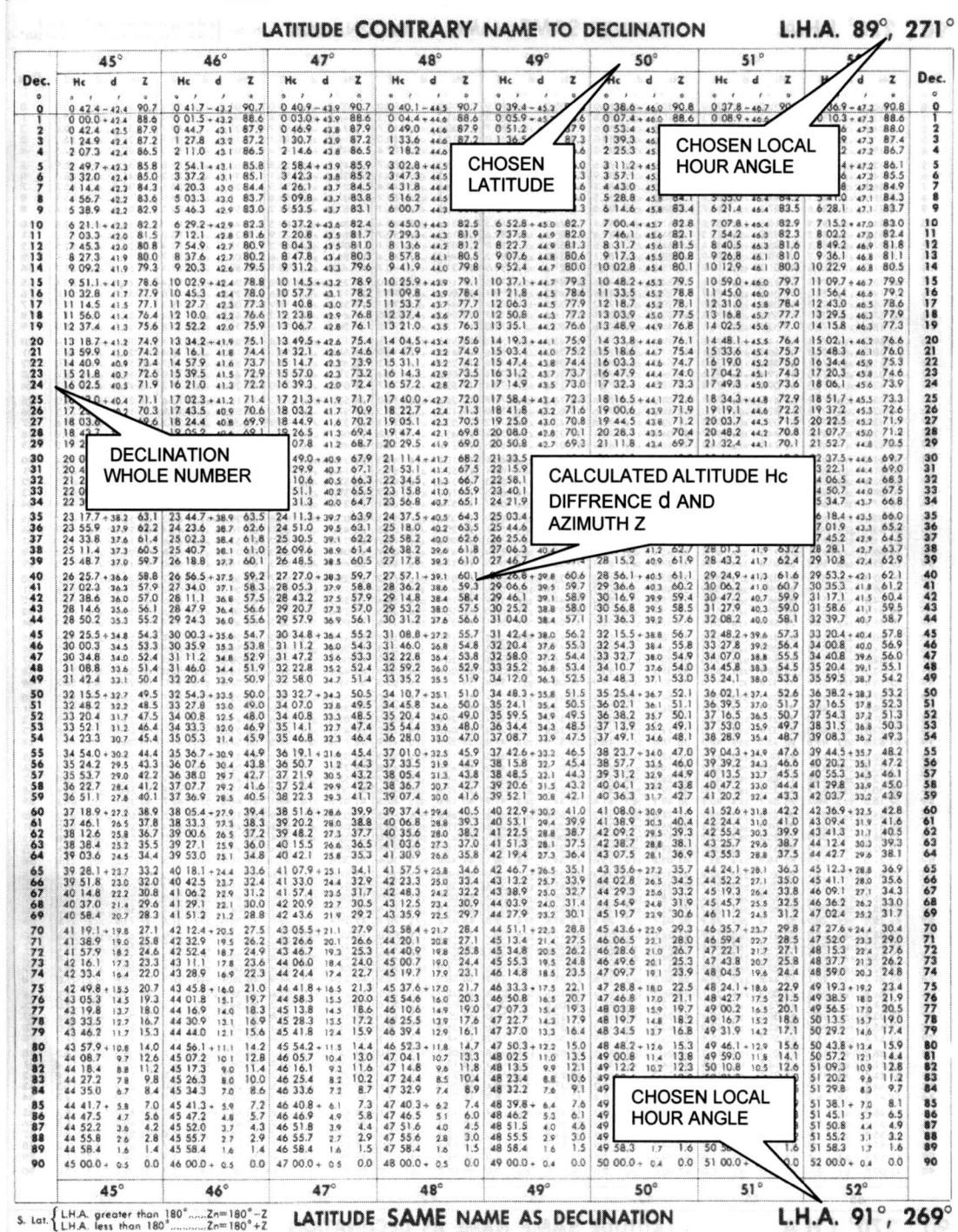

Table 6. Example of a Sight Reduction Tables from Marine Navigation *NP 401 (4)*

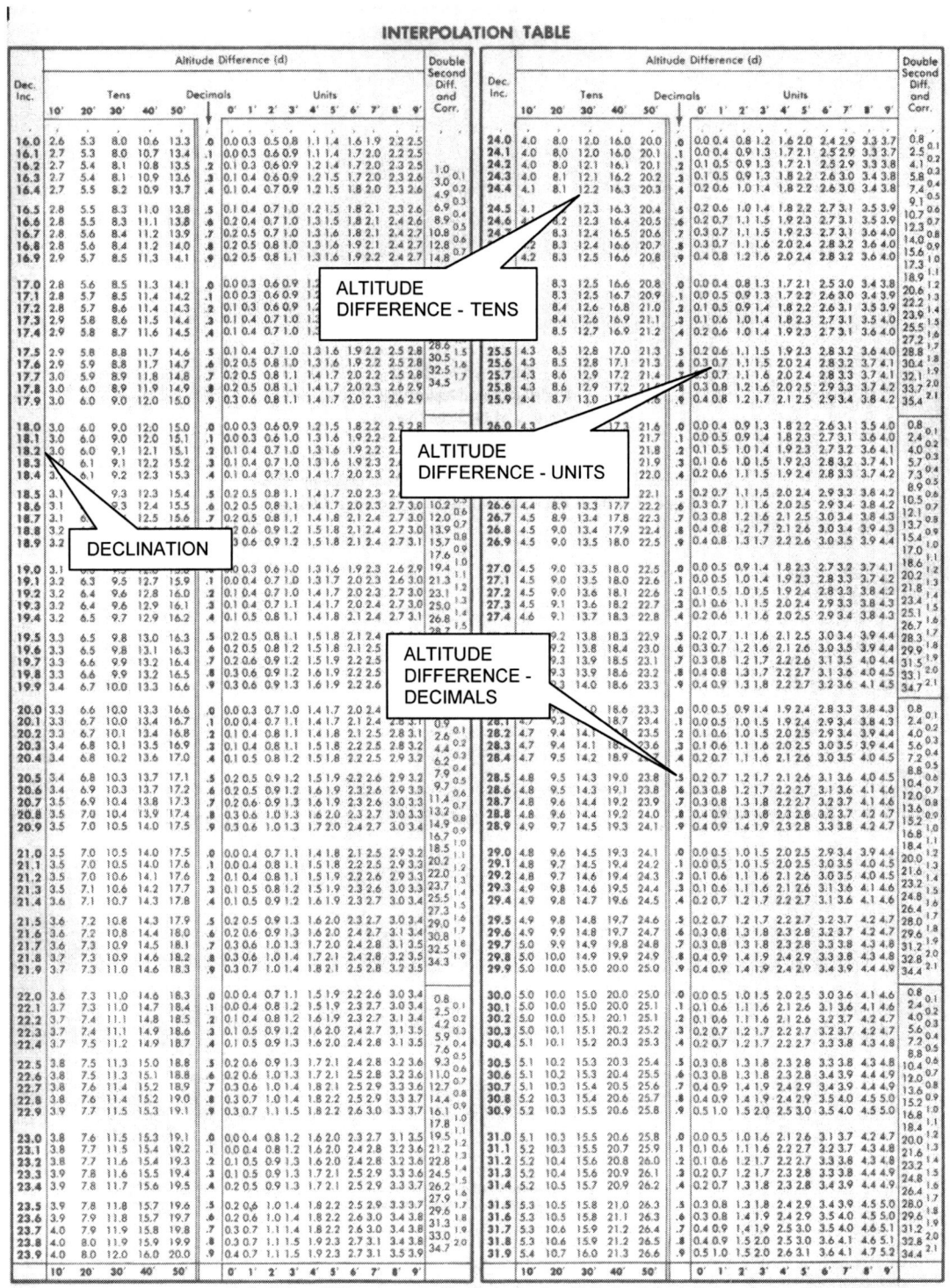

Table 7 NP 401 Interpolation table - minutes of declination - for calculated altitude

– tables, tables, tables………..

The tables, in 360 pages per volume, have each page pair relating to ascending and descending LHA values from 0-360°. The page for the correctly corresponding LHA is dependent on whether the chosen latitude and observed body's declination are in the same hemisphere ("latitude **same** name as declination") or opposite hemisphere ("latitude **contrary** name to declination"), indicated at the head and foot of the page (**Table 6**).

The entry criteria (arguments) for each page are the chosen latitude (arranged to be a whole number) and the sighted body's whole degrees of declination for the moment of observation. Three tabulated figures are found per column, each of which is headed by the latitude whole number. The figures are:-

    ° ′ H_c - the calculated altitude for whole degrees of declination

    ′ d   - the difference between the given altitude at that and the next whole degree of declination (– or +)

    ° Z   - Azimuth Angle

The calculate altitude derived is for the whole number of degrees of declination. The minutes of declination are adjusted for by using the **d** value in an interpolation table (**Table 7**). The figure extracted is added to the whole degree figure of calculated altitude. A complete example is worked in **Section 8 Example 33** p.224.

***Rapid Sight Reduction Tables for Navigation* NP 303** – vols. 1, 2 & 3. Published by HM Nautical Almanac Office, March 2010.

These volumes were originally designed for the navigation of aircraft (published as *AP 3270*), using bubble sextants but became popular with mariners because of their ease of use and acceptable accuracy of 0.5 nautical miles. There are three volumes:-

Vol.1 (ISBN: 978-0-7077-40683 Pages: 334, retail price: £25.90) is arranged such as to enable altitude and azimuth to be extracted rapidly for seven selected stars at all latitudes, using only a whole degree of LHA Aries and a whole degree of latitude as the entry requirements. Star declination is not needed. This volume aims to produce "the optimum selection of stars for a three-star fix" and to an accuracy of 0.5nm. An example of the table type (**Table 32**) is given on p.261. This method is used in **Section 9.15 Example 38** p.259.

Vol.2 (ISBN: 978-0-70-774-0188-Pages: 258, retail price: £25.90) Latitudes 0-39°.

Vol.3 (ISBN: 978-0-7077-4-1109- Pages: 354, retail price: £25.90) Latitudes 40-89°.

Vols. 2 & 3 produce solutions for bodies with declinations from 29° North to 29° South (which includes the Sun, Moon and planets). Both require as entry data the LHA and declination of the observed body from a chosen latitude and give a calculated altitude to 1° and an azimuth to 1°.

The data in NP 303 and NP 401, published by HM Nautical Almanac Office are also published, under different titles in the United States, by the Hydrographic and Naval Oceanographic Offices respectively. NP 303 Vol.3 is used in **Example 27, Section 6.16** p 172.

***The Nautical Almanac* NP 314-11**[4.1] (ISBN 978-0-7077-41062 – retail price £35.85). Published by HM Nautical Almanac Office, the UK Hydrographic Office 2010.

This annual publication, as well as its main content of daily ephemeris tables, also contains a set of concise tables from which calculated altitude and azimuth can be computed from the GHA and Declination of a body from a chosen position to an accuracy of 1nm. They are designed to be used when no more sophisticated method is available with entries of latitude and hour angle in whole degrees. The method used divides the PZX triangle into two right angled triangles, deriving data for each that is then added.

***Brown's Nautical Almanac* -2010** (ISBN 978-1-8492-7016-8: retail price £59.00) by T N Brown, published by Brown, Son & Ferguson, Ltd. (Glasgow) 2010.

This annual publication is aimed at the professional seaman navigator and as well as containing astro-navigational data it includes much sea-port information and inter-port distances for both coastal and deep-sea navigation. It is published each September.

Nories Nautical Tables (ISBN 978-0-85288-945-9 – 568 pages, retail price £24.00) edited by Capt. A G Blance, published by Imray Laurie Norie and Wilson, 2007.

These tables first published in 1803 champion the Marcq St Hilaire Haversine method to give a calculated altitude and contain much data useful to the professional sailor on ports of the world, including altitude correction tables. For attaining calculated altitude, more maths is required than with the publications above, but essentially this is only careful addition and subtraction. This method uses the formula for the solution of a spherical triangle into one using natural and logarithmic haversines and arrives at a zenith distance rather than an altitude (altitude = 90°– zenith distance –**Section 1** p.30). A haversine is half a versed sine (1-sin θ/2) – see **Section 4.8**. ABC Tables for the derivation of a quadrantal azimuth are included. Note that this volume uses the older traditional nomenclature of sextant, observed, apparent and true altitudes.

***Reeds Astro Navigation Tables* 2011** (ISBN 978-1-4081-2336-2: 80 pages, retail price £19.99) by Lt Cdr H J Baker, published by Adlard Coles Nautical 2010.

"*Reeds Tables*" are published each August for the following year by Adlard Coles Nautical and are aimed at the yachtsman. These tables champion the Versine method to arrive at a calculated altitude which, like *Norie's Nautical Tables,* produces a calculated zenith distance. The formula used is: *Versine* CZD = (*Versine* LHA x *cos* Lat x *cos* Dec) + *versine* (Lat ± Dec).

The tables give both natural and logarithmic versine values and logarithmic sine and cosine values, the logs being used to reduce potentially difficult multiplications of 4-figure decimals to simpler additions and subtractions.

The data is well laid out and has a section with worked examples. It fits easily in the chart table! ABC tables are included from which is worked out a quadrantal azimuth see **Section 4.8**. *Reeds* uses traditional altitude nomenclature, calling the observed altitude the "true" altitude.

– tables, tables, tables………...

4.8 *Azimuth Tables* (**ABC tables**) – giving direction to the plot?

In order to plot a position line when calculated and observed altitudes for a given body are compared, the navigator needs a bearing, the direction of the body for which the calculated altitude has been found. The azimuth is that direction and depends upon the local hour angle and declination of the observed body, together with the latitude from which it is observed. The tables are often known as "ABC" tables and are not the easiest of navigational tables with which to get to grips. I find the layout in *"Reeds"* and *"Norie's"* easy to use, the latter being more detailed and so needing less interpolation. The rules for naming A, B and C may vary slightly according to their source.

Table A is entered first with LHA and latitude and a figure extracted (A). Table B is entered with LHA and declination and a further figure extracted (B). A and B are also ascribed either +/– or North/South names (according to the table source) using fixed entry rules. The sum of A ± B is calculated and used with latitude to enter a third table, Table C. From this table an azimuth figure is extracted, in quadrantal notation, e.g. X° East or West of either due North or South, which can then be converted to a true bearing from 0-359°. See **Section 2.18-19** (p.62-63). An example follows in **Section 4.12 – Example 12 p.105**.

Calculator based methods for deriving a quadrantal azimuth are available (**Section 4.14 p.110**). They are much easier and quicker than the longhand method which may require considerable interpolation. But the use of the longhand method should be mastered. **NP 314**, **NP 303** and **NP 401** all produce, from their sight reduction tables, an azimuth for a given body from a chosen position as well as a calculate altitude.

4.9 Tables of Mathematical functions – all the rest put together?

Logarithms, Cosines and Log Cosines, Versines, Log Versines and Haversines.

All mathematical tables associated with astro-navigation have one aim: <u>to make difficult calculations easier.</u>
1. Logarithms are used to transform multiplications and divisions into additions and subtractions. They were widely used before slide-rules and calculators became readily available and are still used in longhand methods. A logarithm is the power, or "exponent", to which one number, the base, must be raised to give another number e.g. $10^2 = 100$, so that the log to the base 10 of 100 is 2. Base 10 is used in "common logarithms". The logarithm to base 10 of $1000 = 3$ and the sum 100 x 1000 in log form is therefore $2 + 3 = 5$, and the antilog of 5 is $10^5 = 100000$. For complicated numbers with decimal places, tables of logs to four or more figures are available.

Example 9: Use logarithms to multiply 3.142 x 6.137 (which by calculator = 19.2825)

$$\text{Log}_{10}\ 3.142 = 0.4972$$
$$\text{Log}_{10}\ 6.137 = 0.7879 +$$
$$1.2851 \quad \text{Antilog } 1.2851 = 19.2825$$

Logarithms of versines and haversines, used with log cosines remain very useful in the longhand methods of calculating altitude.

Cosines, **Sines** and **Tangents** are trigonometric functions; mathematical expressions of the relationship between the sides and the angles enclosed in a right-angled triangle. All can be used in various methods that solve the spherical triangle to arrive at a calculated altitude and an azimuth. They are all complicated numbers, usually to 4 decimal places and logarithmic values are used to simplify multiplication and division into addition and subtraction. The cosine, sine and tangent are considered in detail in the **Appendix** (p.277), which addresses both plane and spherical trigonometry.

Versines and **Haversines** are also "trigonometric functions", related to cosines and are respectively 1-cosine and 1-cosine/2 of an angle and their use evolved to aid in the accuracy of the long-hand methods in the mathematics of astro-navigation. Values of <u>small</u> angles are more accurately reflected by versines or haversines than in corresponding cosine values. Both log and natural (non-logarithmic) versines or haversines are usually presented in the same table, which gives them an initially fearsome look.

4.10 Lists and Indexes – last but not least?

For completions sake, there is one further type of table, that of lists of a star data. There are a finite number of stars useful to the astro-navigator, explored in detail in **Section 9**. Star classification can be based on a number of important variables; brightness or magnitude is often the most defining. NP 314 gives a tri-daily list of 57 of the most navigationally useful stars in alphabetical order, each with their SHA and declination. It also has a list, in order of Sidereal Hour Angle, of 173 stars in total, giving their monthly figure for SHA and Declination, which alters slightly due to the small irregularities in the Earth's orbit around the Sun, discussed in detail in **Section 5**.

Reeds Tables displays a monthly list of the magnitude of 60 useful stars, together with their magnitude, time of meridian transit, declination and sidereal hour angle. By listing stars according to their SHA from 0-360°, gives in ascending order, the true direction as they appear around the observer's horizon.

4.11 Interpolation – a work of the devil?

Some forms of data-containing tables in astro-navigation require **interpolation** and to the newcomer, interpolation can be daunting. The word interpolate is a verb and when used in its mathematical sense means to "fill in an intermediate term of a series". It is simply the method used to fill in gaps in an information table where, for brevity, or compactness, an "interpretation" of the available figures is made to provide an in-between figure that is not listed.

– tables, tables, tables…………

For the uninitiated examples will help, using the following extract of one form of a star altitude correction table which gives the correction figures for apparent altitudes between 30°-40° at heights of eye between 3-6m (**Table 8**):

I have introduced the spaces for clarity. Plainly if the altitude from which the sight was taken was exactly 30°, 35° or 40° and the height of eye had been exactly 3, 6 or 9m then no interpolation is required and the value is simply extracted. Simple interpolation "by eye" is all that is needed for the first example.

Example 10. Find the correction value for an altitude of 37° 30′ at an eye height of 6m?

Usefully, 37° 30′ is half way between 35° and 40°. The correction value required is therefore half way between the corresponding figures for a 6m eye height, namely 8.6 and 10.6. The difference between these two figures is 2.0. This value is then halved and the result added to the lower figure, making the required correction 9.6 as in the amended table (**Table 9**). This is the simplest of interpolations and can be done in the head.

Height of Eye (m)

Altitude (°)	3	6	9
30	4.0	6.0	9.0
35	6.4	8.6	11.2
40	8.6	10.6	14.2

Table 8. Extract from a Star altitude correction table

Height of Eye (m)

Altitude (°)	3	6	9
30	4.0	6.0	9.0
35	6.4	8.6	11.2
37° 30′		9.6	
40	8.6	10.6	14.2

Table 9. Interpolated extract from Star altitude correction table

A more difficult interpolation might require a little maths and can be greatly speeded by the use of a calculator:

Example 11. Find the eye height correction value for a sight taken at an altitude of 33° 45′ at an eye height of 6m.

This is easiest to resolve if minutes of arc are converted to decimals of a degree: 45′ = 45 ÷ 60 = 0.75°. The required value is then found in the following way: The altitude 33.75° is 3.75° greater than 30°. As a proportion of the 5° gap between 30° and 35°, 3.75° = 3.75 ÷ 5.0 = 0.75°. This proportion is then used to multiply the difference between the 30° and 35° values to find the correction at 6m eye height. This difference is 8.6 – 6.0 = 2.6. Multiplying 2.6 x 0.75 = 1.95, which is the required <u>additional</u> correction for 33.75°. This value is then added to the value for 30° to give the result: 6.0 + 1.95 = **7.95**, which is the correction value for a sight at an altitude of 33° 45′ and completes **Table 10**.

– tables, tables, tables…………

Height of Eye (m)

Altitude (°)		3	6	9
	30	4.0	6.0	9.0
	33° 45′		7.95	
	35	6.4	8.6	11.2
	37° 30′		9.6	
	40	8.6	10.6	14.2

Table 10. Extract from a Star altitude correction table with two interpolations

I certainly could not reliably do the maths used above in my head, so a calculator most definitely helps. It is not always necessary to be exactly accurate (although accuracy is something to which navigators must, of course, strive) and rounded figures <u>may</u> suffice in <u>certain</u> circumstances (see **Section 4.12**).

4.12 ABC Azimuth tables and interpolation – a combination too far?

ABC Tables are designed to produce an **azimuth**, the directional bearing to a heavenly body from a chosen position and on which is plotted an intercept and position line. An extract of a typical Table A is shown in **Table 11** and an example will be worked through. It may <u>sometimes</u> be possible with experience to interpolate by eye from ABC tables to obtain an adequately accurate azimuth, because of the fact that it is not feasible to manually plot a bearing to an accuracy of greater than one degree (**Section 4.13** p.110).

Example 12. Find the azimuth for a body with a local hour angle of 5° 30′ and a declination of 37° 30′ from a chosen latitude of 41° North. Table A is typically entered with latitude in the vertical column on the left and LHA in the top <u>or</u> bottom horizontal row, from which the figures have been extracted for interpolation in **Table 11**.

Table A LHA

LAT	5°		6°
40°	9.59		7.98
42°	10.3		8.57

Table 11. Basic extract of table A

The Table A value for 41° is required and in this case the value needs to be interpolated from the given values for 40° and 42°. The LHA value for 5° 30′ needs to be interpolated from the given figures for 5° and 6°. The mathematics in this case is simple, as the chosen values of LHA and latitude are halfway between the given values. By subtracting the smaller from the higher value in each row <u>or</u> column will give a figure that is then divided in two. This figure is then added to the lower of each pair of figures and will enable the gap to be correctly filled. The rows have been completed in **Table 12**.

Table A LHA

LAT	5°	5° 30′	6°
40°	9.59	8.79	7.98
42°	10.3	9.44	8.57

Table 12. Extract of table A with initial interpolations

The figure for 41° can then be worked out in a similar way using the figures for 5° 30′ giving the required result, as in **Table 13**. Interpolation can be quite complicated when fractions of a degree are involved and rounding up or down is usually necessary.

Table A — LHA

LAT		5°	**5° 30′**	6°	
	40°	9.59	8.79	7.98	
LAT	**41°**		**9.12**		
	42°	10.3	9.44	8.57	

Table 13. Extract of table A with completed interpolations

The figure derived from Table A is 9.12. It is given the value of +9.12 as the hour angle was listed at the top of Table A (a somewhat bizarre rule, but nonetheless correct. The rule is usually printed at the top of the table). Had the hour angle in question been listed at the bottom of the table, then the figure derived would have been ascribed a negative value: –9.12.

Table B is then entered with LHA and Declination. It is of a similar arrangement to Table A and interpolation is again necessary. An extract of Table B is shown in **Table 14**, but on this occasion, for illustration, using the columns as the starting point.

Table B — LHA

DEC		5°	**5° 30′**	6°	
	36°	8.34		6.95	
DEC	**37°**	8.65	**7.93**	7.21	
	38°	8.96		7.47	

Table 14. Extract of table B with completed interpolation

The result is arrived at by multiplying the difference between the values for 5° and 6° by 0.5, as 37° is 1° larger than 36° and 1° is half the difference between 36° and 38°: 8.96 – 8.34 = 0.62. 0.62 x 0.5 = 0.31. 8.34 + 0.31 = 8.65. Similarly: 7.47 – 6.95 = 0.52. 0.52 x 0.5 = 0.26. 0.26 + 6.95 = 7.21. This is a fairly straight-forward interpolation that it is just possible to do in one's head. The 37° figure at 5° 30′ is derived in a similar way: 8.65 – 7.21 = 1.44. 1.44 x 0.5 = 0.72. 7.21 + 0.72 = 7.93

This figure derived from Table B (7.93) is negative in this case as both latitude and declination have the same name: they are in the same hemisphere. The rule is usually printed at the top of the table. Had latitude and declination been of different names, one South and the other North then the figure from Table B would be given a positive prefix.

The figures from Tables A and B are then added to produce the entry figure for Table C. In this case: 9.12 + (–7.93) = 9.12 – 7.93 = **1.19**. The second requirement for entry is latitude. An extract follows in **Table 15**.

Table C   C value

	1.10	1.20
40°	49.9	47.4
42°	50.7	48.3

LAT

Table 15.
The basic extract of table C

The interpolation in this extract is slightly more difficult when using the rows, as the required figure for 1.19 is not midway between 1.10 and 1.20. To derive the value for 40°, 47.4 is subtracted from 49.9 which gives a difference of 2.5°. The difference between 1.10 and 1.20 values for C is 0.1. The difference between 1.10 and 1.19 is just 0.09. As a proportion of 0.1, 0.09 is 0.9 (0.09 ÷ 0.1 = 0.9). Multiplying the 2.5° difference between the 40° readings by 0.9 equals 2.25 (2.5 x 0.9 = 2.25). The value at 1.19 is then arrived at by subtracting the 0.9 value of 2.25 from the 1.10 value of 49.9: 49.9 – 2.25 = **47.65**. This value is placed in **Table 16**.

– tables, tables, tables…………

Table C

	1.10	**1.19**	1.20
40°	49.9	**47.65**	47.4
41°		**48.23**	
42°	50.7	**48.8**	48.3

Table 16.
The completed extract of table C

In a similar way the figure of 48.80 is derived from the figures given for the latitude of 42°. The final figure of 48.35 is found for the latitude of 41°, by subtracting the value for 40° from that for 42°, dividing it by 2 (as 41 is midway between 40 and 42) and adding it to the value for 40° (48.8 – 47.65 = 1.15 ÷ 2 = .0.575 + 47.65 = 48.23).

Here comes the crunch: plotting sub-divisions of a degree by hand is not feasible therefore the rounded down figure of 48° would be adequate. **This is the result - 48°** and not attained that easily! It is lot of work, to just round up or down to a whole number.

The overall result of the addition of the values derived from table A and table B can be positive or negative, depending on the entry data to the table, derived in turn from the sight parameters. It is the sign that indicates either North or South in the final quadrantal notation. If positive then the figure derived from Table C, the azimuth determining value, is South in the northern hemisphere and if negative it is North in the northern hemisphere. The sign is not relevant when entering Table C itself, only for its result. The LHA of the sight indicates the eastern or western aspect of the azimuth in quadrantal notation. If the LHA is less than 180° it indicates the azimuth is West and if greater than 180° it indicates it is East.

Some tables, notably *Norie's* give the A and B figures a compass direction rather than a sign. A is named opposite to the latitude (North or South) except when the LHA is between 90-270°; it is then named the same as latitude. B is always named as the declination. The seemingly odd rules stem from the orientation changes that the PZX triangle undergoes when LHA passes 180° (**Section 2.12** p.55) and the relationship imposed between azimuth angle and azimuth.

In the worked example above, the result of A + B was positive so that the azimuth is South. The LHA was less than 180° so the azimuth was also West. So the overall quadrantal azimuth was **South 48° West**. In 360° notation this will be 180 + 48 = **228°**.

4.13 The problems of interpolating Azimuths by eye – a guesstimate?

Having worked through the paragraphs above the navigator can see that considerable work is required for a result that is, when finally reached, rounded up or down to a whole degree. It is possible to do the same work by eye and in the above example and fairly readily arrive at the figure of 9 in Table A and 8 in Table B.

The problem comes in Table C, where the decimal divisions of A ± B become important in choosing the correct column from which to extract the C value. If the rounded figures of 9 and 8 are subtracted, giving the C entry figure of 1, then the resulting azimuth value for an observation from latitude 41° North is 53°. This is a 5° difference from the mathematically interpolated value of 48°, which is an unacceptable error.

4.14 *Practical alternatives to ABC Tables* – smoothing the troubled ABC brow

Whilst **Section 4** is devoted to an understanding of the use of astro-navigational tables and not formulae, it is nevertheless useful at this point to look other ways of achieving an **azimuth**. There are two practical alternative solutions to using the laborious ABC Tables and both use formulae. One formula, derived from the cosine formula, is easily and quickly performed on a scientific calculator. It uses the **LHA** and **Dec** of the observed body together with its **calculated altitude** (H_c) from the latitude and longitude of the chosen position, to give the **Azimuth**:

$$\sin \text{Azimuth} = (\sin \text{LHA} \times \cos \text{Dec}) \div \cos \text{Calculated Altitude}$$

The result is in **quadrantal notation** so that some recourse to the ABC tables is necessary for directional determination. There are worked examples in later sections (**Section 7.13 Example 30 p.198 & Section 9.14 Example 37 p.251**).

The second formula, really four formulae used in sequence and also applicable to a calculator is called the **ABC Formula**. It requires inputs of observed body LHA and Dec together with the Lat for the chosen position, but does not require the calculated altitude (H_c):-

$$A = \tan \text{Lat} \div \tan \text{LHA}$$

$$B = \tan \text{Dec} \div \sin \text{LHA}$$

$$C = A \pm B$$

$$\tan \text{Az} = (1 \div C \div \cos \text{Lat})$$

A is given the name North/South according to the LHA and latitude. If LHA is either 0-90° or 270-360°, A is given the name opposite to the name of latitude. If the LHA 90-270°, A is given the same name as the name of latitude.

B is given the same name as declination of the observed body: North or South.

C = A ± B: added if names of A and B are the same i.e. both North or South

– tables, tables, tables…………

: subtracted if names of A and B, i.e. North and South, are different.

The figures obtained are absolute and any negatives are ignored.

C is named North or South according to the sign of C after the sum: C=A±B. If C is positive, it is assigned the prefix South in North latitudes and North in South latitudes. If C is negative, it is assigned the prefix North in North latitudes and South in South latitudes.

The Azimuth is prefixed North or South, the same as the prefix of C and then suffixed East or West according to the LHA value. Azimuth is suffixed East if LHA > 180° and West if LHA < 180°.

A worked example is shown in **Section 7.13 Example 30** p.198.

The ABC formulas are relatively complex and there are circumstances where reasonable and safe short cuts can be made. A forenoon sight of the Sun will always be on an azimuth East of South and an afternoon sight West of South when viewed in the northern hemisphere from a latitude above the tropics (23° North). When viewing from the southern hemisphere, the reverse is true; below 23° South the forenoon Sun will always be East of North and the afternoon west of North. When taking sights in the region of the tropics plainly great care needs to be taken as to whether the Sun is sighted North OR South of the observer.

For sights of bodies other than the Sun these simple rules may not apply and it might be a reasonable course of action to take a bearing of the body in question, which will indicate within which quadrant a sight is being made. Sights close to azimuths of 90°, 180°, 270° and 360/000° will prove problematic and it <u>may</u> then be necessary to resort to the formal equations.

Section 5: The Sun and Time – their important link

This section describes the Sun and its intimate relationship to time. It explores the movement of the Earth in its solar orbit and explains its irregularities and their effects on time. There is a short diversion into the physics of planetary motion, from Newton but not quite to Hawking and the latest theories of time and space[5.1]. The relevance to the navigator of the useful inventions of Greenwich Mean Time and Zone Time are explained.

By the end of Section 5, the following KEY FACTS should be understood:-

- Apparent solar time, which varies cyclically, is the passage of time recorded by observing the Sun's movements to the West.

- Mean solar time is an artificial calculated and exact average of the passage of time, evening out the Sun's cyclical variations.

- The Equation of time is the daily difference between solar and mean time, found by subtracting solar time from mean time.

- Apparent solar time varies cyclically because of the summative effects of Earth's elliptical solar orbit and its fixed axial obliquity.

- Earth's elliptical orbit is caused by the Sun and the Earth's gravitational attraction for one another and which causes a variation in speed of the Earth in its orbit. The speed changes occur in a predictable fashion.

- Earth's fixed obliquity produces a change in the aspect of the Earth facing the Sun as the Earth orbits it over a year. This change, perceived on Earth as a cyclical changing of the Sun's declination, involves the Sun's movement in a direction other than that perceived as a westerly movement of the Sun and so affects time.

- There are other very small effects on Earth's motion such as precession and nutation.

- World time is divided into 24 time zones each of 15° of longitude in breadth, based on the Greenwich meridian and each country elects to use a specific mean time within its zone.

- When crossing to an adjacent time zone, an hour is subtracted from ship's time when travelling East and added when travelling West.

- Radio time signals are based on Universal Coordinated Time, measured by precise clocks that are related and adjusted to Universal Time which is itself based on mean solar time measured and recorded at Greenwich.

– the Sun and Time

5.1 Our Sun – what do we know about it?

As sailors and navigators we hugely rely on our Sun (Old English *sunne*, Latin: *sol*). Its age is 4.5 billion years and it probably has a 10 billion year life-expectancy. Termed a yellow dwarf star by astronomers, the Sun is not very bright when compared to other stars in the heavens when viewed from afar, but of course we see it from a very privileged position.

The Sun is unbelievably enormous, with a diameter of 1.4 million kilometres (over a hundred times that of Earth) and a volume of $1.4 \times 10^{27} m^3$ (1.3 <u>million</u> Earths). Its vast mass is of the order of 2×10^{30} kg (300 000 x Earth) and most surprisingly, it contains more than 99% of the <u>total mass</u> in our solar system. Its shape is almost a perfect sphere, rotating once every 25 days and orbiting around the centre of our galaxy just once in 250 million years.

The Sun consists primarily of the gases hydrogen (75%) and helium (24%), the two lightest gases and simplest atoms known, together with mere traces of heavier elements, revealed by spectroscopy from Earth. The gases exist as plasma; highly energized ionised atoms where, in the dense core at huge pressure, hydrogen ions fuse together producing helium. This reaction releases unimaginable amounts of energy, some 4×10^{23} kilowatts per second, generating a temperature of 13 million degrees.

The outer part of the Sun that we see is called the "photosphere" and it has much lower temperature of around 6000°. It is from here that energy is released as infra-red (heat) and light wavelengths, along with other more refined particles, which then radiate across the solar system and beyond into infinity. External to the photosphere lays the corona, visible from Earth during an eclipse of the Sun. This is a further region of great heat, reaching a temperature of a million degrees, but the processes producing this are incompletely clear.

By the time the Sun's released energy reaches Earth's upper atmosphere, having travelled 150 million kilometres, the incident power is 1.4 kw/m². This is then further attenuated by absorption and reflection in the atmosphere, reaching the surface at an approximate 1 kw/m², the equivalent of one bar of an old electric fire.

The Sun's radiation reaching the Earth is broad spectrum from the shorter ultra-violet and blue wavelengths, through green, yellow and orange, to the longest red and infra-red wavelengths. The human eye would ordinarily see this mix as white light but the atmosphere differently affects the spread of wavelengths. The longer wavelengths, green through to red, tend to pass unhindered straight through to the Earth's surface, but the shorter blue wavelengths tend to be absorbed then re-emitted unchanged by gas molecules, largely nitrogen (N_2) and oxygen (O_2). This emitted light is scattered widely through the atmosphere, causing the sky to appear blue to us as observers on the Earth. This is called "Rayleigh scattering", after the physicist Lord Rayleigh, who described it[5.2].

The speed of light is an enormous 3×10^8 m/sec and the Earth is 150×10^6 kilometres from the Sun. By dividing the distance by the speed, one finds that it takes 500 seconds, a little over 8 minutes for light to travel to the Earth. Having reached the Earth, man has from time immemorial, used the light from the Sun as his timepiece.

5.2 *Apparent and Mean Solar Time* – what are they and how are they related?

Time kept purely by observing the movements of the Sun is called **apparent solar time**. It is by some authors called true solar time. It is quite simply the passage of time that is apparent to us on Earth by watching the movements of the Sun across the celestial sphere. In the language of navigation, time is the hour angle of the Sun, its celestial longitude from a given meridian. It is the local time as determined by the Sun at that meridian.

The passing of time culminates in a solar day, that is, the interval between successive crossings by the Sun of a local meridian. A solar day is measured from midday to mid-day and its progress and passage would be indicated by the shadow cast on a sundial.

Careful observers, as far distant as the Babylonians and Ancient Greeks, found that there were irregularities in the Sun's movement, suggesting, quite correctly, that it did not keep perfect time[5.3]. Observing sundials and perhaps using instruments such as hourglasses and water-clocks, the Sun was seen not to pass across an observer's meridian at exactly the same moment each and every day. It seemed, for some periods of the year, that the Sun crossed an observer's meridian early and at other times it was late.

The amount of variation is quite considerable, often many seconds per day and these losses or gains are cumulative, with "short" or "long" days tending to occur in runs. Over the course of a year we will see that the Sun can be up to 14 minutes "early" and at other times up to 16 minutes "late". This would plainly play havoc with modern life. In 2011 for example, having picked a month at random, there is an 8 minute difference in the time of the Sun's crossing of the Greenwich meridian between the first and last days of March. On March 1st, the Sun's meridian crossing occurs at 1212 and on the 31st it happens at 1204.

Enlightened astronomers, adhering to the heliocentric rather than the geocentric view of our solar system (outlined in **Section 1.13** p.22) realised that this meant that the Earth's passage around the Sun was irregular and not the Sun itself. They also observed it was a "regular irregularity", the variation occurring at similar times each year and was not randomly changing.

5.3 *Mean Solar Time* – how does our world cope with our Sun's irregularities?

The modern world, for very good reasons, needs to run on uniform time and therefore man has invented the concept of **mean solar time**. This is a calculated artificial time that keeps to a fixed and very exact 24 hour cycle of an imaginary regular Sun, with each day being exactly the same length and running from mid-night to mid-night [5.4]. Every mean solar day is, to the nearest second, 86400 seconds long, the equivalent of 1440 minutes or 24 hours. It is solar time but with all the variations ironed out. In fact the mean Sun's movements tally exactly with those of the real Sun on only four occasions each year.

Mean solar time familiar to navigators as **Greenwich Mean Time** (GMT) [5.5], is properly called UT1 (**Universal Time** 1), based on an averaging of the Sun's movements that are tracked by the Royal Greenwich Observatory. In 1972 by international agreement, **Universal**

– the Sun and Time

Coordinated Time (UTC) was generally introduced, recorded by special Caesium-containing atomic clocks and to which the whole world's time is referenced. UTC can and often is adjusted to UT by the addition of "leap seconds" in June and/or December. GMT is, strictly speaking UT1 +12 hours, as UT is measured from midnight to midnight. When an accuracy of just 1 second is required, then it is reasonable for the navigator to equate GMT, UT and UTC.

It is useful that the term GMT is retained in the world of seamanship, given its time-honoured relationship with all things navigational. At Greenwich, the "mean Sun" is presumed to cross the Prime Meridian at exactly 12hrs 00 minutes and 00 seconds of each and every day, measured by clocks which do not follow the vagaries of the real Sun. Clocks and watches certified for accuracy are termed chronometers, which themselves can be corrected by reference to atomic clocks, originally invented in 1955 and held as standards which measure the passage of time to very many decimal places[5.6].

Until 1833, nautical almanacs listed all times as apparent solar times, as shipboard time was based exclusively on Sun observations. After 1833, almanacs were published using mean solar times as ship's time was by then, determined by the recently perfected chronometer, registering Greenwich Mean Time[5.7].

5.4 *The Equation of Time* – bridging the gap in time?

The difference between apparent solar time and mean time is the sum of various influences. This difference is calculated by subtracting mean solar time from apparent solar time and the resulting figure is called the **equation of time,** its daily value being listed in the astro-nautical almanac (**NP 314**), given as a time related to mean noon at Greenwich (see foot of **Table 3 Section 4 p.89**).

It is important for navigators to be aware of the daily figure as a correction for the difference will need to be made when planning and timing Sun sights in GMT/UT. The equation of time undergoes a sinusoidal change over a year and these changes can be readily illustrated in graphical form in **Figure 66**, which shows the daily difference in minutes, between Solar and Mean time through 2011, but any year gives an identical graph.

In the early weeks of the year the apparent solar time lags mean time and subtracting one from the other results in a negative number. This deficit increases, reaching a maximum of 14 minutes in February, indicating that the apparent Sun (that which we see) is slow when compared to the mean Sun. At this point it crosses the Greenwich meridian at 1214 hours Greenwich Mean Time, some 14 minutes after midday in our "mean" modern world. As the year moves on, the deficit decreases until mid-April when the equation is zero, apparent and mean solar time briefly coinciding.

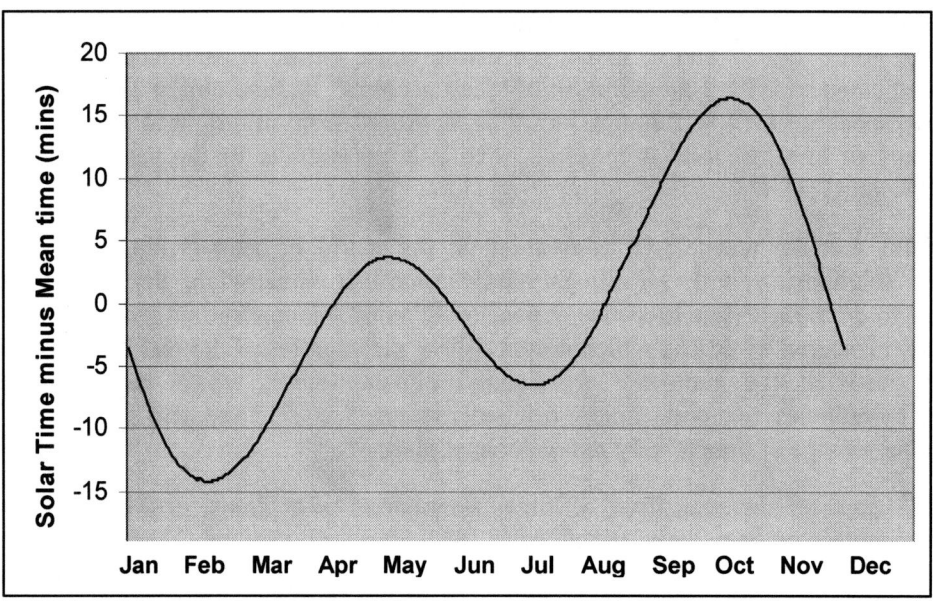

Figure 66. Graph of the Equation of time in 2011 and all years

The two time frames then diverge again, but this time the apparent Sun leads the mean Sun, the graph line becoming positive and peaking at 4 minutes in mid May. This is followed by a further reversal, the apparent Sun falling behind again, lagging by a maximum of 6 minutes in early August. Through September and October the situation is again changed with the apparent Sun catching and then leading the mean Sun up to a maximum of 16 minutes. The difference then decreases again into December.

The changes are cyclical but not cumulative beyond a year's span and the aspiring navigator needs to explore further understand their source.

5.5 *Apparent and Mean solar time* - what causes the time difference?

There are two main causes for the odd behaviour of apparent solar time; the elliptical shape of the Earth's orbit round the Sun and the fixed obliquity of the plane of the Earth's equator to the plane of that orbit. To begin, we must first look in more general terms at the movement of the Earth and other planets around the Sun, as this is crucial to an understanding of time on Earth. To achieve this requires a further delving into history and some basic physics.

Remember that our view of the solar system is a very biased one and is based on our less than central position on a relatively small rotating planet and where much that we see is moving relative to us and the Sun. For an understanding, it is easiest to look down from above our solar system and examine how the planets react to the solar engine that drives the system.

5.6 How our planets orbit the Sun - what makes them to move as they do?

Johannes Kepler (1571-1630)[5.8], a German astronomer and mathematician, described how our planets rotate around the Sun, having studied the very careful observations of fellow astronomer Tycho Brahe (1546-1601) and from which he formed three main conclusions, now known as Kepler's laws of planetary motion. Kepler saw from the data he examined, that planets must move around the Sun not in circles, but ellipses. The mathematical definition of an ellipse is relatively complex, involving an oblique section of a cone, but the detail need not unduly concern us.

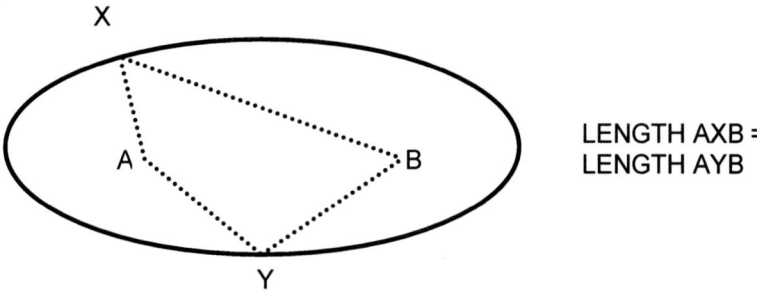

Figure 67. A simple ellipse showing its two focal points

Essentially an ellipse is a figure having two focal points A and B, as in **Figure 67**. The total length of lines joining the foci to any point on the circumference of the ellipse is constant. As the foci move towards each other, the ellipse becomes more circular and if they merge into one, the shape is an ellipse no longer, but a circle. In our solar system, the Sun is situated at one of the two focal points of the elliptical path of the Earth (**Figure 68**) and at one of the two focal points of the elliptical orbits of each of the other planets.

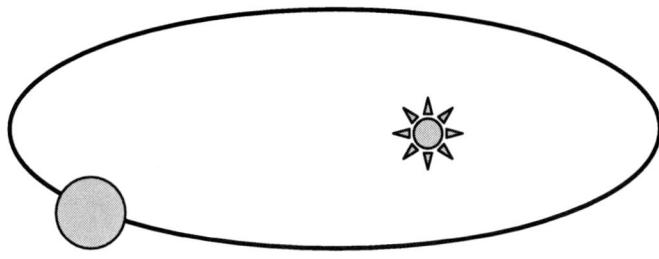

Figure 68. The Sun at one focus of the Earth's orbital ellipse

Kepler's second finding was that a line joining one focal point of an elliptical orbit to the circumference of the ellipse sweeps over equal areas in equal times, depicted in **Figure 69**, where the area of segment "A" equals the area of segment "B". This means, very significantly, that with the Sun at one focal point, a planet rotating about it must move faster while close to the Sun and <u>slower</u> when farther from it; a very important observation.

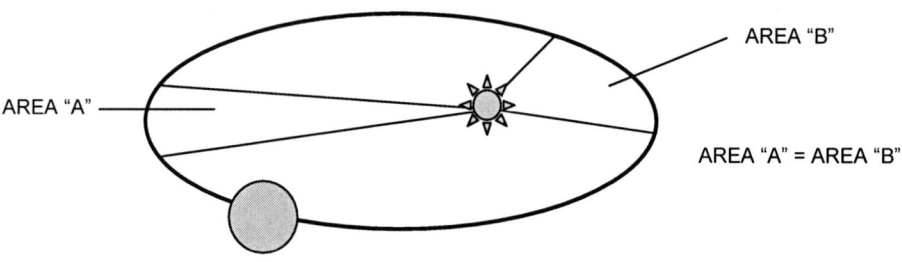

Figure 69. Kepler's second law of planetary motion

Thirdly and finally, Kepler found that larger orbits, such as those of Mars, Saturn and Jupiter, take longer to complete and that the speed of a planet in a larger orbit is lower than that of a planet in a smaller orbit. Just as Kepler predicted, the speed of the Earth does change through its yearly path which is approximately 58×10^{12} miles in length. Its <u>average</u> speed is an enormous 30 kilometers per second and this varies by up to 3%, approximately 1 km/sec., dependent upon just where the Earth is in its yearly cycle, but generally tending to speed up as its path moves more towards the Sun and slowing down as it path moves away.

5.7 Describing orbits by their Eccentricity - what does it measure?

The elliptical shape of an orbit can be described by a mathematical term called <u>eccentricity</u> (abbreviated to ε - *epsilon*, fifth letter of the Greek alphabet) [5..9]. Eccentricity can vary from 0 to 1, the bigger its value, the "longer" the ellipse. An ellipse has two axes, the major longer one and the minor shorter one (**Fig. 70**). The axes are related to eccentricity (ε) by the following equation:

$$\varepsilon = \sqrt{1 - \frac{b^2}{a^2}}$$

where **a** = half the length of the long axis, called the semi-major axis and **b** = half the length of the short axis, called the semi-minor axis.

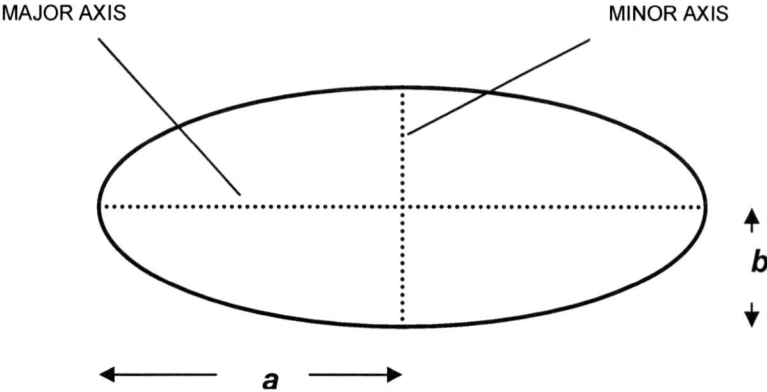

Figure 70. Defining an ellipse by its two axes: major and minor

The eccentricity of the Earth's path is low at 0.0167, indicating that its path is not a pronounced ellipse. Only Venus and Neptune have lower values but the highest is that of Mercury, which has an eccentricity of 0.2 and consequently, a much more pronounced elliptical movement around the Sun. The major diameter of our Earth's ellipse is 299.19 million kilometres and the minor diameter is only slightly less at 299.15 million kilometres; these two extremes differ by only 0.01%. The two foci of Earth's ellipse are only 5 million or so kilometres apart, and at one sits the Sun, which itself has a diameter of 1.4 million kilometres. So in reality the Earth's orbit is an *almost* circular ellipse.

5.8 Explaining planetary motion – why does the speed of a planet vary?

Isaac Newton (1643-1727) the great natural philosopher and mathematician, in his universal theory of gravitation, produced a vitally important scientific explanation of the findings of Kepler[5.10]. Newton very significantly described and explained the situation in which every object in the universe attracts every other object with a force (**F**), proportional to the sum of the mass of each body (**M$_1$** and **M$_2$**) and inversely proportional to the square of the distance (**d**) between them:

$$F = G (M_1 + M_2) / d^2$$

This means that the closer the masses, this bigger the force of attraction. Indeed, when the distance between the masses halves, the value of the attractive force is squared. **G** is the gravitational constant (6.67×10^{-11} Newtons/m^2/kg) and the unit of force, the Newton, is that force which imparts to a mass of 1kg an acceleration of 1m/sec^2.

G is a very tiny number but when multiplied by masses as huge as the Sun or that of a planet, the attractive force resulting from the above equation becomes very much more significant. Its existence can explain the planetary motion we see. The Sun's mass is 300 000 times that of the Earth and causes it to exert a very powerful gravitational pull on each planet.

This attractive force pervades space and is akin to that of the magnetic field around a magnet, which has an effect on iron containing objects some distance away from it and strengthening the closer the object gets to the magnet. Likewise, the closer the planet is to the Sun, the greater the attractive gravitational force. As a planet begins to approach the Sun, the increasing gravitational force accelerates the planet, its velocity (a vector which has both speed and direction) changes, increasing in speed and altering its direction as the planet moves in a tighter arc, more towards the Sun. This is the explanation of the elliptical course which leads eventually to the planet reaching its closest point to the Sun, called the **perihelion**.

Having passed this point, the distance between the Sun and the planet then begins to increase and accordingly, the gravitational pull decreases and the planet decelerates. Its speed and direction slowly change and the planet slowly veers away from the tightest part of its curving course, continuing on its elliptical path.

Ultimately the planet reaches its farthest point from the Sun, the **aphelion** and curving round as the gravitational pull begins to increase again, as the whole cycle repeats itself. So in summary it is gravity that causes the changes in the speed and direction of the Earth as it orbits the Sun.

5.9 How Earth's orbit affects apparent solar time – what causes this effect?

The Earth orbits the Sun in 365 days, moving through the full 360° at an average of just under one degree of arc per day. The rotation of the Earth about its <u>own</u> axis remains essentially unchanged through each of these periods, but in addition to its daily rotation, the movement of Earth in its solar orbit also imparts a small amount of rotation of the Earth around its axis, in the same direction as the eastwards axial spin. The <u>average daily</u> orbital angle moved is 0.99° (360/365 = 0.99°), or 59.2´ of arc. The additional rotation is illustrated in **Figure 71**, much exaggerated for clarity.

– *the Sun and Time*

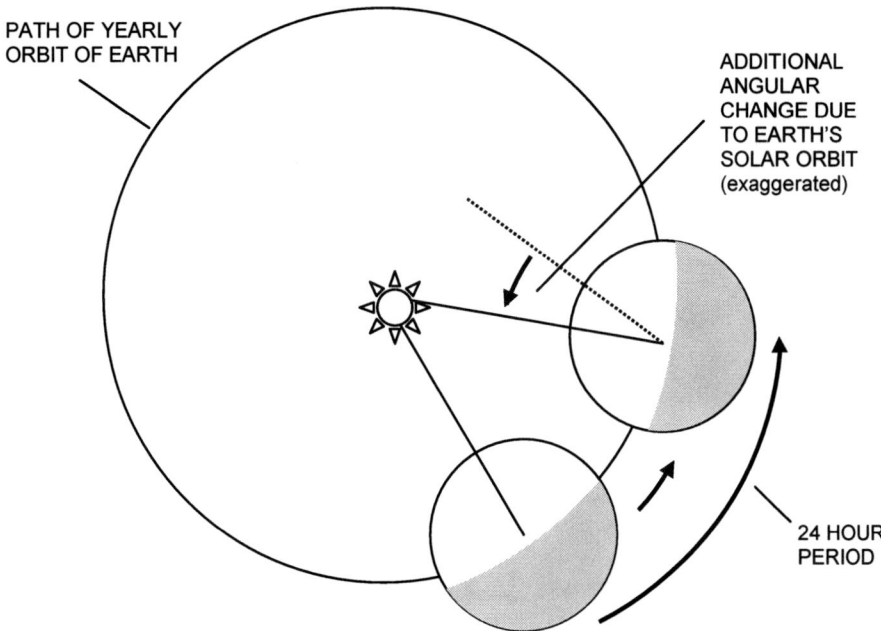

Figure 71. Earth's daily orbital movement, exaggerated to show the angular change

In the period where the Earth's speed is <u>increasing</u> it will move <u>slightly</u> <u>more</u> than 0.99° around its solar orbit in each day, as in **Figure 72**. Conversely, when the Earth is moving away from the perihelion towards the aphelion and slowing down, it moves slightly less than 0.99° of the way round its yearly orbit, in each earthly day, as in **Figure 73**.

Both of these speed changes directly affect the angular relationship with the Sun when viewed from Earth, so they must both affect apparent time as this depends upon the position of the Sun as seen from Earth.

The times of the year at which the Earth reaches perihelion and aphelion in its solar orbit, change slowly over very long time periods, but currently, the Earth reaches perihelion in early January and aphelion in July. This results in an increase in the speed of the Earth in its orbit in the latter months of our year peaking in late September.

As the distance of the Earth from the Sun decreases, the attractive gravitational force increases, causing the speed of the Earth to increase. This has the effect of turning the Earth a little more to the East, which in turn, when viewed from Earth, has the effect of making the Sun appear to move in a <u>westerly</u> direction and this apparent movement is seen as an increase in the passage of solar time.

Astro-navigation from Square One

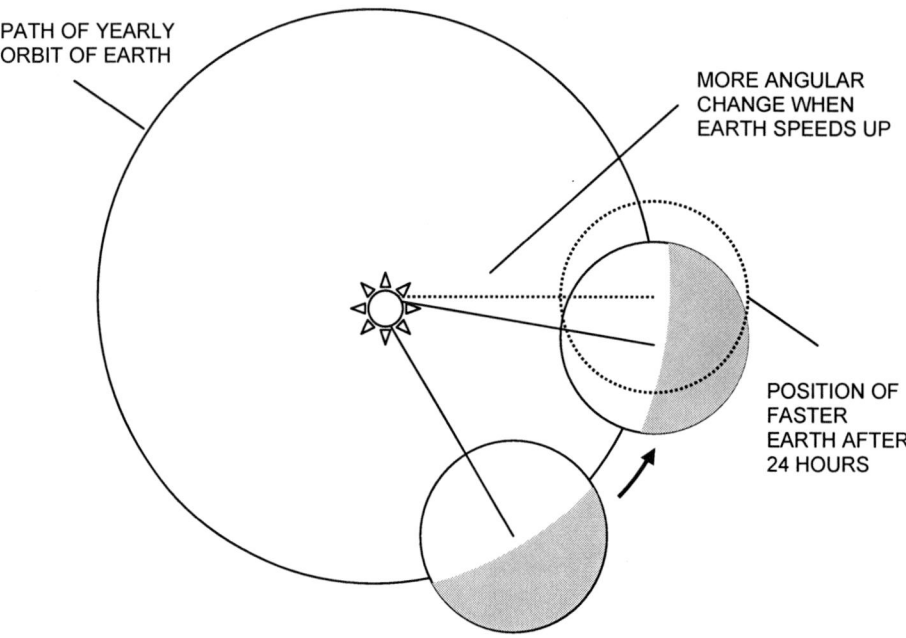

Figure 72. Showing the angular effect of Earth speeding up

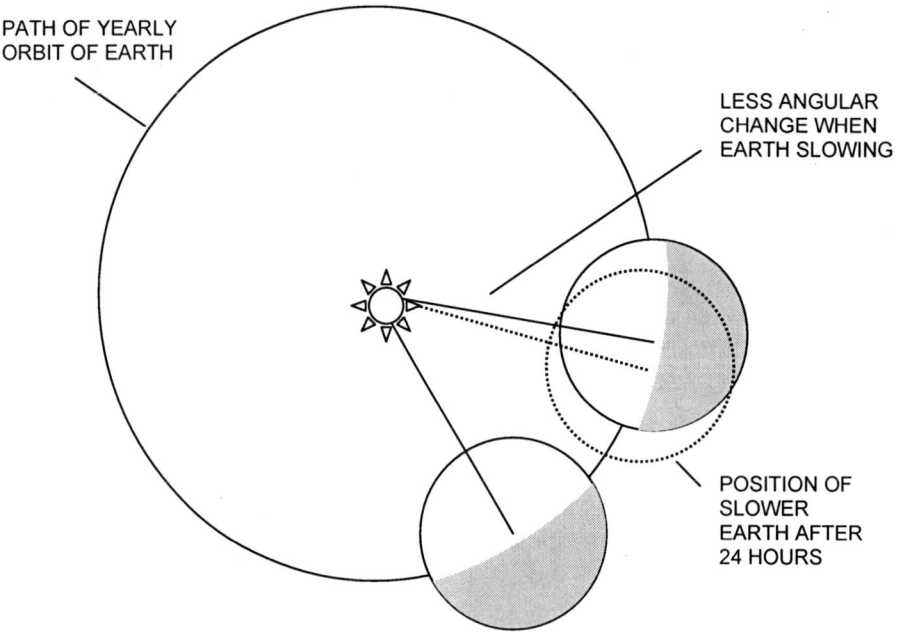

Figure 73. Showing the angular effect of Earth slowing down

– the Sun and Time

Conversely, Earth's speed decreases in the months after perihelion, the nadir (lowest point) being reached in early April. In this period solar orbital movement each day is less, so the Earth rotates to the East a fraction less, so the angular change in the relationship to the Sun is not as pronounced, resulting in less apparent movement of the Sun to the West. This has the effect of slowing solar apparent time when compared to the regular plod of mean time.

The average speed of the Earth's orbit is approximately 30 kilometres per second and during its yearly cycle this varies from 29.5 to 30.5km/s. The effect of the changing speed slows or hastens time by a figure averaging a few seconds a day. This is cumulative and can be plotted on a graph as in **Figure 74**. The resulting curve follows the shape of a sine wave and the daily incremental change results in a lag on mean solar time totalling almost 8 minutes in early April and lead of a similar amount in late September.

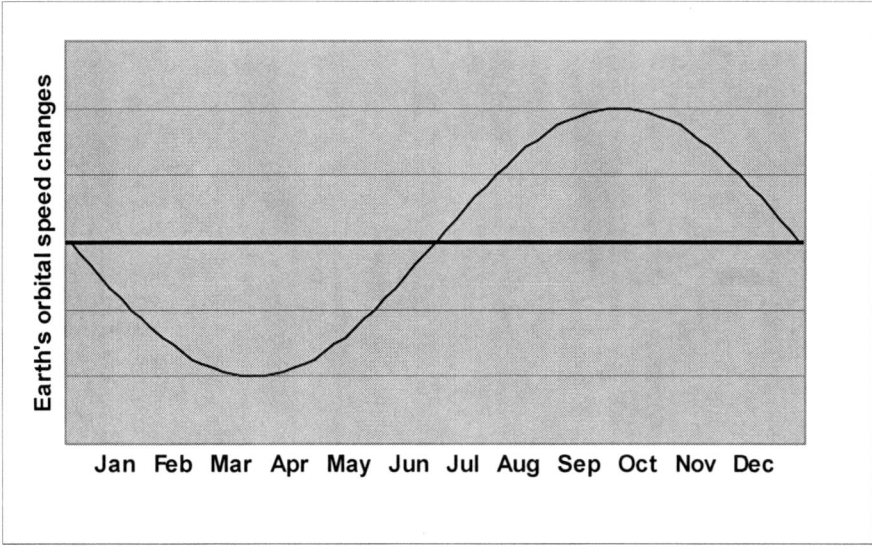

Figure74. The yearly variation of Earth's orbital speed

5.10 How Earths axial obliquity affects apparent solar time

The Earth's axis, around which it spins once in every 24 hours, is not vertical but has a fixed incline. The equatorial plane makes an angle, called the **obliquity**, of 23° 27′ with the plane of the Earth's orbit around the Sun. It is the presence of the obliquity that directly influences apparent solar time because it alters, from day to day, the apparent height of the Sun in the sky that we measure as continual changes in the Sun's declination.

As introduced in **Section 1.12** (p.20), it is the presence of the obliquity that results in the seasonal variation that we see, with the Sun rising higher in the sky in the summer months than those in winter. As Earth moves round the Sun over a year, the constant obliquity of the

Earth's equator to the path of the Earth's orbit causes a change in the aspect of the Earth presented to the Sun.

It is important to grasp that the angle of obliquity <u>does not change</u>, but its mere presence means that there is a change in the aspect or face of the Earth presented to the Sun as the Earth progresses around it on its yearly orbit, seen by the observer on Earth as a change in the Sun's declination. It is worth reviewing the detail of **Figure 17** from **Section 1.12** p.20.

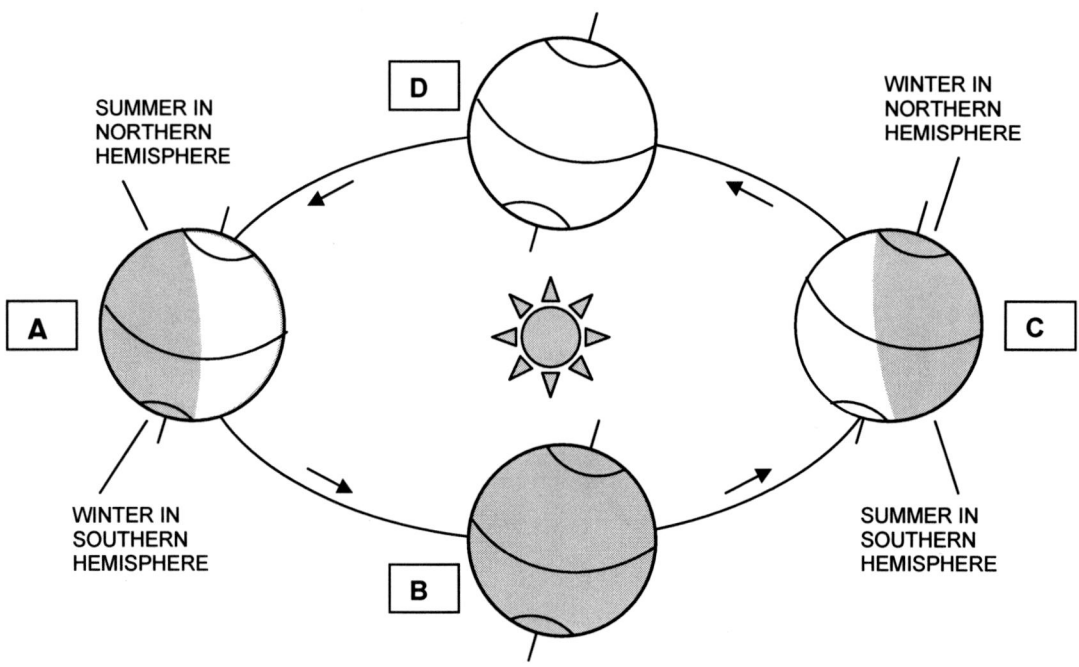

Figure 17. Earth's rotation round the Sun with the resulting seasons

From our viewpoint on Earth, we see an annual cycle of the motion of the Sun in the sky which can be resolved into either horizontal or vertical movement. On a daily basis, the rise and fall of the Sun in the sky is caused by the rotation of the Earth on its axis as detailed in **Section 1.14** (p.24), but the vertical height that the Sun reaches is a function of the changing <u>declination</u> of the Sun due to Earth's solar orbit. The horizontal movement of the Sun is measured by its westward march, which is time.

The last two sentences are <u>critical</u> to the understanding of this whole section. Taken together, they mean that the part of the Sun's apparent movement that we see as the vertical change in declination, does <u>not</u> contribute to the perceived passage of time from observing the Sun. In **Section 1.13** (p.22), I described the Sun's movement in the astro-navigator's geocentric world and illustrated it as **Figure 20**, which is repeated below.

– the Sun and Time

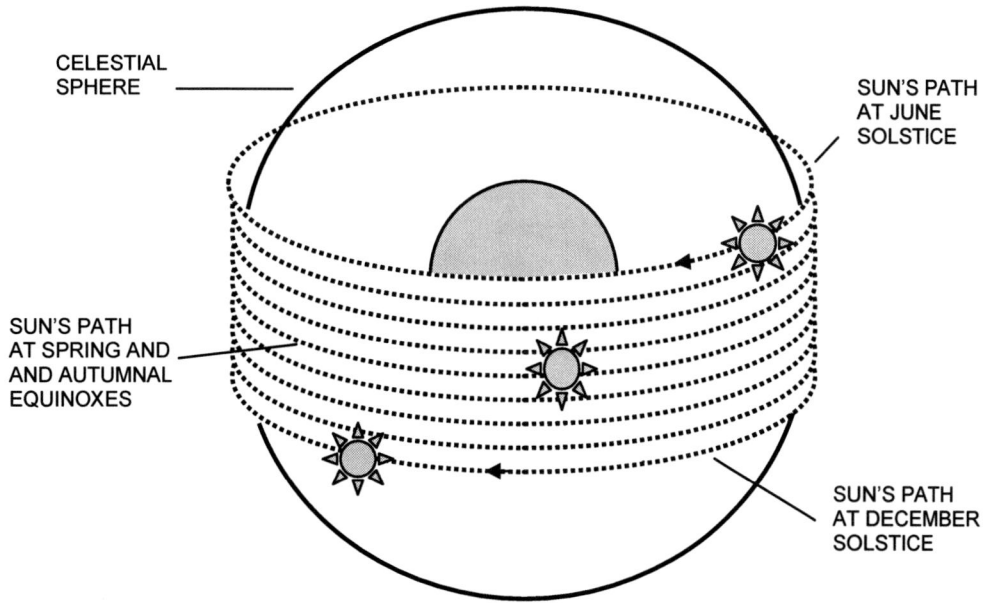

Figure 20. The complex apparent motion of the Sun in the geocentric system

The Sun's movement across the celestial sphere is in a <u>horizontal plane,</u> parallel to the celestial equator, only at the <u>extremes</u> of its yearly cycle; at the solstices in June and December. At all other times, its movement is not purely horizontal but has a small <u>vertical</u> element. The relative amounts of horizontal and vertical movement change with the position of the Earth in its orbit and can be illustrated by **Figure 75**.

The degree of vertical movement is zero at the solstices and increases after them, reaching a maximum at the equinoxes and then declines again, becoming zero at the succeeding solstice. It is possible to break down the daily movement of the Earth round the Sun into relatively simple figures that represent the apparent horizontal and vertical elements of the Sun's movements as seen by the observer on Earth. Ignoring for the sake of understanding, the speed changes associated with the Earth's elliptical orbit, the Earth could be assumed to move through an <u>average</u> of 360°/365 x 60 = 59.2′ of arc of the 360° of her annual orbit in each period of 24 hours.

Each day would experience an <u>average</u> daily change in declination of 15′. But in reality, at the solstices, all the Sun's movement is in the horizontal plane, declination being zero, where at this time, the daily movement in the orbital plane would be 61.1′. At all other times, the horizontal element of the movement is less than 61.1′ and in proportion to the increase in the vertical movement of daily declination. As one increases the other decreases as the seasons pass.

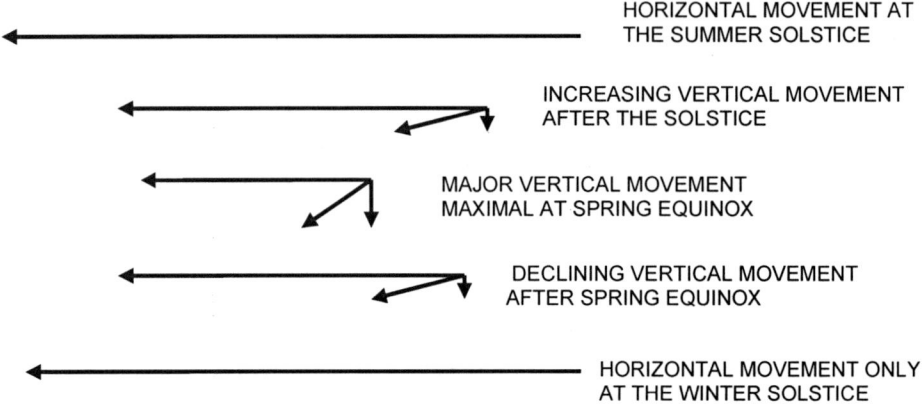

Figure 75. Resolving the Sun's apparent movement between June and December

The daily <u>average</u> change of 15´ in declination produces a horizontal arc deficit of just less than 2 minutes of arc equivalent to a little more than 7 seconds of time per day. The maximum daily change of 23´ produces a deficit of around 20 seconds of time per day and this occurs at the equinoxes. These deficits in horizontal movement of the Sun converted to the vertical movement of changes in declination cause the apparent time (based wholly on our observation of the Sun moving to the West) to slow after the solstices as the Sun's daily declination increases and reach a maximum at the equinoxes.

After each equinox is passed, vertical displacement (seen as declination change) decreases and horizontal displacement consequently increases. This causes the passage of apparent solar time to speed up. These daily deficits in apparent time are cumulative over the cycle from solstice to equinox and reach a total of 8 minutes or so at the March and September equinoxes. **Figure 76** is a graph of the daily change in the Sun's declination for 2011, although they are the same in every year. The peaks relate to the comparatively rapid changes in declination before and after the equinoxes and the smoother curves are at the solstices when daily changes are at a minimum.

As declination increases, apparent solar time slows as the Sun's movement is slightly less westward. Likewise, as declination decreases apparent solar time increases. This means that from the December solstice to the March equinox solar time slows by an increasing amount (a few seconds per day). At the March equinox solar time stops slowing and begins a daily increase, which continues until the June solstice at which time it begins to slow again until the September equinox when it again begins to speed up until the December solstice. This change in declination, when converted to a change in solar time, can be represented in a graph of solar time minus mean time (**Figure 77**).

– the Sun and Time

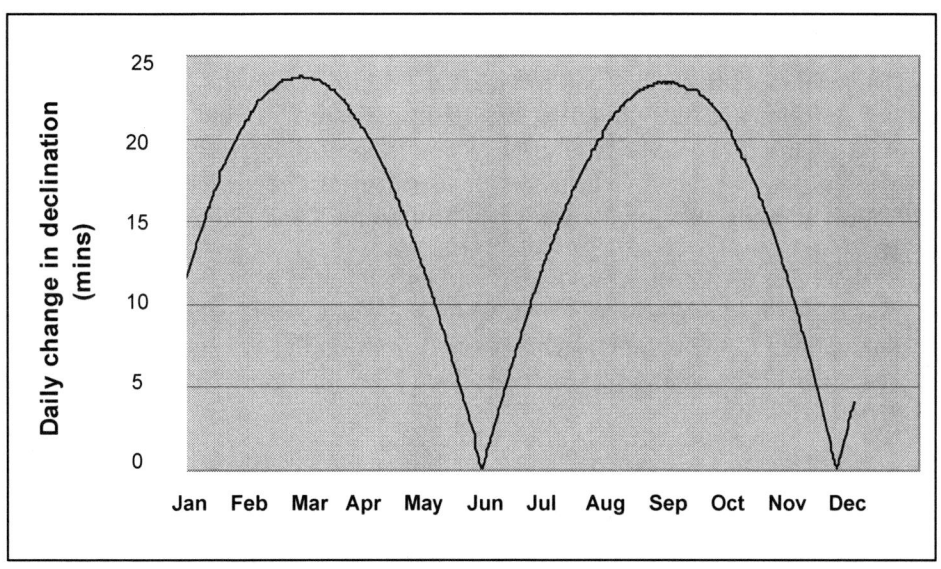

Figure 76. Graph of the daily change in declination for 2011

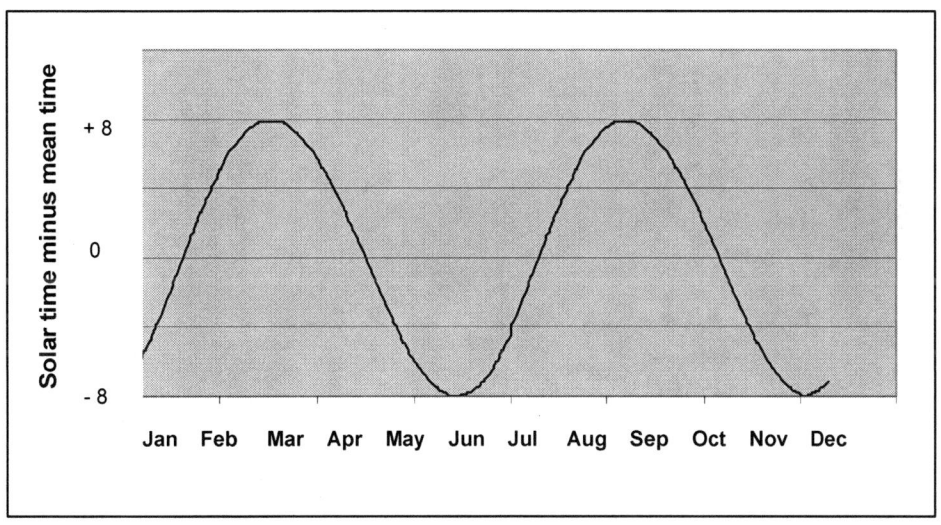

Figure 77. The effect of obliquity on solar time minus mean time over a year

The graph curve of **figure 74**, the apparent time effect of orbital speed changes and **Figure 77**, the effects of declination, are additive and it is this sum of the two that produce the major contribution to the graph of the **Equation of Time (Figure 66** p.116).

5.11 Visualizing the effect of speed changes and obliquity - mind spinning?

The changes in apparent time are difficult to visualize because it requires both the Earth's real motion around its orbit and the model of the static geocentric Earth at the centre of the moving celestial sphere, to be considered in the same train of thought. I have found that it is helpful to imagine viewing the Sun from the 0° meridian at Greenwich, peering through a narrow slit that just allows you to see the Sun's whole diameter if and when it passes.

Let us first consider that the Earth is moving around its solar orbit at a <u>constant</u> speed; neither speeding up nor slowing down and at a time when there is no vertical element of the Sun's apparent movement due to the obliquity when viewed from Earth. At exactly 1200 Greenwich Mean Time, the Sun will appear and <u>fill</u> your narrow slit view, illustrated in **Figure 78**.

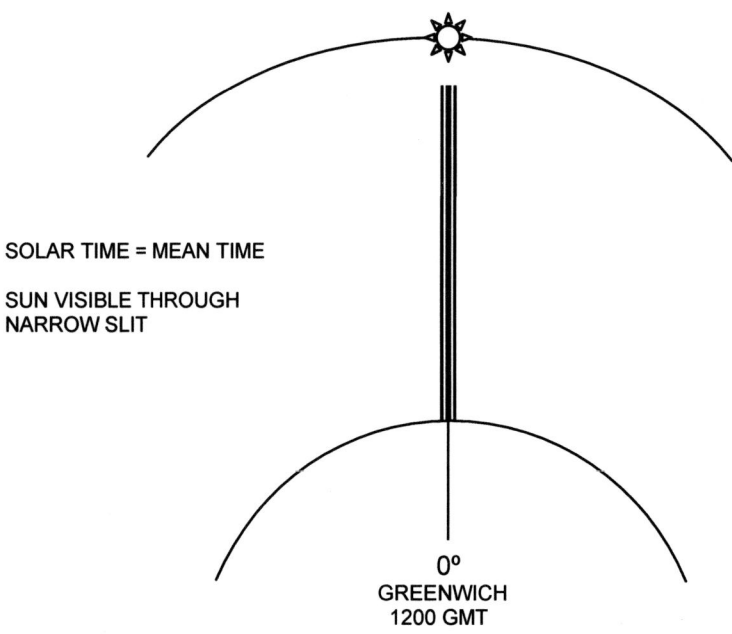

Figure 78. Visualizing the difference between mean and solar time (1)

Why? Because under these circumstances, mean and apparent time coincide. A lead or lag in apparent time (seen as more or less westward movement of the Sun) only occurs with orbital accelerations and decelerations of the Earth, or with the slow changing of the aspect of Earth facing the Sun as a consequence of vertical movement due to the fixed obliquity. Remember, what you are observing is <u>apparent</u> solar time (based solely on your view of the Sun from Earth).

Now consider the situation when the Earth is travelling faster, in a "speeding up" phase. As the Earth completes its own 24 hour rotation, it has moved along slightly more than $1/365^{th}$ of its solar orbit in that 24 hour period, because it is going a little faster. At 12-midday Greenwich Mean Time, looking through your narrow slit along the 0° meridian, you will find that the Sun is not visible, as in **Figure 79**. It has already passed your slit view and you have missed its meridian passage.

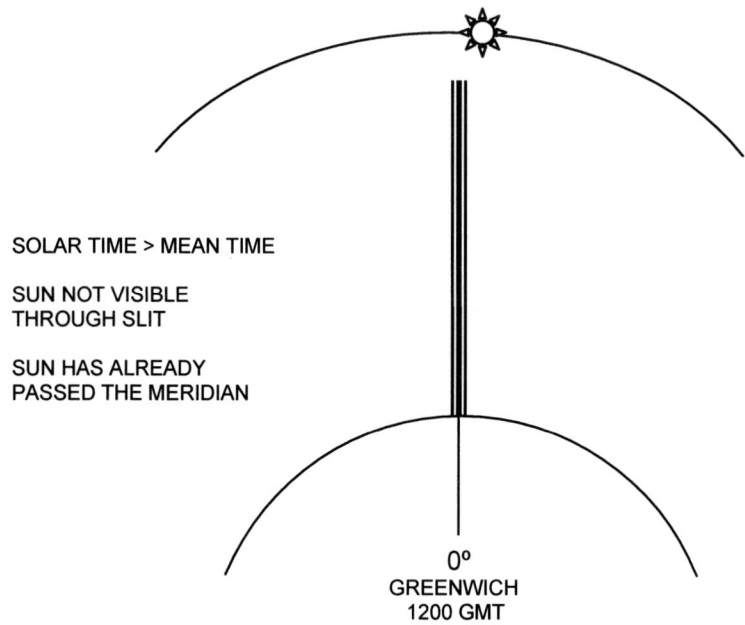

Figure 79. Visualizing the difference between mean and solar time (2)

Why? Because the Earth's orbital position has been advanced, the angle at which the observer views the celestial sphere has rotated slightly to the East, causing the Sun to appear to move slightly to the West. Apparent solar time has moved on with the Sun. If you had been observing for a little prior to 1200 GMT/UT, then you may have caught the passing Sun.

Exactly the reverse situation occurs if the Earth is slowing down or the obliquity orientation is causing a declination increase that detracts from the Sun's horizontal movement. As it completes its 24 hour rotation, the Earth has moved along slightly less than $1/365^{th}$ of its solar orbit in that 24 hour period. At 1200 Greenwich Mean Time, looking through the narrow slit along the 0° meridian, the observer will find that the Sun is again not visible. It has yet to make its meridian passage and pass the view through the slit (**Figure 80**).

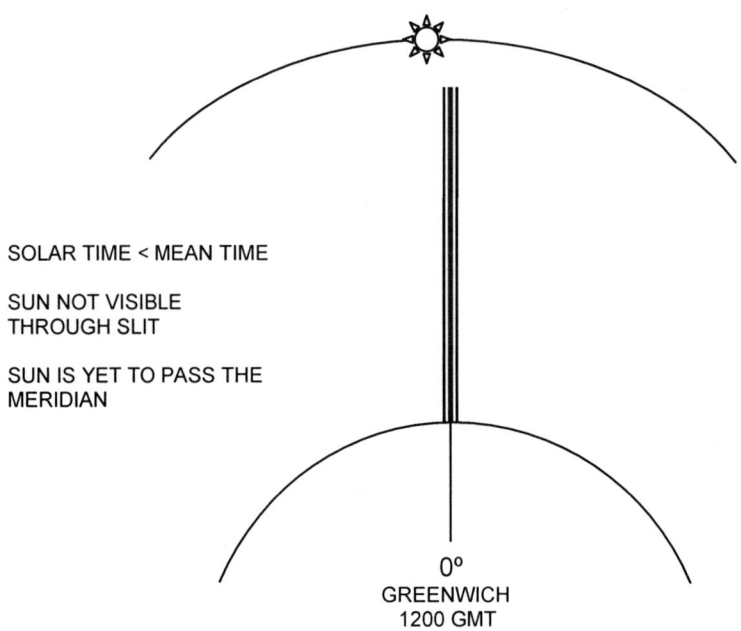

Figure 80. Visualizing the difference between mean and solar time (3)

Why? Because the Earth's orbital position has been retarded, the observer's angle of view of the celestial sphere is slightly less to the West, causing the Sun, relative to the observer to be also slightly more to the East. Apparent solar time has slowed with the Sun. The observer should continue to observe for a few minutes after 1200 GMT/UT and should then see the Sun passing the meridian.

In summary: if the Earth is accelerating on its journey around the Sun, it causes the apparent time as seen from Earth to advance and if the Earth is slowing down, it causes the apparent time to be retarded. Likewise, if the orientation of the Earth due to its obliquity is producing decreasing declination of the Sun (after the equinoxes), then apparent time will advance, but if declination is increasing (after the solstices) then apparent time will be retarded.

5.12 Other causes of orbital variation – further complications?

For completeness, there are other causes of orbital variations of the Earth (and the other planets), such as **precession** and **nutation**, but these have <u>negligible</u> daily effects but do make a measurable difference over very much longer timescales. They arise because the Earth's axis itself is not rock-steady; it describes a very slowly evolving helical pattern. This natural process is called **precession** and is a gravitational effect. **Figure 81** attempts to depict this movement of Earth's axis.

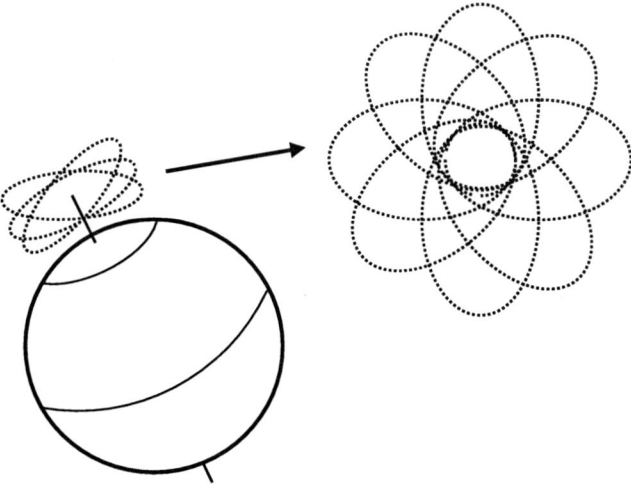

Figure 81. The precessional path of the Earth's axis

Precession is caused by the fact that the Earth is not a perfect sphere but an oblate spheroid, meaning that it bulges slightly around its middle and this bulge influences the way that the Sun attracts the Earth, according to Newton's law, causing a little eccentricity in its rotation.

Superimposed upon this is a further small but regular nodding oscillation in the precessional movement called **nutation**. Lastly and not at all relevant to navigators, clocks or sun-dials, but just worthy of note, is that the Earth's axial rotation is very gradually slowing but not at a rate in any way relevant to the astro-navigator.

So in conclusion to this part of this section, the evidence shows that our Sun does not keep perfect time and the reasons for this are explicable if not easily calculated, in relatively simple terms. Fortunately for the navigator, these variations are allowed for in the tables computed for use in reducing Sun sights, but their existence and effect need to be understood.

If you look at any daily page in the astro-nautical almanac (e.g. **Section 4.1 p.89**), you will find that the Sun's **Greenwich Hour Angle** (its celestial longitude), does not change by exactly 15° per hour but a few tenths of an arc-minute more or less than 15°, depending upon whether the Sun is in a phase of increasing or decreasing speed.

5.13 The relevance of each day's equation of time – does it matter?

It will be appreciated that when using a meridian passage Sun sight to enable latitude to be calculated, that knowledge of the equation of time for the specific day is vital in planning the timing of your sight. Solar time may well be ahead of mean time so that 10 minutes to twelve is too late to start looking for the almanac. Depending on the equation of time for that day which is generally listed to the nearest whole minute, you may find that the Sun has already passed your meridian. Get used to looking for the meridian passage time at the relevant daily page in the almanac when <u>planning</u> your day's sailing. Seize any and every opportunity to take sightings of the Sun. Be especially careful when UT is adjusted to BST etc.

5.14 *GMT/UT and Time zones* – "Round the World in 80 days" explained?

Greenwich Mean Time, introduced in **Section1.16** (p.26), now officially called Universal Time (UT) is used as the standard reference point for timekeeping across the world and for its division into 24 time zones, demarcated by specific meridians of longitude. Each zone spans 15° (360 / 24 = 15) and is based on the central meridian in each zone, extending 7.5° to the East and West from the central point. At the central meridian in any zone, 1200 hours local time coincides with 1200 hours mean solar time, the point at which the mean Sun is assumed to be at its highest in the sky and crossing the zone's central meridian. Within each zone a specific standard time is adopted such that man's affairs can be regularized.

Passing to the East from Greenwich (zone 0), the zones are –1 to –12, and to the West from Greenwich, zones +1 to +12. But, life is complex and zones –1 to –12 are each given a letter, from A to M in the usual order but missing out J. The zones from +1 to +12 are also ascribed a letter; from N-Y. Zone 0, in which the UK resides has the term Z applied (hence the term Zulu Time, from the phonetic alphabet). Each country decides its time zone or zones. All of the aforementioned is best illustrated in the formal HM Nautical Almanac Office publication, the World Map of time Zones (**Figure 82**).

If at sea and travelling across time zones, the local time in a zone can be quite straightforwardly converted to Greenwich Time (UT). Simply add or subtract the zone number to or from the local time in the zone.

Example 13: if you are in zone – 4 and the local time (LT) is 0600 hours, then 0600 – 4 hours = 0200 hours. It is 2 o'clock in the morning on the same day on the Greenwich meridian, which lies on the centre line of Zone 0. The time is 0200 GMT/UT.
Care needs to be taken when adjusting local time to GMT/UT when midnight is crossed in the calculation. Simply remember that the Earth turns in an easterly direction (perceived by man as the Sun moving West), so easterly longitudes are always ahead of more westerly ones.

Example 14: if you are in zone +9 and the local time is 2000 hours then the time at Greenwich is 2000 +9 hours = 0500. It is 5 o'clock in the morning at Greenwich. Will this be the day before or the day after? Greenwich is to the East of zone +9 and since the Sun passes from East to West, Greenwich sees the Sun earlier than the + zones West of it. It is therefore 0500 GMT/UT on the next day.

— *the Sun and Time*

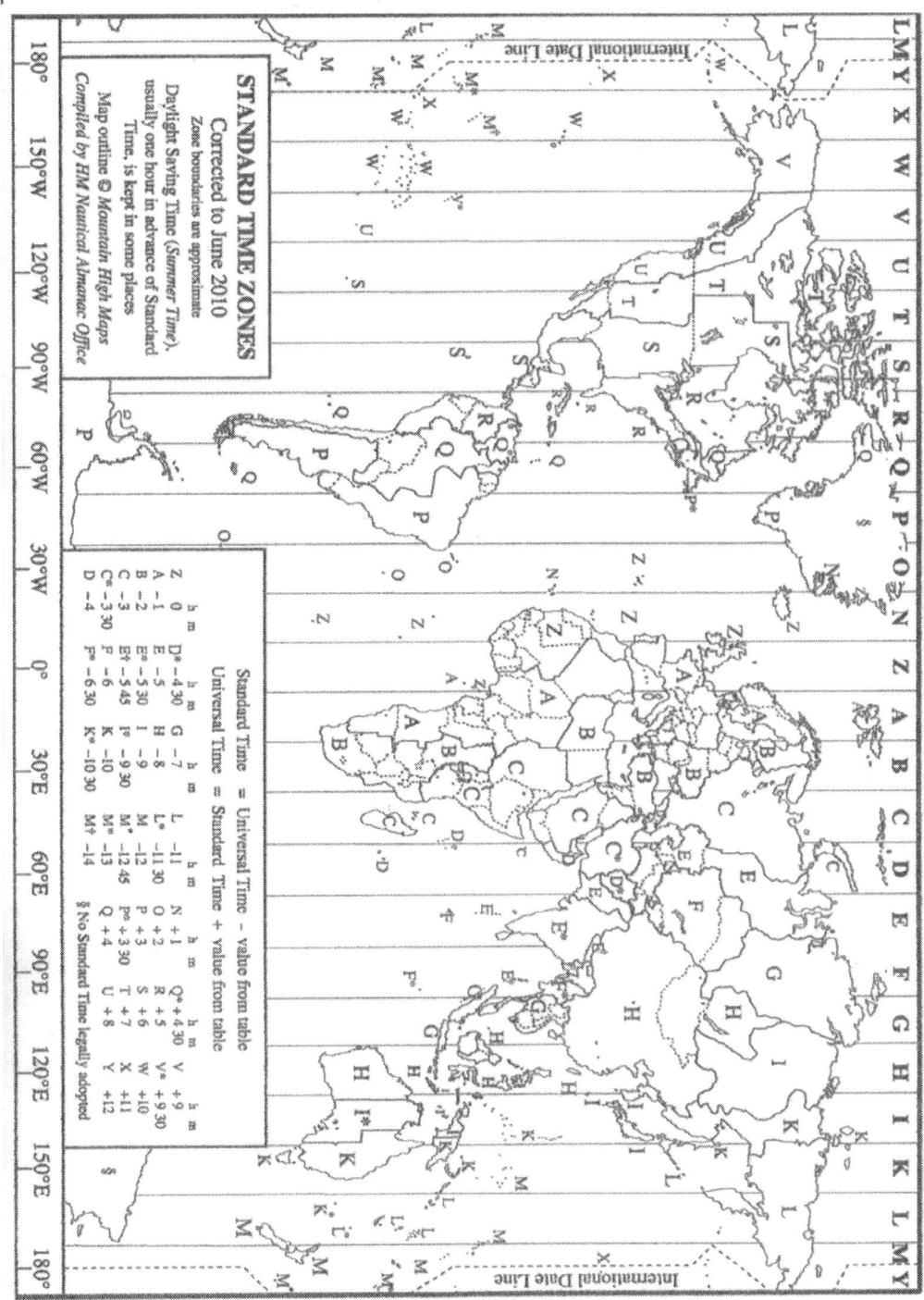

Figure 82. The current World Time Zones centred on Greenwich: Zone 0

At the transition lines between zones, 7½° either side of its central meridian, the recorded time is advanced by one hour when travelling in a westerly direction or retarded by one hour, if going East. The +12 and −12 zones overlap one another where the 180°E and 180°W meridians are one and the same.

The central meridian of these overlapping zones is called the International Date Line. Passing across the Date Line in an easterly direction a whole calendar day is lost and when passing from West to East a whole calendar day is gained (remember "Round the world in 80 days", by Jules Verne) [5.11]. Because the Earth rotates from East to West, when moving round the world to the East, each time a new zone is entered, an hour is subtracted to match local time. This saves an entire calendar day in an East-going circumnavigation. When travelling to the West, the reverse is true; an entire calendar day is lost in a West-going circumnavigation.

Where a country spans more than one zone, straying only marginally into adjacent zones, then usually a single time zone is selected for the whole country. If a country is large and spans several zones, an agreement is made to use more than one zone across the country e.g. USA. At the International Date Line, adjustments are also made for regional convenience as certain pacific islands are deliberately retained within the time of a particular zone, although geographically they may lie on opposite sides of the Date Line.

5.15 *Keeping up with the times* – useful time checks?

The navigator is potentially beset by several different "times", depending on his location. **Apparent solar time**, the time that is apparent from observing the Sun or a sundial and is due to the peculiarities of Earth's movement round the Sun and linked to mean time by the equation of time. **Greenwich Mean Time** (UT) is relatively straight forward, it is Apparent Solar Time smoothed of its irregularities making each day exactly 24 hours. It is the time at the reference "Time Zone 0", based at the Greenwich meridian. BBC World Service usefully broadcasts a time signal exactly defining the start of each hour. This is UTC-based on the atomic clock.

For navigators requiring time accuracy greater than 1 second, a correction can be obtained to adjust UTC (based on the atomic clock) to UT (based on mean solar time). The BBC World Service is broadcast on various short-wave frequencies through the day and night. Some of the useful frequencies are: 6195, 9410, 12095, 15308, 15400, 15480, 15575, 17830 kHz, but check on www.bbc.co.uk/worldservice/schedule/frequencies.html as changes are fairly frequent and there is a wide regional variation. There is little excuse for not knowing the GMT/UT wherever you may be. Time signals taken from DAB radios may not reflect true GMT/UT.

On long sea voyages an accurate and reliable chronometer should be adjusted to and remain on GMT/UT and its error recorded. **Local Time** (LT) is the official time kept in any of the defined Time Zones, or their agreed variations and this will include the daylight saving time adjustments twice per year. The locally broadcasting radio and TV stations will indicate this time.

– the Sun and Time

To the navigator, Greenwich Mean Time is of paramount importance since all astro-navigational tables relate to GMT/UT. The equation of time is the second most important piece of temporal information as without it one cannot easily relate apparent solar time to GMT/UT and the meridian passage of the Sun for that all-important sight. Local Time should be largely reserved for indicating when the Sun is over the yard arm, necessitating a retirement for suitable refreshment.

5.16 Summary of section 5

The section began by describing the Sun and the relationship between apparent and mean solar time, linked by the equation of time. Earth's elliptical orbit, its eccentricity and the significance of Kepler's and Newton's laws were explained. The methods of recording time: UT, UTC and GMT/UT and their similarities and differences were explored. The effect upon apparent solar time of both Earth's orbit and axial obliquity was detailed, the latter requiring the resolution of the Sun's declination into both horizontal and vertical elements. A means of visualising the differences between mean and apparent time was presented and the other less immediately important effects of precession and nutation were described. The world time zone system was explained with examples of converting LT to GMT/UT and radio time signals detailed.

Section 6 – Navigating with the Sun

In **Sections 1.5** and **1.6** (p.9-10) the terms **Greenwich hour angle** (GHA) and **declination** (Dec) were introduced as the means of fixing a body's position on the **celestial sphere**. GHA, Dec and the celestial sphere are all inventions of man, now somewhat historic but still wholly appropriate devices that have stood the test of time and enabled the navigator to extract some useful meaning from the relative position of the objects in our sky. In this section, the details of Greenwich Hour Angle and its relationship to the observer's position will first be considered, to be followed with details on Declination and then Altitude. The section then moves on to the practicalities of navigating using the Sun which is the most important of the objects which continuously move across the celestial sphere. In the two subsequent sections the subjects will be the Moon and planets, looking at their particular peculiarities and similarities when compared with the Sun.

By the end of Section 6, the following KEY FACTS should be understood:-

- **The Greenwich Hour Angle of the Sun, increasing at an average of 15° per hour, is the angular distance on the celestial sphere measured at the North Celestial Pole, in a westerly direction from the Greenwich meridian.**

- **The Local hour angle of the Sun is the angular distance on the celestial sphere, measured in a westerly direction from the meridian of the observer.**

- **The Declination of the Sun, measured North or South of the celestial equator, varies cyclically over the year. The Sun's hourly declination value for a given observation is adjusted by a *d* value whose size corresponds to a mean (average) declination change over 3 days.**

- **The apparent passage of the Sun is a parabolic curve, reaching its peak height at the Sun's crossing of the meridian of longitude of the observer and which defines local noon on that meridian.**

- **The altitude of the Sun at meridian passage, together with its declination at that moment in time enables simple computation of the observer's latitude using one of three specific equations, dependent upon the relative position of Sun and observer.**

- **The change in altitude of the Sun in the minute centred upon its meridian passage is less than the resolution of the sextant's micrometer scale giving a significant opportunity to obtain an accurate and reliable altitude measurement.**

- **Non-meridian Sun altitude sights must be timed accurately to the second such that both the Sun's GHA and Declination for the moment of observation can be found from the day's ephemeris table.**

- **The Sun's Local Hour angle is computed for the chosen longitude of the** observation which with the chosen latitude and tabled declination permits a calculated altitude and azimuth to be found for the moment of observation.

- **The Sun's calculated altitude and azimuth from the observation point is found from Rapid (Air Navigation) or the standard Marine Sight Reduction tables, ABC tables/equations or adaptations of the cosine formula for longhand use or with a scientific calculator.**

- **The intercept is derived from the Sun's observed and calculated altitudes and plotted on the Sun's azimuth drawn through the chosen position to give the position line at the moment of observation.**

- **Sequential Sun sights can be used through the day to produce a "running" position fix by transferring position lines forwards or backwards in time on an assumed course.**

6.1 *The Greenwich Hour Angle* – the Sun's relative position?

The term Greenwich Hour Angle is an explanation in itself. It is the <u>A</u>ngular distance in a westerly direction at a given <u>H</u>our, between the meridian on which lays a heavenly body and the meridian of <u>G</u>reenwich. Here is the vital linkage to time. The GHA of a body lasts but for a moment of time, which is why the majority of the pages in an astro-nautical almanac are taken up by lists of Greenwich Hour Angles over 24 hour periods, for each and every second of every day of the year.

We know that in reality the Earth is rotating 360° in 24 hours, at an average of 15° per hour, after which period a given point on Earth is orientated in the same way with respect to the Sun. **Table 17** shows the NP 314 Sun GHA and Declination table for April 10th – 12th 2011 where the GHA is given hourly. When closely examined, the Sun table will be seen NOT to increase at the expected 15° per hour. This is because the tables <u>include</u> corrections for the variation in Earth's rotation round the Sun due to its elliptical orbit, the inclination of the ecliptic due to Earth's fixed obliquity and the small effects of **precession** and **nutation**.

The GHA increments for minutes and seconds within each hourly period are given in separate tables and the 8 and 9 minute increment and correction tables, used in later examples are shown in **Table 18**. *Reeds Astro Navigation Tables* has a 2 hourly Sun GHA/Dec and Aries GHA table, together with an increments table spanning the 2 hourly readings.

The GHA is the cornerstone for solving the PZX spherical triangle (**Section 2.9** p.47) that enables the navigator to work out a calculated altitude (H_c) to compare with an observed altitude (H_o) obtained from an observation. This produces the all-important intercept on the calculated azimuth of the observed body at which point is drawn the position line. From the known Greenwich Hour Angle and observer's longitude we can determine the **Local Hour Angle** (LHA) for the place of observation at the same moment as the observation.

2011 APRIL 10, 11, 12 (SUN., MON., TUES.)

Table 17. NP 314-11: April 10-12th 2011 Sun & Moon GHA, Dec & rising/setting

Annotations on the table:
- EXAMPLE 26 p.167 — SUN GHA 1000 HRS 329° 38.6′ DEC 7° 53.6′ NORTH & INCR.
- EXAMPLE 27 p.172 — SUN GHA 0900 HRS 314 42.4′ DEC 8° 14.8′ NORTH & INCR.
- EXAMPLE 27 p.172 — SUN GHA 1500 HRS 44° 34.4′ DEC 8° 20.3′ NORTH & INCR.

– *navigating with the Sun*

Table 18. NP 314 -11: 8 and 9 minutes increment and *v* and *d*

Astronavigation from Square One

6.2 *Determining Local Hour Angle* -East's not least and West's not best?

The Local Hour Angle is the <u>A</u>ngular distance in a westerly direction at a given <u>H</u>our, between the meridian of the <u>L</u>ocality of the observing navigator and the meridian on which the body lays.

To clarify GHA and LHA it is easiest to draw a simple diagram, based on a plan view of the Celestial Sphere from above its Northern Pole, with the Earth central, its Greenwich meridian at the 6 o'clock position, using the **polar view**. This places the Easterly direction to the right and Westerly to the left, remembering that the Earth is central and still, with the celestial sphere rotates from East to West as in **Figure 83** and introduced in **Section 1.8** p.13.

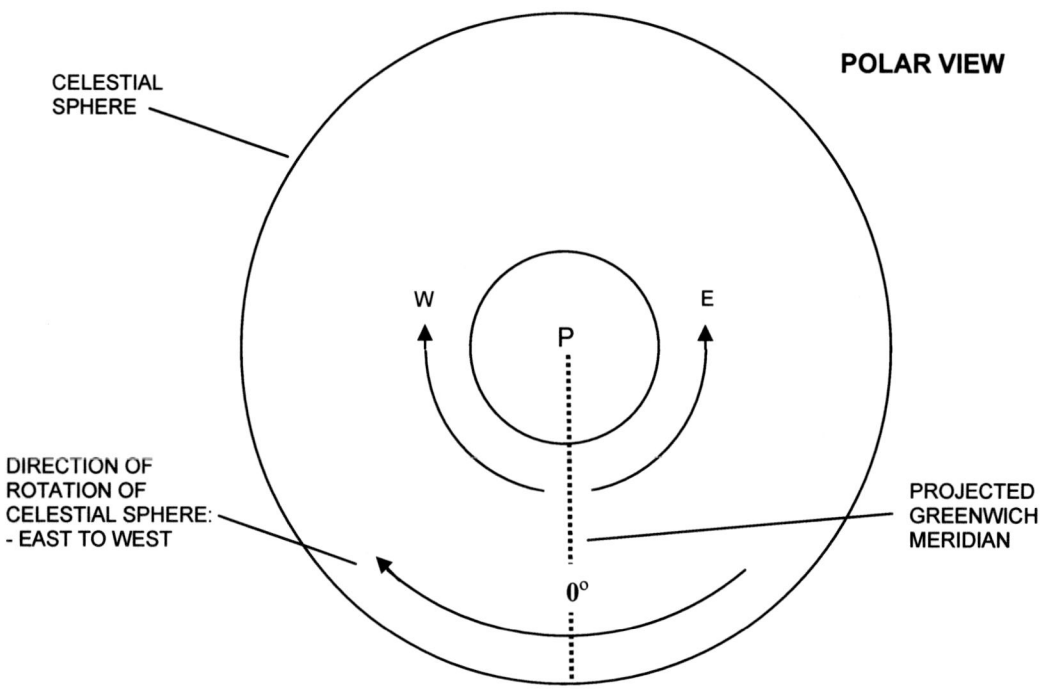

Figure 83. Polar view: an orientation from which to consider GHA and LHA

Two major rules apply for working out LHA from GHA, dependent upon whether the observer is in a Westerly or Easterly longitude. If the observer is in a Westerly longitude then that longitude must be <u>subtracted</u> from the GHA to give the LHA, as in **Figure 84** and explained in **Example 15**.

– navigating with the Sun

Example 15. Putting simple figures into the above example: if the observer's position is 50° West and the Sun's Greenwich hour angle at a <u>specific time</u> is given as 135° then the Local hour Angle of the Sun from the observer's position <u>at that same moment</u> is:

LHA = GHA – Longitude West: LHA = 135° – 50° = 85°.

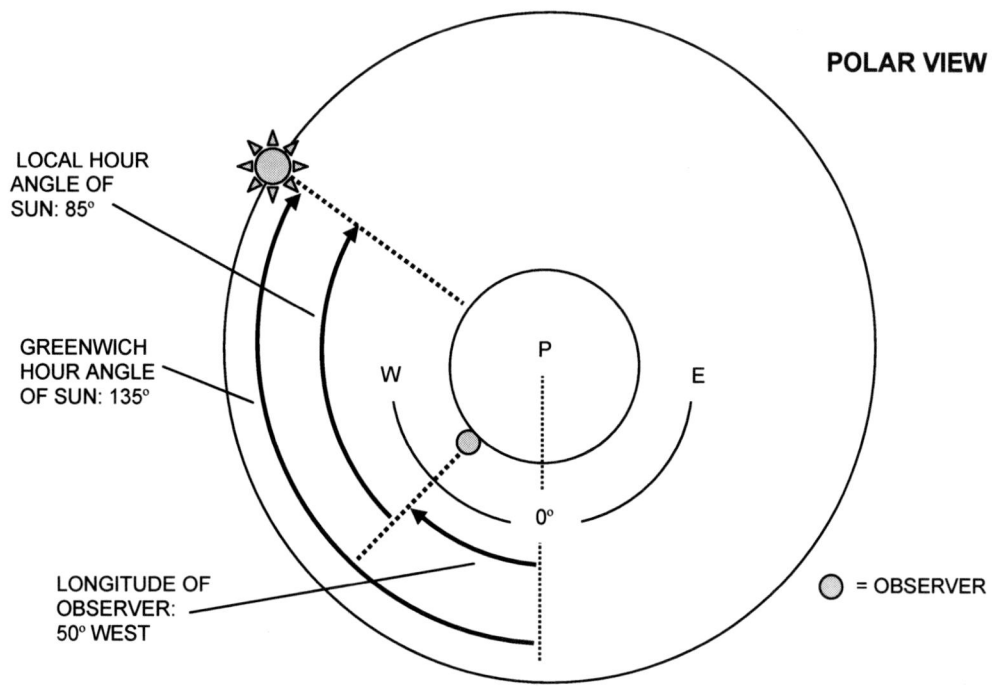

Figure 84. Rule for finding Local Hour Angle: LHA = GHA minus Long. West

Example 16. Using **Tables 17 and 18** above, what is the LHA of the Sun on Sunday 10th April 2011 at 0808 and 15 seconds at a longitude of 14° 31.2′ West?

GHA at 08 hrs		299° 38.3′
Increm. for 8m 15s		2° 03.8′
GHA 0808 15s		301° 42.1′
LHA = GHA – long. West		14° 31.2′ –
LHA	=	**287° 10.9′**

If the observer is in an Easterly longitude then that longitude must be <u>added</u> to the GHA of an observed body to give the LHA of that body, as in **Figure 85**.

141

Astronavigation from Square One

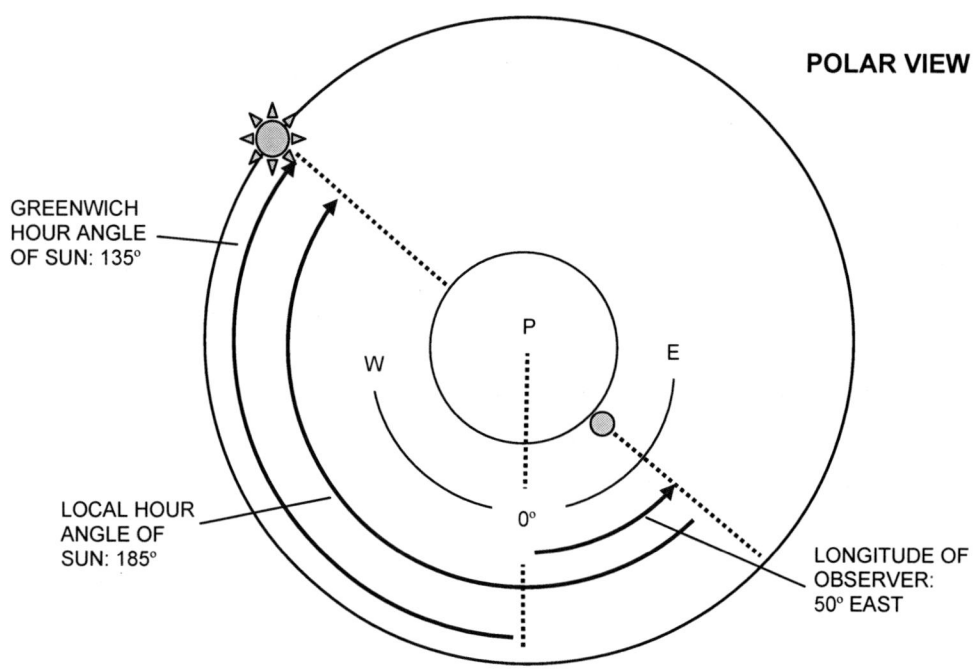

Figure 85. Rule for finding Local Hour Angle: LHA = GHA plus Long. East

Example 17. Again putting simple figures into the above example: if the observer's position is 50° East and the Sun's Greenwich hour angle at a specific time is given as 135° then the Local Hour Angle of the Sun from the observer's position at that same moment is:

LHA = GHA + Longitude East: LHA = 135° + 50° = 185°.

Example 18. Using **Tables 17** and **18** above, find the LHA of the Sun on Tuesday April 12th 2011 at 1709 and 9s, from a longitude of 43° 21.6′ East.

GHA at 17 hrs	74° 47.7′
increm. for 9m 9s	2° 17.3 ′ +
GHA 1709 9s	77° 05.0′
LHA = GHA + long. East	43° 21.6′ +
LHA =	**120° 26.6 ′**

Remember, GHA and LHA are always measured <u>westwards</u> from 0-360°. Draw the diagram and try a few calculations. You will find that if the observer's longitude West is greater than the GHA of the observed body then you will end up with a negative number as in **Figure 86**. In these cases just add 360° and you will have the correct answer as in **Example 19**. Again, LHA is always measured to the <u>West</u> of the observer.

– *navigating with the Sun*

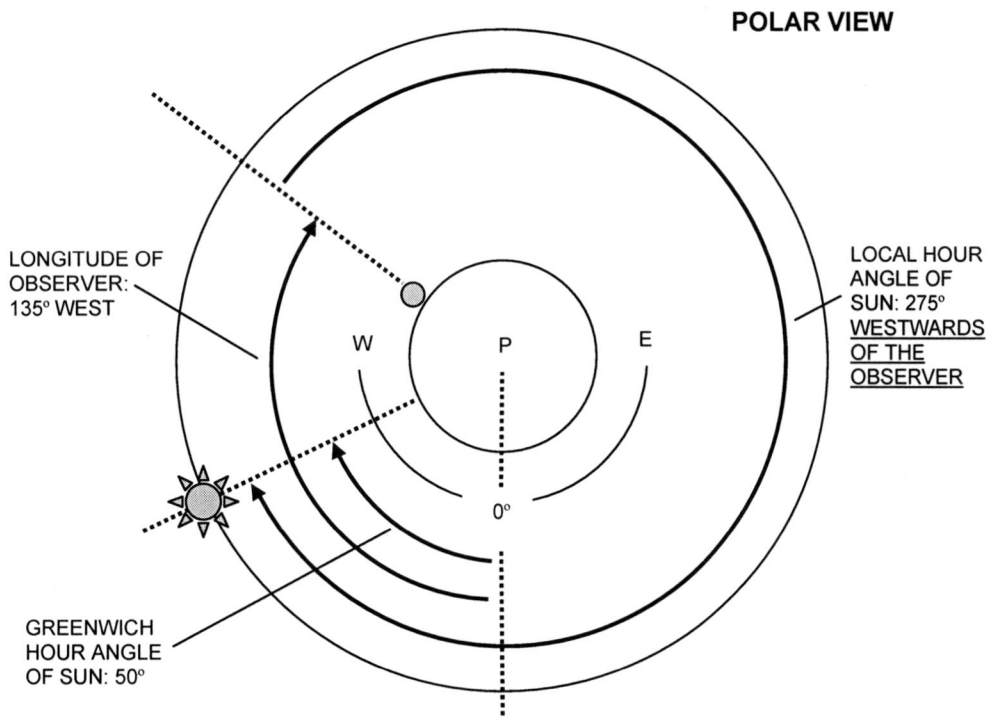

Figure 86. Rule if observer's longitude West is greater than Sun's GHA

Example 19. Putting simple figures into the above example: if the observer's position is 135° West and the Sun's Greenwich hour angle at a <u>specific</u> time is given as 50° then the Local Hour Angle of the Sun from the observer's position <u>at that same moment</u> is:

$$\text{LHA} = \text{GHA} - \text{Longitude West:}$$
$$\text{LHA} = 50° - 135° = -85°$$

Add 360° to this negative: $\text{LHA} = -85 + 360 = 275°$

In a similar fashion, when adding easterly longitudes to GHA, you may produce a number greater than 360° as in **Figure 87**, simply <u>subtract</u> 360° to get the correct answer.

Astronavigation from Square One

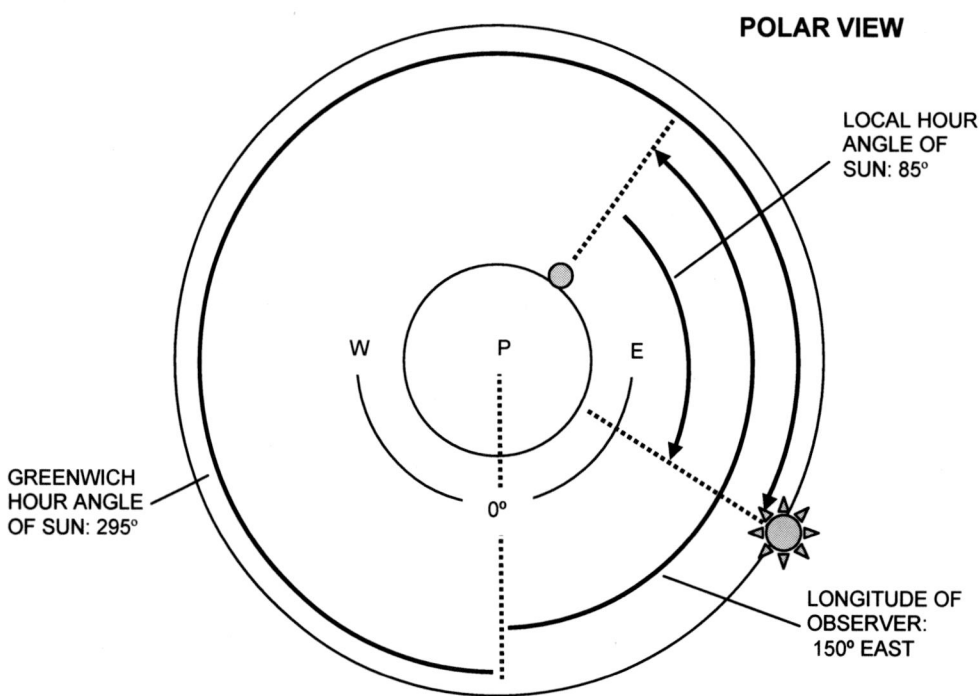

Figure 87. Rule if observer's longitude East + Sun's GHA = >360°

Example 20: If the observer's position is 150° East and the Sun's Greenwich hour angle at a <u>specific</u> time is given as 295° then the Local Hour Angle of the Sun from the observer's position <u>at that same moment</u> is:

LHA = GHA + Longitude East: LHA = 295° + 150° = 445°

Subtract 360° from 445° and the LHA is: 445 − 360 = 85°.

The key to calculating LHA from GHA and longitude is to draw a simple polar view diagram. The angles do not have to be exact. Then draw in the arcs of GHA and longitude followed by the arc of LHA drawn <u>westwards</u> from the observer's longitude. This will help considerably in orientating your thinking and will help to avoid simply committed but potentially catastrophic errors.

6.3 *The Sun's changing declination* – a variable feast?

We have seen in **Section 5** that the declination of the Sun changes continuously from 0–23′ per day, being at a maximum during the equinoctial periods and zero at the solstices. This variable rate of change has to be allowed for when calculating the Sun's declination at a precise moment

144

– navigating with the Sun

in time. The almanac figures given for declination are those at <u>Greenwich</u> i.e. at one specific meridian.

NP 314 gives hourly figures for the declination of the Sun on each tri-daily page and gives a *d* value at the foot of each page, to the right of the Sun's semi-diameter value (**Table 17** p.138). This *d* value is an average <u>rate of change</u> of declination for each and every hour of those three days. It is used to enter increments and corrections page for the relevant minute of the hour for which the declination is required (**Table 18** p.139) to give a correction value (really an incremental value) which is used to adjust the whole hour value to give a declination value to the nearest minute.

The *d* value for the Sun, its average change per hour, is always relatively small, varying from 0.0′ at the solstices, to 1.0′ at the equinoxes. Do not be alarmed at the apparent large range of the *d* values (up to 18.0) on each page of the table; the larger figures relate to the Moon, covered in the next section and whose movements are much more varied and rapid when compared to the Sun.

A further important point to note and to be found by inspecting the changing hourly figures for declination on a daily page, is whether declination for the Sun is increasing (from the solstices to the equinoxes) or decreasing (from the equinoxes to the solstices). If increasing, the *d* correction figure is <u>added</u> to the previous whole hour figure, but if it is decreasing then it is <u>subtracted</u>.

Most meridian sights are not taken whilst sitting astride the Greenwich meridian so that the declination listed at Greenwich in the almanac needs to be adjusted to the Greenwich time of local noon at the observer's meridian. This is done by converting the observer's longitude to time in the usual way, adding for westerly longitudes and subtracting for easterly ones, as easterly longitudes see the Sun before Greenwich and the westerly ones after. If necessary the Arc to Time table is used, shown in **Table 1**, **Section 1.16** p.28, or longitude is converted to time by dividing it by 15, as detailed in **Section 4.3** p.90.

Example 21. Find the declination of the Sun at its meridian passage at a longitude of 57° 35′ East on Monday April 11th 2011.

First convert longitude to time using the Arc to Time Table

```
   57°    =   3hr 48m 00s
   35′    =       2   20
              3hr 50m 20s
```

The longitude is East of Greenwich and therefore the Sun's meridian passage will occur <u>before</u> that at Greenwich at: 12hr 00m 00s
                                        3   50   20 −
Greenwich time at local noon: 8hr 09m 40s GMT/UT

Then using the information in **Tables 17** and **18**:

Sun's declination at 0800 GMT/UT	08° 13.9′	(increasing)
d value 0.9	0.1′ +	(9m page)
d corrn. is added as Dec increasing:	08° 14.0′	

Declination of Sun at time of meridian passage at 57° 35′ East = **08° 14.0′**

It is easy to forget to adjust the Sun's declination to the longitude of the observation and simply take the declination at 1200 GMT/UT from the daily table. If coastal sailing around the UK then this error makes little practical difference because of the relative proximity of Greenwich, but when crossing seas some distance from the Greenwich meridian then a real and significant error creeps in.

6.4 Navigating with the Sun – what makes it so important?

Of all the heavenly bodies used for navigation, the Sun is the most important and **Section 5** explored its intimate relationship to time and the fact that the Sun is time. The second critical factor that places the Sun above all other heavenly bodies in its usefulness is its production of a visible horizon through the entire length of the day (weather permitting), consequent upon its own light. Navigation using all the other heavenly bodies depends on the Sun's light to produce a useful horizon in the day for the Moon or at dawn and dusk for all bodies. The "horizon" often visible on a brightly moonlit night cannot be relied upon to be the true horizon.

The Moon as we will explore later in the section is not just a twilight tool but is useful in a good proportion of days. But if the navigator is allowed only one heavenly body from which to find his position then the Sun would be unfailingly chosen. The following paragraphs explain the use of the Sun in obtaining sights and will begin with the all-important noon sight to obtain latitude.

6.5 The Sun's motion across the sky – what shape does its path describe?

The Sun's apparent motion, the motion we see from Earth, was introduced in **Section 1.13**.(p.22). It rises from the horizon and crosses our sky, not in a smoothly symmetrical curve but that of a parabola (**Figure 88**), a curve that turns tightly at one particular point. In simple terms this means that it is not rising and falling in our sky at a constant rate but a variable one. The rate of rise and fall both vary, not only with the season, being lower for the winter Sun than in summer, but also varying from hour to hour and near the meridian passage, from minute to minute.

There is a very significant change in the rate of rise and fall in the hour around local midday. In the last few minutes prior to midday its rise in the sky slows and slows, then ceases, pinpointing exact local midday on the observer's particular meridian of longitude. During its rise of course, the Sun continues moving westwards, so even though its rise stops, it is never really still. It then begins its inevitable descent, gaining pace into the afternoon, its rate of fall increasing and then levelling off.

Rising in midsummer at the latitude of Great Britain, the Sun edges over the horizon at the somewhat ungodly hour of 0350 GMT/UT and rises at a spirited average of 9° per hour up to mid-morning at 1000 or so. Following this there is a noticeable slowing of its rise, halving in the next hour to a rate 4.5° or so and similarly halving to a rate of less than 2.5° between 1100 and 1200, finally reaching a total of more than 61° of altitude at meridian passage. It then begins its descent, slowly at first then gaining momentum, reaching identical rates in its descent to that of its rise, finally setting at nearly 2030 GMT/UT, having been above the horizon for over 16 hours.

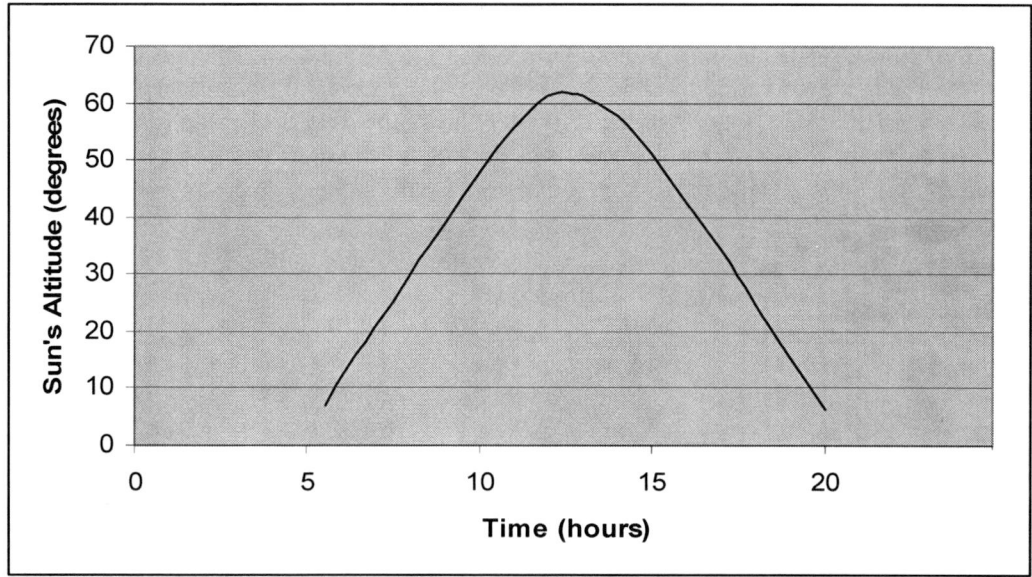

Figure 88. Sun altitude parabolic curve - mid-summer's day, latitude 52° N

Contrastingly in mid-winter, the Sun rises much later, after 0800 GMT/UT and its initial rise is a gentle 5.5° per hour, producing a relatively shallow parabolic curve, noticeably slowing between 1000 and 1100 to just a little over 3° per hour, slowing to 1° in the last hour's run to the meridian. After noon the winter Sun fades and sets before 1600, having spent less than 8 hours above the horizon. Remember, these varying periods that the Sun is above the horizon are caused by the Earth's fixed axial obliquity and the orientation that this imposes with respect to the Sun, as detailed in **Section 1.12** p.20.

6.6 *The Sun at meridian passage* – does it really stand vertically still?

We have established that the Sun rises in the sky after dawn, ascending in a parabolic curve in which the rate of climb decreases markedly in the half-hour or so prior to that moment when it stops ascending and begins descending. For how long is the Sun vertically still? The true answer to this is almost no time at all, for it does not stand still for any measurable period. Its rate of rise slows and slows, stops for a small fraction of time and then it begins to fall.

The standard calculated altitude tables can be used to illustrate the details of the Sun's movement over this critical period. For example, using 4-figure tables and calculating altitude every 2 minutes on the 21st of June at the latitude of Greenwich, the peak altitude reached is 61° 26′ and this appears not change over a period of 4 minutes; a situation plainly not borne out by sextant observation. This apparent anomaly is due to mathematical limitations in the tables when very small angular changes are examined. If 8-figure tables are used to increase the accuracy, it can be shown that the altitude does still change within this period, continuing to rise and then fall by a very small amount.

In the 2 minutes prior to transit the Sun still continues to rise at an <u>average</u> of 0.5 arc-minutes per minute of time. But in the half minute before and after transit, the change in altitude is much <u>less</u> than 0.2′, the resolution limit on the vernier scale of a typical sextant; the equivalent of one-fifth of a nautical mile. The graph in **Figure 89** shows the details of these tiny changes at the meridian passage, where in the 30 seconds either side of the meridian passage the change in altitude of the Sun is only 0.012′, a figure too small to be significant on a sextant's micrometer.

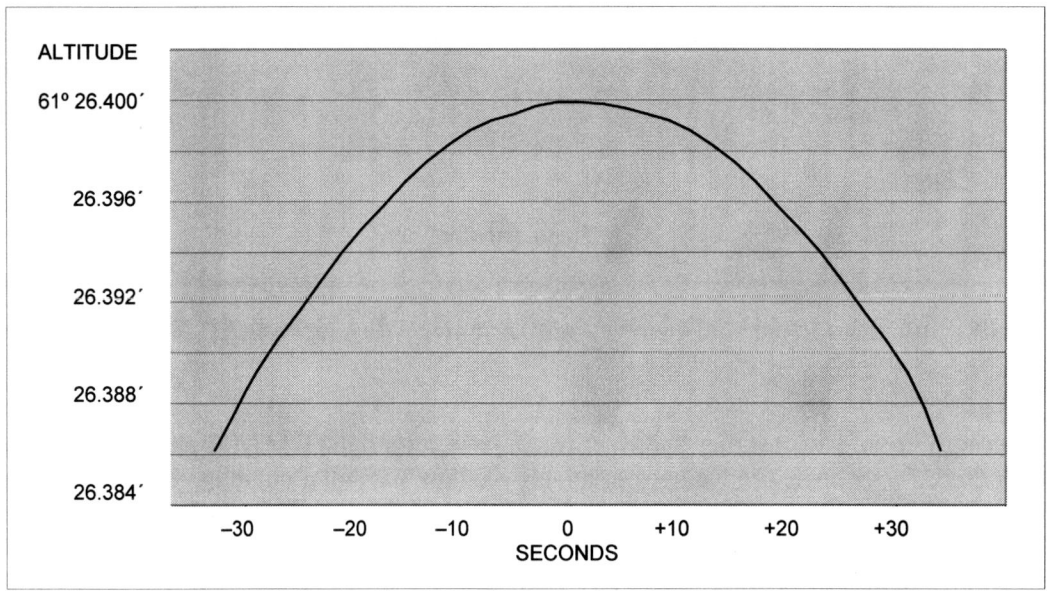

Figure 89. Changes in the Sun's altitude in the 60s around Meridian Passage

This gives us some useful evidence of the Sun's altitude changes in its last moments of ascent and indicates that sextant practice will make a perfect sight. A sight taken every 30-60 seconds around the expected noon should enable the Sun's peak altitude to be obtained within 0.2′ of the true peak. The weak link is always likely to be in the navigator's decision that the

– navigating with the Sun

Sun is aligned correctly on the horizon; at best a refined form of guesswork on the moving sea, but improving with experience.

The graph above uses mid-summer figures of altitude; when the Sun reaches its maximum declination and is thus highest in the sky at that latitude for the whole year. Plainly, transit observations on other days and at other latitudes will produce different values, but the basic points; that the Sun is never quite still at the transit but that the changes in the altitude are very small in the minute around the meridian, will hold good.

6.7 *Timing <u>non</u>-meridian Sun sights* – why it is important?

For a forenoon sight of the Sun to produce a position line, the actual time in the morning at which the site is taken is not critical but its timing is; meaning the recording of the exact moment in time at which the sight is taken, is vital for accurate navigation. The weather and cloud cover may of course limit the available options. The noon sight is effectively independent of time, its aim being to find the peak height of the Sun at the observer's particular latitude.

Sun sights at times other than meridian passage are time <u>critical</u> as the observed altitude (H_o) resulting from reducing the sight must be compared with a calculated altitude at the specific time the sight was taken. The forenoon sight may be planned as a single sight or more often as the first of a series of 2 or 3 sights in a Sun-run-Sun series, using a forenoon sight, the noon sight for latitude and perhaps an afternoon sight. The position line from the forenoon can be run up to the noon latitude line and the position line from the afternoon sight can be run back to the noon sight, to give a "cocked hat" arrangement for a daily position fix of both latitude and longitude.

The forenoon sight should if possible, be taken 3 hours or more prior to midday and the afternoon sight 3 hours or more after midday, such that the angle of cut of the position lines is acceptable, similar to the technique used when taking bearings for position fixing in coastal pilotage. GHA changes at an average 15° per hour so that a 3-4 hour gap between sights gives a 45-60° horizontal angular difference, giving a reasonable angle of cut.

If the time between the sights is overlong however, there is more room for error to creep into the estimation of position, due to current, tidal and steering influences. Plainly it helps if the course and speed of the vessel are steady so that an accurate estimation of distance run can be made by carefully noting the log reading when taking sights. An example of a Sun-run-Sun plot was shown in **Figure 55** of **Section 2.25** p.71 and an example is worked through at the end of this section.

When the sighting is made, its timing to the nearest second is noted on the deck watch, which can then be compared with and corrected from the ship's chronometer kept safe below. My chronometer is a quartz wrist watch that receives broadcast time signals such that it is corrected to UTC. I can check it against time signals from BBC radio (see **section 5.3** p.114 for GMT/UT/UTC equivalence). But note, these time signals do not necessarily cross oceans.

Remember, 4 seconds of time equate to a distance of 1 nautical mile, so it's a general rule to be as accurate as you can by taking care with the things you can control. A good technique is to take several sights over a short time period, perhaps a few minutes and then to plot them on graph paper. A general trend will be seen and any obviously erroneous reading can be discarded.

6.8 *Sun Sight Reduction* – the keys to celestial navigation?

The Sextant Altitude (H_s) is first corrected for any known sextant index error and then for dip to become the Apparent Altitude (H_a) which in the Sun's case, when corrected for semi-diameter, parallax, and refraction becomes the Observed Altitude (H_o), as described in **Section 2.2-2.7** (p.38-45). The observed altitude (H_o) is that which would have been found if the observation had been from the centre of the Earth. The diagram in **Figure 90** aids in thinking about correcting for the limb of the Sun, it being usual to sight the lower limb.

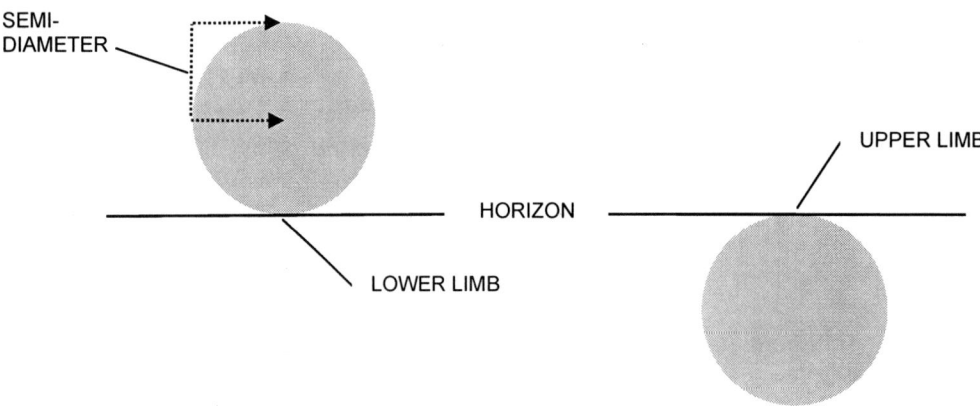

Figure 90. Comparing sights of the Sun using lower and upper limb

The Sun altitude correction table in **NP 314** requires that index error is corrected and dip is subtracted to enter the table with apparent altitude (H_a). A dip correction table is included within the overall page. The table itself, introduced in **Section 4 (Table 5** p.94) is to the novice rather disconcerting given its detail and the fact the page on which it appears encompasses the altitude corrections for Sun, stars and planets. The page is also reproduced as the bookmark for NP 314.

The Sun table is itself split into two for use between October to March and April to September. This compensates approximately for the small effect on the Sun's altitude of the minor changes in the Sun's diameter when seen from Earth, associated with its elliptical orbit. The value for correcting either the upper or lower limb altitude measurement is then read off.

The corrections are organized as "critical" tables (as is the Dip table), where within a given interval of apparent altitude, all entry values are corrected with the given figure e.g. in October to March, all apparent altitude figures between 20° 03′ and 20° 46′ are corrected by a 13.7′ correction for a lower limb sight, or subtracting 18.6′ for an upper limb sight. This single correction adjusts the apparent altitude (H_a) of the Sun sight, for semi-diameter, parallax and refraction.

NP 314 gives the actual values for the Sun's semi-diameter at the foot of the "Sun GHA and Dec" column on each tri-daily page (see **Table 17** p.138). It varies from 16.3′ in January at **perihelion**; Earth's nearest point to the Sun, to 15.8′ at **aphelion** in July; Earth's furthest point from the Sun. The values vary proportionately between these months. Having adjusted the Sun's apparent altitude (H_a) the resulting figure is the observed altitude (H_o). This completes the procedure of reducing the Sun sight to one that would have been found had the sight been taken at the Earth's centre.

NP 314 also gives guidance and a table with which to correct for effects on refraction of non-standard temperature and pressure but this is not a critical factor to the average navigator. There is also a table for correcting altitudes of less than 10°, but as indicated previously, sights at these altitudes can be assumed to be unreliable and are not generally recommended.

Example 22:
On April 11th 2011 from an eye height of 3m, a sextant with an error of 2.2′ on the arc gave an altitude reading (H_s) of the lower limb of the Sun of 23° 36.6′. What is the value of the observed altitude? What would the sextant altitude of the upper limb have been at the same moment? Use **Tables 5** p.94, **17 & 18** p.138-9.

Apparent altitude (H_a) = Sextant altitude (H_s) ± sextant error, – dip angle
= 23° 36.6′
2.2′ – (sextant error "on the arc" is taken off)
= 23° 34.4′
3.0′ – (dip correction, always subtracted: eye 3m)
Apparent altitude (H_a) = 23° 31.4′

Observed altitude (H_o) = App. altitude ± corrn for semi-diameter, parallax and refraction

Apparent altitude (H_a) = 23° 31.4′
± Main correction = 13.8′ + (+ for lower limb sights)

Observed altitude (H_o) = 23° 45.2′

Table 17 gives the semi-diameter of the Sun on the day in question to be 16′. The whole diameter of the Sun's disc is therefore 32′. The sextant altitude of the upper limb would therefore have been 32′ greater than the figure for the lower limb: at 24° 17.2′.

Astronavigation from Square One

6.9 *The Noon Sight* – how is it used it to find latitude?

The Noon Sight, of any sight, is one of the most important and useful. It was briefly outlined in **Section 1.18** (p.31) and **Figure 27** is repeated below. In essence, the noon sight measures the altitude of the Sun as it crosses the observer's meridian; i.e. at that fleeting moment when the Sun's Greenwich Hour Angle (celestial longitude) coincides with that of the observer's Earthly longitude i.e. they both lie on <u>exactly</u> the same **meridian**.

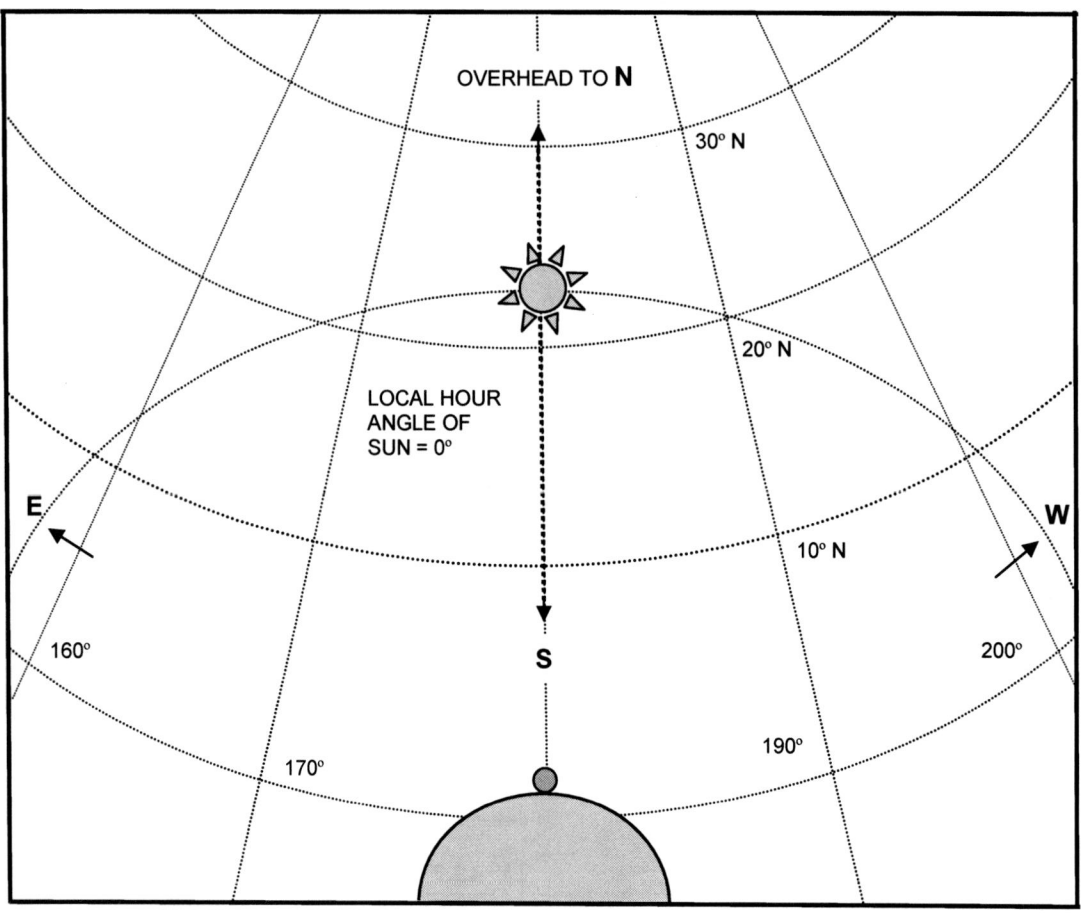

Figure 27. The special features of the Sun's position at local noon

– navigating with the Sun

The Local Hour Angle of the Sun at meridian passage is <u>zero,</u> the Sun lies on the same North-South axis as the observer. From this sight the observer's latitude can be calculated, due to these special positional circumstances.

There are 4 variables that are linked together at the key moment, one relates to the Sun and three to the observer. If three are known then the fourth can be worked out with simple arithmetic. The variable for the Sun is its Declination; the angle a line from the Sun makes with the plane of the celestial equator at the centre of the Earth.

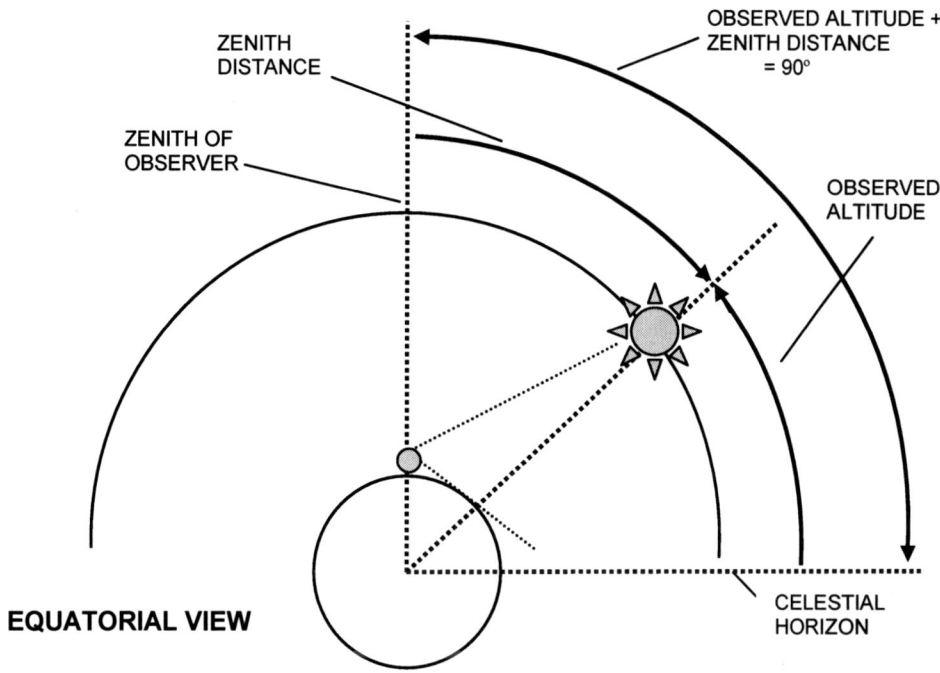

Figure 91. A right angle is formed between lines to the celestial horizon and zenith

The three variables relevant to the observer are the observed altitude of the Sun, found by correcting the sextant altitude from the observer's position, the zenith distance; that is the angular distance between the observer's zenith and the Sun. and lastly the observer's latitude.

Notice that time does not feature critically as one of the 4 variables involved in a noon sight, as the meridian passage of the Sun is the variable that allows us to fix a specific moment in time, that which we call **local apparent noon**.

The zenith is that theoretical point on the celestial sphere directly above the navigator's head, shown in **Figure 91**. This figure and subsequent ones in this section use <u>not</u> the **polar view**, but the **equatorial view** of the celestial sphere and Earth, as if looking along the same plane as the celestial equator.

Remember that latitude is the angle that a line from the observer to the Earth's centre makes with the Earth's equatorial plane. The Sun's declination is provided in the daily table in the astro-nautical almanac. The length of the arc of the zenith distance can be expressed in terms of altitude, based on the observer's sextant skills, as zenith distance and observed altitude always add up to 90°. Why does the zenith distance and observed altitude always equal 90°? It is the right angle formed between two lines arising at the Earth's centre. One line runs up through the observer to the observer's zenith and the other runs out to the celestial sphere at which point lays the celestial horizon for the observer's position.

The arc-length of the zenith distance need never be worked out, but simply expressed as 90° minus the observed altitude, the latter being found by correcting the sextant altitude such that it gives the result that would have been measured from the Earth's centre, if that were possible.

Summarizing; declination is found in the relevant table, zenith distance is expressed in terms of observed altitude. Sextant altitude is measured and corrected to give observed altitude. Latitude is the only unknown quantity and is found from the other three.

6.10 *The 3 equations used to produce latitude* – why more than one?

The equations used for the solution are 3 in number, each for a different circumstance, depending on the relative positions of the Sun and the observer; whether they are in the same or opposite celestial hemispheres and if in the same, whether the Sun passes to the South or North of the observer. The basis for each situation is best explained in diagrammatic form, as introduced in **Section 1.18** p.31.

Each equation relates the latitude of an observer to the altitude and declination of the Sun at a given moment in the set of specific positional circumstances at meridian passage. In all three equations, zenith distance is treated in exactly the same way:

zenith distance + observed altitude = 90°

therefore: **zenith distance = 90° – observed altitude**

The first of the three equations for the calculation of latitude from the Sun's meridian altitude is:

Equation 1: **latitude = zenith distance + declination**

This equation is used when the Sun is in the same hemisphere as the observer, after the spring equinox in the northern hemisphere and when the latitude of the observer is greater than the Sun's declination, as illustrated in **Figure 92**.

— *navigating with the Sun*

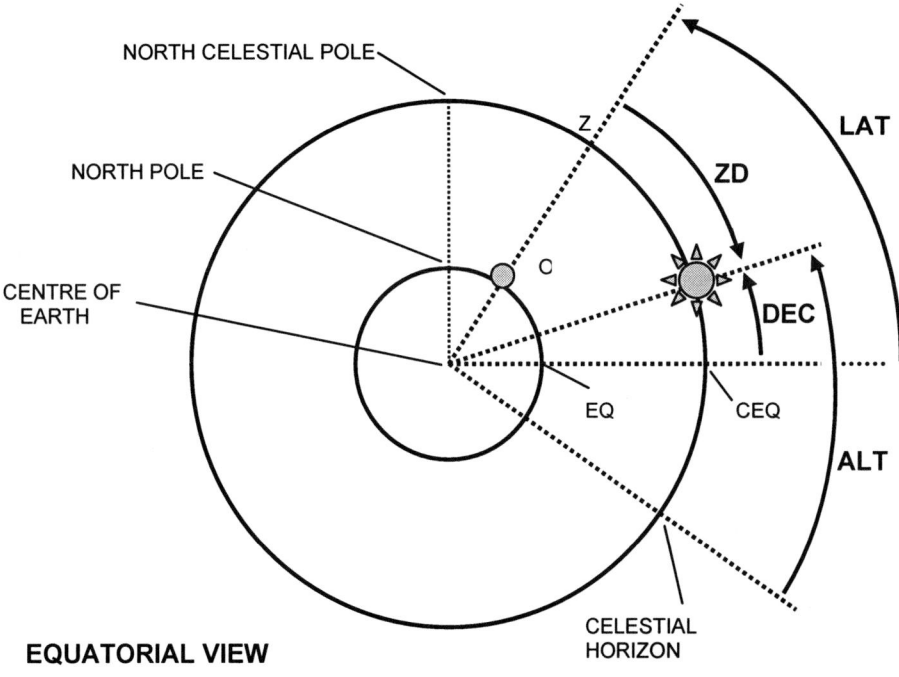

Figure 92. Equation 1: observer and Sun in <u>same</u> hemisphere

In this instance the arrangement is straightforward. It should be easily seen by looking at the angles that the following is true:

latitude = zenith distance + declination

Then substituting for zenith distance with: **zenith distance = 90° − altitude**, it follows that:

latitude = (90° − altitude) + declination

Example 23. On June 7th 2011 in the northern hemisphere, a noon sighting of the Sun was corrected and gave an observed altitude of 53° 41.5′. At what latitude was this sight taken given that the Sun's declination at the point of the site on the day in question was North 22° 47.5′?

Substituting: **zenith distance = 90° − altitude**

into: **latitude = zenith distance + declination**:

latitude = (90° 00′ − 53° 41.5′) + 22° 47.5′

When subtracting a figure from 90° 00′ it is all too easy to commit a simple mathematical error and produce the following sum:
90° 00.0′
53° 41.5′ −
36° 58.5′

155

Astronavigation from Square One

The error is forgetting that there are only 60′ and not 100′ in a degree. To avoid this potential catastrophe it is best to get into the habit of always writing 90° 00′ as 89° 60′, therefore the correct sum appears as:

$$89° \; 60.0′$$
$$\underline{53° \; 41.5′} \; -$$
$$36° \; 18.5′$$

Continuing with **Example 23**:

latitude = (89° 60′ − 53° 41.5′) + 22° 47.5′

latitude = 36° 18.5′ + 22° 47.5′

latitude = **59° 06.0′ North**

$$36° \; 18.5′$$
$$\underline{22° \; 47.5′} \; +$$
$$59° \; 06.0′$$

The second of the 3 equations for finding latitude by the Sun's meridian passage is:

Equation 2: **latitude = zenith distance − declination**

This equation is used when the Sun is in the <u>opposite</u> hemisphere to the observer, after the autumnal and before the spring equinox in the northern hemisphere and this is illustrated in **Figure 93**.

Again by looking at the angles, it is straightforward to see that in this case:

latitude = zenith distance − declination

Then by substituting for zenith distance using:

zenith distance = 90° − altitude

It follows that:

latitude = (90° − altitude) − declination

Example 24. On December 7th 2011, in the northern hemisphere, a noon sighting of the Sun was corrected and gave an observed altitude of 33° 41.5′. At what latitude was this sight taken given that the Sun's declination at the moment of the sight on the day in question was South 22° 39.1′?

Substituting: **zenith distance = 90° − altitude**

into **latitude = zenith distance − declination**:

latitude = (90° 00′ − 33° 41.5′) − 22° 39.1′

156

— *navigating with the Sun*

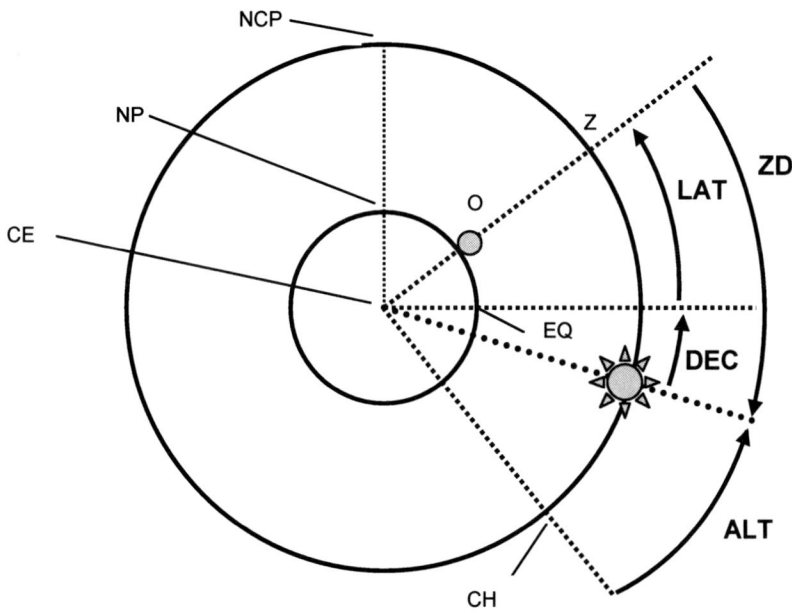

Figure 93. Equation 2: observer and Sun in <u>opposite</u> hemispheres

Again taking care with subtracting from 90° 00′ and when subtracting one figure from another it is again best to add 60′ to the minutes by dividing up one degree into minutes:-

$$\begin{array}{r} 89°\ 60.0' \\ 33°\ 41.5' - \\ \hline 56°\ 18.5' \end{array}$$

latitude = 56° 18.5′ − 22° 39.1′

This sum then becomes, for safety: $\begin{array}{r} 55°\ 78.5' \\ 22°\ 39.1' - \\ \hline 33°\ 39.4' \end{array}$ by converting 56° 18.5′ to 55° 78.5′

latitude = 33° 39.4′ North

The third of the 3 equations for finding latitude by the Sun's meridian passage is:

Equation 3: latitude = declination − zenith distance

This is a special case, used when Sun is in <u>same</u> hemisphere as the observer, but latitude is <u>less</u> than declination. This case may occur when the observer is in tropical waters, perhaps heading towards a first crossing of the equator. The Sun is in the same hemisphere but its declination my be <u>greater</u> than the observer's latitude; indeed the Sun may rise, cross the observer's meridian and then set, passing <u>North</u> of the observer and not the usual South, as in **Figure 94**, all being dependent on the relative sizes of latitude and declination.

Astronavigation from Square One

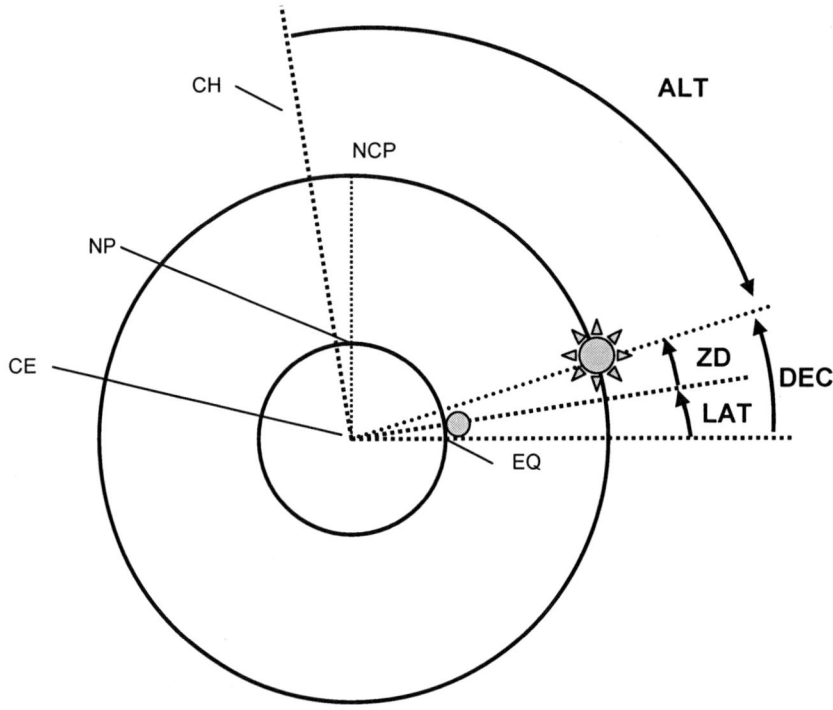

Figure 94. Equation 3: latitude less than Sun's declination (same hemisphere)

The slightly peculiar arrangements of the same 4 angles: declination, altitude, zenith distance and latitude are less obvious than in the preceding diagrams. As the Sun is to the <u>North</u> of the observer, then so is the direction of the celestial horizon from which the sextant altitude is measured and then corrected to give true altitude.

It should still be seen that as:	**zenith distance + altitude = 90°**
and also, as usual:	**zenith distance = 90° – altitude**
it can then be seen that:	**latitude = declination – zenith distance**
and therefore by substitution:	**latitude = declination – (90° – altitude)**

The next example will be done from "square one". For consistent results the navigator must practice, practice, practice. Working things out on a scrap of paper is not ideal and leads to an increased amount of avoidable mathematical error. It is best to record all sights in a sight book, ruled for date, time and altitude.

– navigating with the Sun

Example 25. On Tuesday April 12th 2011 a meridian passage noon sighting of the lower limb of the Sun was timed at 1208 12s local time. The sight was made from an estimated latitude of 1° North, longitude 32° 10.0′ West. The sextant altitude was 82° 07.6′, the Sun appearing to the North of the observer, at an eye height of 3m and with an off the arc sextant index error of 2.6′. From what latitude was the sight taken, given the Greenwich Meridian passage was at 1201? Use **Tables 5** p.94, **17** & **18** p.138-9.

To find latitude, two variables need to be defined: the altitude of the Sun at meridian passage and the declination of the Sun at the same moment. The equation used will be **Equation 3**, as the Sun in April, after the Spring equinox, will be in the same hemisphere as the observer, but the Sun was to the North of the observer at meridian passage. Look back at **Figure 94** which will help in orientating you to the navigator's situation.

1. Find the **observed altitude**

The sextant altitude (H_s) was 82° 07.6′ which needs to be converted to an observed altitude (H_o), that which would have been obtained at the Earth's centre. The sextant altitude (H_s) is first corrected for index error and then the angle of dip correction is subtracted to give the apparent altitude (H_a), which is the altitude of the body from the sensible horizon at the level of the eye of the observer.

The sextant altitude (H_s) =	82° 07.6′
Index error: off 2.6 ′	2.6 ′ + (added as error off arc)
	82° 10.2′
Height of eye 3m	3.0′ – (Dip 1.76 x √3 = 3.0 or **Table 5** p.94)
Apparent altitude (H_a)	82° 07.2′

The apparent altitude is then amended to give the observed altitude by correcting for semi-diameter, parallax and refraction, all included in the single, main correction derived from **Table 5** (arranged as a critical table). In April, the apparent altitude of 82° 07.6′ lies between 79° 43′ and 86° 31′, for which the correction value given for a lower limb sight in April is +15.8′.

The **observed alt.** (H_o)	= apparent altitude (H_a) + main correction	
The apparent altitude (H_a)	=	82° 07.2′
		15.8′ + (main correction)
The **observed altitude** (H_o)	=	**82° 23.0′**

2. Find the **Declination** of the Sun at time of sight

The GMT/UT of the sight must be found to work out the Sun's **Declination** (which is tabled in the almanac for Greenwich) at the moment of the sight (**Section 6.3** p.144).
The sight was actually made at 1221 12s LT.
The estimated longitude is 32° 10.0′ West which is in Time Zone +2 (22.5°-37.5° W)
The Greenwich Time at the moment of the sight is therefore: Local Time + Zone time

Astronavigation from Square One

Local Time of sight	=	1208 12s
Zone time		0200 00 +
GMT/UT of sight	=	1408 12s

Declination of the Sun at 1400 hours: 8° 41.4′ North (**Table 17** – daily page)
d value (page foot) 0.9   0.1′ +(incr). (**Table 18** - 8 min page)
Declination of Sun at time of sight: 8° 41.5 ′ North

3. Find the **latitude** of the observation

Using Equation 3:   **latitude = declination – zenith distance**

where as usual: **zenith distance = 90° – altitude**

it then follows that: **latitude = declination – (90° – altitude)**.

$$\text{latitude} = 8°\ 41.5' - (90°\ 00.0' - 82°\ 23.0')$$

$$= 8°\ 41.5' - 7°\ 37.0'$$

latitude = 1° 04.5′ North – just North of the equator.

```
        89° 60.0′
        82° 23.0′
         7° 37.0′

         8° 41.5′
         7° 37.0′
         1° 04.5′
```

While it is not mandatory to have the exact GMT/UT time of the meridian passage as the equations from which is derived latitude do not need an external time input. It is important however, in finding the declination of the Sun at the meridian passage. Knowing GMT/UT at the moment of transit is also a check on longitude, since the GMT/UT of local meridian passage, when adjusted for the day's equation of time, can be converted to arc angle either by the use of the Arc to Time (**Table 1** p.28) or dividing the time by 15 (longitude in time **Section 4.3** p.90).

Given that the Sun's altitude changes little in the minute either side of the meridian passage and the time recorded for the passage might vary considerably within that minute, it is <u>not</u> a highly accurate check on longitude. One minute of time equates to a 15nm distance in longitude, hence the potential inaccuracy.

6.11 *Using a proforma* – avoiding errors or undermining understanding?

As with any systematic recording, to "reduce" or "work" the sight, some seasoned yachtsman and professional navigators use a printed proforma, filling in the various readings aiming to reduce procedural errors. A proforma <u>may</u> also help the novice as it clearly marks each step and makes checking for errors more straight-forward, but a proforma should <u>NOT</u> be a substitute for understanding the process and each important step of sight reduction.

– *navigating with the Sun*

Sun Meridian Passage proforma

Date	

Estimated position:	Latitude	° . ′	N / S
	Longitude	° . ′	E / W

	Day	Hour	Minute
Meridian Passage at Greenwich			
Longitude deg. °			
− E / + W mins. ′			
Local time meridian passage (UT):			

Declination	Hours	° . ′	N / S inc/decr.
d value __	Min page corrn.	° . ′	+ / −
Declination at observation		° . ′	N / S

	Sextant Altitude (Hs)	° . ′	
	Index Error on − / off +	. ′	
		° . ′	
	Height of Eye m dip	. ′	
	Apparent Altitude (Ha)	° . ′	
	Correction + LL / − UL	. ′	+ / −
	Observed Altitude Ho	° . ′	N / S
		89 ° 60. 0′	
Change Name -	Zenith Distance	° . ′	− N / S
	Declination	° . ′	+/− N / S
	Latitude	° . ′	N / S

	° . ′ N / S	Same hemisph. Sth:-	Lat = ZD + Dec
	° . ′ N / S	Diff. hemisph:-	Lat = ZD − Dec
Lat	° . ′ N / S	Same hem. but Nth:-	Lat = Dec − ZD

Certainly in my hands, doing without a proforma concentrates the mind to the specific task, which is the logical step-wise process of sight reduction. I am though, not the tidiest navigator and a proforma often aids in keeping things in order.

Some professional trainers deliberately restrict proforma use for relative newcomers to the subject, a stand to which I would offer some support. In essence, reducing sights for all heavenly bodies follows the same logical steps, with relatively minor differences. The similarities of data handling greatly outweigh the differences between the bodies.

The remaining sight reduction examples in this book have been arranged to follow 8 logical steps for in order to enhance the understanding of the similarities and differences´ when reducing sights of all bodies. Examples of proformas will be shown in the appropriate section. One for a meridian Sun sight is shown in **Figure 95**.

6.12 Summing up the meridian passage – hemispheric relevance?

It might seem confusing to have to know three equations for working out latitude but it is very likely that for most if not all one's sailing life, only two equations will apply; one when the Sun is in the navigator's hemisphere – in the spring and summer and one when the Sun is in the opposite hemisphere – in the autumn and winter. Only if the European sailor heads very markedly South and enters the tropics, south of 23° 27´ North, may the third equation begin to become relevant, depending upon declination and latitude

The trick is being able to explain why three equations are needed and I find that this is most easily done by being able to roughly draw the previous three diagrams; all that is needed is an old school compass, a pencil and persistence. Consistent results <u>may</u> be partly assured by using a proforma into which the data is placed BUT an understanding of why and how the particular parameters are used is <u>much</u> more important.

6.13 Taking the meridian Sight – what are the realities?

In the best of conditions, it can take an experienced observer 15-30 seconds or more, to be reasonably sure that the Sun is squarely on the horizon, the smoother the sea the easier the task. In the real world a careful sight and a record of altitude and time can be reasonably made once every minute or so.

In bad conditions with large peak to trough wave heights, even a crack navigator such as Francis Chichester, in his single-handed round the world epic in 1966-7, describes that just one single sight, because of the prevailing conditions in the Southern Ocean, could take many minutes to achieve [6.1]. Bearing down on Australia as a lee shore in 60 knot winds was not a position in which to make mistake, having been out of sight of land for 12000 miles, in the pre-electronic era.

However, in the minute or so prior to transit, we have seen that the calculated altitude figures used in **Section 6.5** p.146. suggest that the Sun climbs only a further 0.5´ in the minute before meridian passage at temperate latitudes (a little more towards the equator), the equivalent of one half of a nautical mile in the final position line calculation. So the navigational accuracy of a sight taken in the minute or so around the transit should be pretty good. If I could guarantee that all my sights gave me an accuracy of 0.5 nautical miles, I would be a very happy man. Like the scout: be prepared.

Organize yourself at least 10 minutes before the expected transit time having remembered to check the GMT/UT of the equation of time for the day in question (**Section 5.4 p.115**). The last few pre-passage minutes are not the time to find, check and adjust your sextant; that should have been done much earlier. Get the crewmate ready, if you are fortunate enough to have one; with pencil, pad and deck watch. Get into a comfortable position. I tend to sit on the cockpit back-rest and wedge my body against a mizzen lower shroud.

Start taking some readings and get into a rhythm. You should be able to follow the Sun up and the crewman may be able to give you a commentary on the difference between adjacent readings, but the careful recording of both time and reading is far more important. The moment comes when you think that the appearance has not changed from the last view some seconds ago. Note the time. Then, wait 30 seconds or so and see if the Sun bites into the horizon without further adjustment. If it does, then you have achieved your aim in capturing the altitude of the Sun at meridian passage and that time is generally held to indicate the Sun's meridian passage.

If you have made an error and the Sun has continued to rise and you are still adjusting to bring it down to the horizon, then keep your nerve and continue observing every 30 seconds or so until you are sure, or as sure as you can be that you have got the sighting. It can be very instructive to later draw a simple graph of your measured altitudes versus time and see the developing parabolic curve (or not). It helps your confidence as an observer and builds faith in your own eyes and abilities and will also give you a useful curve from which to deduce the moment of transit and the Sun's altitude at that point.

In any event, you will work out a sighting regime that suits yourself and your circumstances, so just get out there and do it! Ignore the odd looks you get when you are seen peering through your sextant, whilst sitting on a local sea-wall or beach!

6.14 *The horizon and height of eye* – the difference of a foot or two?

It is important for accuracy and consistency to measure with care your height of eye, best done of course in flat calm weather (usually associated with a hazy horizon!). An example of a height of eye/dip graph was given in **Section 2.2 Figure 31** p.40. Notice that the graph is not linear, especially at the lower end. At the left hand end of the graph, an error of 0.5m in eye height creates an error of up to 0.5´ in altitude measurement. It is worth when in port, measuring from the waterline up to eye-level for sight taking and marking a nearby shroud with tape.

We cannot always rule out the errors due to swell height etc., but those we can control should be kept to a reasonable minimum. In significant swell the true horizon can only be seen from the peak of a wave. The peak to trough vertical distance needs to be estimated, then halved and added to the observer's eye height to give the effective eye height at the moment of taking the sight.

In the world's greatest small open boat journey in 1916, Shackleton, with his navigator Worsely, crossed the Southern Ocean in the whaler "James Caird", from Elephant Island to South Georgia, in order to get help for his stranded crew after his failed attempt to cross Antarctica. Worsley records how he tended to keep low in the boat to foreshorten the horizon distance in order to combat persistent poor distant visibility and so maximize the accuracy of his sights[6.2]. This is a valuable lesson regarding eye height to us (hopefully) less crisis-stricken navigators

6.15 *Non-Meridian Sun sights and their handling*.

At times other than local noon the Sun does not lie exactly on the observer's meridian. In these situations the solution to sight reduction lies in solving the PZX triangle to find a calculated altitude (H_c) and Azimuth (Z_n) for the Sun from a chosen position, for the same moment at which the Sun's altitude is measured. From these two altitudes, an intercept is calculated which is then drawn on the azimuth of the body to construct a position line, as outlined in detail in **Section 2.22**.p.66.

The azimuth for the Sun from the observer's position can be derived either from sight reduction tables, ABC azimuth tables, the ABC formula or the calculator formula, as outlined in **Sections 4.8** (p.101) and **4.14** (p110). For clarity, the 8 stages are set out below:

1. Decide upon a **Chosen Position** from which to find a calculated altitude for the Sun. This may be based on a dead reckoning position, an estimated position (having allowed for tide and current) or one picked for the convenience of round numbers, if using sight reduction tables.

2. Find the **Observed altitude** of the Sun from the Sextant altitude at the recorded time.

    Sextant altitude of Sun taken and time of sight recorded, corrected to GMT/UT. Sextant sight (H_s) corrected for sextant error (IE) and Dip to give the apparent altitude (H_a).

    Apparent altitude (H_a) corrected for semi-diameter, parallax and refraction (main correction) to give the Observed altitude (H_o).

    Summarized: $H_s \pm IE - Dip = H_a$ + main correction = H_o.

3. Find the **Greenwich Hour Angle** (GHA) of the Sun at the GMT/UT of the observation.
    Convert the GHA to the **Local Hour Angle** of the using the chosen longitude.
    LHA = GHA + longitude East or − longitude West.

4. Find the **Declination** (Dec) of the Sun at the GMT/UT of the observation.

5. Find the **Calculated altitude** of the Sun at the same time as the sextant sight.

    Using Lat, Dec and LHA work out the Calculated altitude (H_c) of the Sun at the same time as the Sextant sight using one of the following methods, all of which solve the PZX triangle:

a). **Cosine formula**: $\cos a = \cos b \cos c \pm \sin b \sin c \cos A$

which when appropriately substituted reads:

$90° - \text{Alt} = [\cos(90° - \text{Lat})\cos(90° - \text{Dec})] \pm [\sin(90° - \text{Lat})\sin(90° - \text{Dec})\cos\text{LHA}]$

b). **Scientific Calculator formula**:

$\sin \text{Calc.Alt} = (\cos\text{Lat} \times \cos\text{Dec} \times \cos\text{LHA}) \pm (\sin\text{Lat} \times \sin\text{Dec})$

- Derived from the cosine formula, see **Appendix**: paragraph **A10** p.293

c). **Sight Reduction Tables**

Rapid Sight Reduction Tables for Navigation NP 303/AP 3270 or *Sight Reduction Tables for Marine Navigation* NP 401, *The Nautical Almanac* NP 314,

d). **Versine formula** – favoured by *Reeds Astro Navigation Tables*

$\text{vers CZD} = \text{vers}\text{LHA} \times \cos\text{Lat} \times \cos\text{Dec} + \text{vers}(\text{Lat} \pm \text{Dec})$

CZD = calculated zenith distance = 90° − calculated altitude

e). **Haversine formula** – favoured by *Norie's Nautical Tables*

$\text{haversine CZD} = \text{hav}\text{LHA} \times \text{hav}\text{Lat} \times \text{hav}\text{Dec} + \text{hav}(\text{Lat} \pm \text{Dec})$

f). **Programmed Calculator** - e.g. Star Pilot-89 navigation calculator.

g). **Computer programme** e.g. NavPac software (UK Hydrographic Office) or AstroCalc software (www.pangolin.co.nz)

6. Find the **Intercept** using the observed (H_o) and calculated (H_c) altitudes

observed altitude − calculated altitude = intercept

If negative then intercept is AWAY from the chosen position, if positive it is TOWARDS the body from the chosen position.

7. Find the **Azimuth** of the Sun at the moment of the observation from the latitude and longitude of the chosen position using:

a). *Rapid Sight Reduction Tables for Navigation* NP 303/AP 3270 or *Sight Reduction Tables for Marine Navigation* NP 401, *The Nautical Almanac* NP 314,

b). ABC Azimuth Tables in *Norie's Nautical Tables* and *Reeds Astro Navigation Tables*

c). ABC Formula (see **Section 4.14** p.110) $A = \tan \text{Lat} \div \tan \text{LHA}$

$B = \tan \text{Dec} \div \sin \text{LHA}$ $C = A \pm B$ $\tan \text{Azimuth} = 1 \div C \div \cos \text{Lat}$ (Az in quadrantal)

d). Calculator formula: $\sin \text{Az} = (\sin \text{LHA} \times \cos \text{Dec}) \div \cos H_c$ (Az in quadrantal)

f). Computer program - NavPac, AstroCalc.

8. Plot the **chosen position**, **azimuth**, **intercept** and **position line**.

The next example will use a form of the 3-dimensional diagram developed in **Section 2**, intended to help the reader <u>understand</u> the 3-dimensional nature of the problem.

Example 26 is drawn in **Figure 96**. The PZX triangle, the Greenwich hour angle of the Sun, the Local hour angle of the Sun and the azimuth angle are highlighted in bold. The GHA and LHA angles are adjusted for clarity and not angular accuracy.

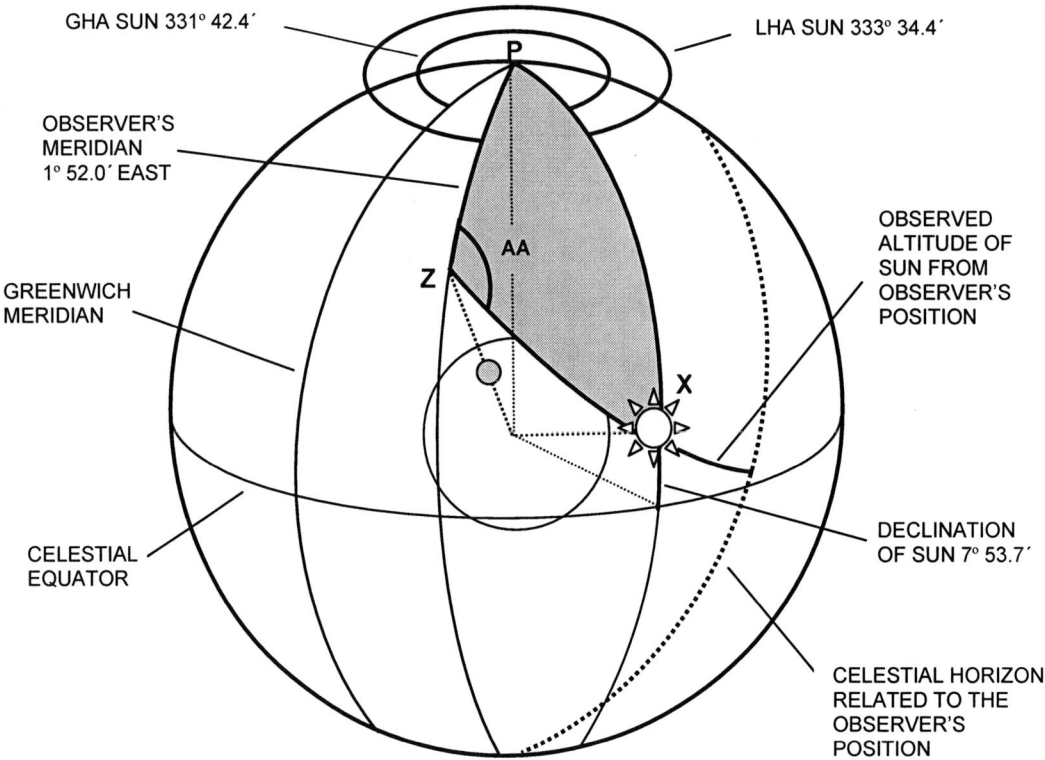

Figure 96. Visualising the 3-dimensions of Example 26

166

– *navigating with the Sun*

Example 26. A sight of the Sun was taken on Sunday 10th April 2011 at 1008 hours and 12 seconds GMT/UT from an assumed position of latitude 51° 33.5′ North and longitude 01° 52.0′ East, with a 3.2m height of eye. The deck watch was 3s slow. The sextant altitude recorded was 41° 01.4′ with an index error of 5′off the arc. Using NP 314 GHA and declination data in **Tables 17** and **18** (p.138-9) and the Sun Altitude correction data in **Section 4: Table 5** (p.94), reduce the sight and derive the calculated altitude, intercept, azimuth and position line.

Using stages 1-8 outlined above, **Example 26** will be worked through and in this instance the cosine formula will be used to solve the PZX triangle to arrive at the calculated altitude (H_c) and the calculator formula will be used to find the azimuth..

1. **Chosen position**: Latitude 51° 33.5′ North.   Longitude 01° 52.5′ East

2. **Observed altitude** (H_o)

   $H_s \pm IE - Dip = H_a + $ main correction $= H_o$

   Sextant altitude (Hs) = 41° 01.4′

   Sextant Index error $\underline{\quad 5.0′\quad}$ + (off the arc, so added on)
   41° 06.4′

   Dip correction 3.2m $\underline{\quad 3.1′\quad}$ – (**Table 5** p.94- always subtracted)

   Apparent alt.(H_a)   41° 03.3′

   Main corrn. $\underline{\quad 14.9′\quad}$ + (**Table 5** p.94 - added for lower limb)

   Observed alt.(H_o)   41° 18.2′

3. **Greenwich Hour Angle** of Sun: daily page of almanac:- convert to **LHA**

   Time of sight 1008 12s + 3s watch correction = 1008 15s

   GHA at 1000 hrs GMT/UT April 10th = 329° 38.6′   (**Table 17** p.138)
   Increment for 8m 15s   = $\underline{\quad 2°\ 03.8′\quad}$ (**Table 18** p.139)

   GHA of Sun at 1008 15s   = 331° 42.4′

   Longitude of observation $\underline{\quad 1°\ 52.0′\quad}$ + (added for East)

   Local Hour Angle of Sun   = 333° 34.4′

4. **Declination** of Sun: daily page of almanac:

   Declination at 1000 hrs.   = 07° 53.6′
   d value (page foot) 0.9 (8m page) $\underline{\quad 0.1′\quad}$ + (**Table 17**) Dec. incr.

   Declination of Sun 1008 15s   07° 53.7′

5. **Calculated altitude**: using the cosine formula to work out calculated altitude from the given observer's latitude and Sun's Local Hour Angle and Declination:

   cos a = cos b cos c + sin b sin c cos A   -which after substitution reads:

Astronavigation from Square One

$cos(90°- \text{Alt}) = [cos(90°- \text{Lat}) \times cos(90°- \text{Dec})] + [sin(90°- \text{Lat}) \times (sin(90°- \text{Dec}) \times cos\text{LHA}]$

$90-\text{Lat} = \begin{matrix} 89° \ 60.0 \\ \underline{51° \ 33.5} \\ 38° \ 26.5' \end{matrix} = 26.5/60' + 38 = 38.4417°$

$90-\text{Dec} = \begin{matrix} 89° \ 60.0 \\ \underline{07° \ 53.7} \\ 82° \ 06.3' \end{matrix} = 6.3/60 + 82 = 82.1050°$

$cos(90-\text{Alt}) = (cos\ 38.4417)(cos\ 82.1050) + [(sin\ 38.4417)(sin\ 82.1050)(cos\ 333.5733)]$

$cos(90-\text{Alt}) = (0.7832 \times 0.1374) + [0.6217 \times 0.9905 \times 0.8955]$

$cos(90-\text{Alt}) = 0.1076 + 0.5514$

$cos(90-\text{Alt}) = 0.6590$

$90-\text{Alt} = arccos\ 0.6590 = 48.7763°$

$\text{Altitude} = 90° - 48.7763° = 41.2237° = 41° + 0.2237 \times 60' = 41° \ 13.4'$

Calculated altitude = 41° 13.4'

6. Intercept: observed altitude (H_o) – calculated altitude (H_c) = intercept

$H_o = 41° \ 18.2'$
$H_a = \underline{41° \ 13.4'} -$
$+ 4.8'$

Intercept = + 4.8'

The intercept is positive as H_o is greater than H_c and it therefore lies <u>towards</u> the geographical position of the Sun. The figures are illustrated in **Figure 97**.

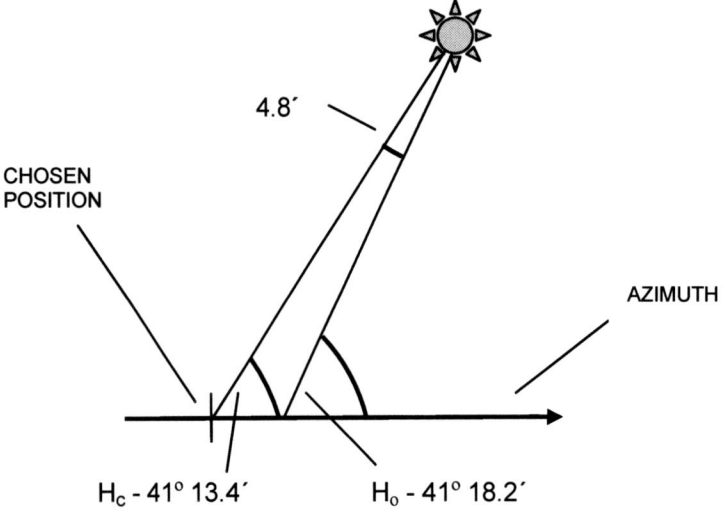

Figure 97. The observed (H_o) and calculated (H_c) altitudes drawn to find the intercept

— *navigating with the Sun*

7. **Find the Azimuth**: Using the Calculator formula: *sin* **Az** = (*sin* **LHA** x *cos* **Dec**) ÷ *cos* **H**$_c$
(Azimuth is in quadrantal notation)

LHA = 333° 34.4′ = (34.4/60) +333 = 333.5733°

Dec = 7° 53.7′ = (53.7/60) +7 = 7.8950°

Alt (Hc) = 41° 13.4′ = (0.2233) x 60′ + 41 = 41.2233°

sin **Az**	= (*sin* **LHA** x *cos* **Dec**) ÷ *cos* **H**$_c$
sin **Az**	= (*sin* 333° 34.4′ x *cos* 7° 53.7′) ÷ *cos* 41° 13.4′
sin **Az**	= (*sin* 333.5733 x *cos* 7.8950) ÷ *cos* 41.2237
	= (0.4451 x 0.9905) ÷ 0.7521
	= 0.4409 ÷ 0.7521
sin **Az**	= 0.5862
Az	= arccos 0.5862
Az	= 35.8878° = 35° + 0.8878x60 = 35° 53.3′
Azimuth	= **35° 53.3′**

The sight was taken in the northern hemisphere, above the tropics; the azimuth plainly South of the observer. The sight was taken in the forenoon when the Sun is plainly East of South. Converting the quadrantal azimuth (Z) of 35° 53.3′ to a true azimuth (Z_n) is straightforward:

Z_n = 180 – Azimuth East.

= 180 – 35° 53.3′

Z_n = **144° 06.7′** (rounded to 144°)

8. **Plot the azimuth and intercept**: plotted in **Figure 98**.

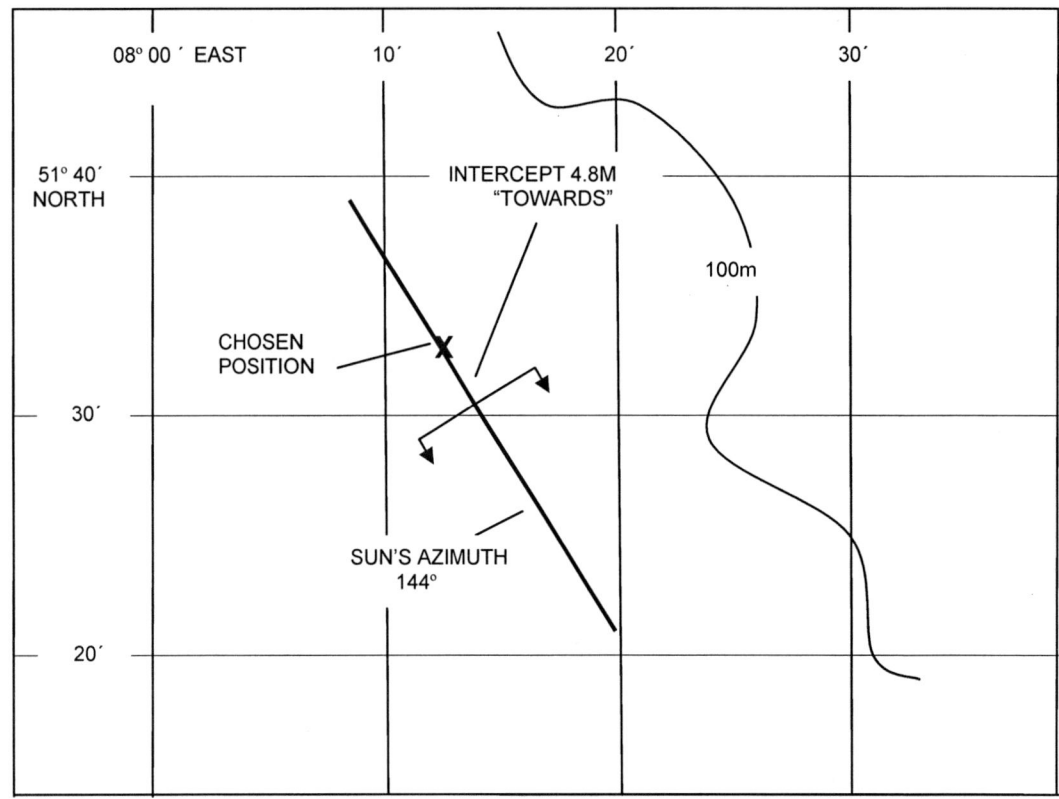

Figure 98. Plot of azimuth, intercept and position line for Example 26

A Non-meridian Sun sight proforma could be used as in **Figure 99** below, subject to the caveats detailed in **Section 6.11** p.160. The proforma shown is for use with sight reduction tables such as NP303 and NP401. The former, Rapid Sight Reduction Tables for Navigation NP303, also known by the code AP3270, are used for the next example, **Example 27**.

– *navigating with the Sun*

Sun Sight proforma

Date								
					− 30′			
Estimated position	Latitude	° . ′	N / S	Longitude		° . ′	W / E	
					+ 30′			

	Day	Hour	Min
Zone time			
Zone			
GMT			
GMT / UT			

GHA Hours ___	° . ′
Min ___ Sec ___	° . ′
GHA:	° . ′
Chosen longitude −W+E	° . ′
LHA:	° . ′

	Day	Hour	Min	Sec
Watch time				
Watch error				
GMT / UT				

Declination Hrs ___	° . ′
d ___ + / − Min page ___	° . ′
Declination: N / S	° . ′

Sextant Altitude (Hs)	° . ′
Index Error on − / off +	. ′
	° . ′
Height of Eye m dip	
Apparent Altitude (Ha)	° . ′
Correction + LL / − UL	. ′
Observed Altitude Ho	° . ′

Chosen Lat. N / S	
Declination N / S	
SAME / CONTRARY	
Chosen Lat	° N / S
Azimuth Z	°
Zn	°

	Ho	° . ′
	Hc	° . ′ −
	Intercept	. ′

Hc	° . ′
d +/− ___ incr	. ′ +
Hc	° . ′

Towards/away

Chosen position	N / S	° W / E	° . ′
Intercept		towards / away	
Azimuth		° T	

Figure 99. An example of a Sun Sight proforma for NP 314 data

171

Astronavigation from Square One

6.16 *Sun-run-Sun position fixing* – the gold standard in daylight navigation?

Using serial Sun sights for fixing position is a valuable technique to master; moving the sights forwards or backwards in time, combining morning and evening Sun sights with or without the noon latitude estimation. The principles were outlined in **Section 2.25** p.71. The three sights are reduced to give the observed altitude in the standard way.

Two positions are chosen, based on dead-reckoning or estimated positions; one at the time of the morning sight and one at the time of the evening sight. Calculated altitudes for the Sun are then worked out at the exact times of morning and evening sightings and compared to their respective observed altitudes to arrive at two intercepts.

Careful recording of the course and speed during the period between the sights is important so that the chosen positions well approximate the actual but unknown positions, with the aim of producing small intercepts.

The intercepts are then plotted on the azimuths and position lines drawn. The noon latitude position line, if available, is also plotted. Using the known speed and course of the vessel between the sights, the morning sight is transferred forward and the afternoon sight transferred backward to intersect with the noon latitude line. This produces (with good sights and careful maths) a position fix in the "cocked hat" formation. An example follows:

Example 27. During Monday 11th April 2011 during a sail across the North Sea the weather was favourable with a light and following north-easterly breeze. The boat was steered on a course of 250°T and the log recorded an average speed of 4 knots through the day. Three lower limb Sun sights were achieved; in the morning, at local noon and in the afternoon. The index error was 2.2′ off the arc and height of eye 3.6m for all three sights

Estimated positions were recorded at the times of the morning and afternoon sights, the details are listed below. Use the sights to produce a position fix at local noon, using the altitude correction table and daily ephemerides: **Tables 4** (p.92) **5** (p.94), **17** (p.138), **18** (p.139), **19** and **20**.

1. Decide upon the **chosen positions**:

	Morning	**Noon**	**Afternoon**
Estimated lat.	54° 03.0′ N		53° 58.0′ N
Estimated long.	7° 12.0′ E	~ 7° 03′E	6° 54.0′ E
Chosen Latitude	54° 00.0′ N		54° 00.0′ N

 - a whole number is chosen for Rapid Sight reduction Air tables.

Chosen Longitude - see GHA/LHA calculation below.

2. Find the Observed altitudes:

	Morning	Noon	Afternoon
Sextant alt. (H_s)	35° 27.6′	44° 04.2′	27° 15.4′
Index error (2.2′ off)	2.2′ +	2.2′ +	2.2′ +
	35° 29.8′	44° 06.4′	27° 17.6′
Dip 3.5m	3.3′ −	3.3′ −	3.3′ −
Apparent alt. (H_a)	35° 26.5′	44° 03.1′	27° 14.3′
Main corrn.	14.7 +	15.0′ +	14.2′ +
Observed alt. (H_o)	35° 41.2′	44° 18.1′	27° 28.5′

3. Find the GHA and adjust the LHA of Sun for the chosen positions:

Time of sight (GMT/UT)	0908 20s	1133	1509 11s

		Morning		Afternoon
GHA Sun (hr)	09hrs	314° 42.4′	15hrs	44° 43.4′
+ min/sec incr.	8m 20s	2° 05.0′ +	9m 11s	2° 17.8′ +
GHA Sun at sight		316° 47.4′		47° 01.2′
+ longitude East		7° 12.6′ + E		6° 58.8′ + E
Local hour angle Sun		324° 00.0′		54° 00.0′

Note that the chosen longitudes are carefully arranged so as to obtain a whole number for the LHA for entry into the Air Navigation tables and are within 30′ of the estimated position.

4. Find the Declination of the Sun at the times of the sights:

Declination (hr) inc.	09hrs	8° 14.8′ N	11hrs	8° 16.7′	15hrs	8° 20.3′
d value (0.9)	8m	0.1′ + inc.	33m	0.5′	9m	0.1′ +
Declination at sight time		8° 14.9′ N		8° 17.2′ N		8° 20.3′ N

5. Find the calculated altitudes and azimuths from Rapid Sight Reduction tables:

	Morning sight	Afternoon sight
Chosen Lat	54° N	54° N
LHA	324°	54°
Dec	8° 14.9′ N	8° 20.4′ N

Table 19 is then entered with **LHA** and **Dec** (whole number) at the <u>correct page</u> for the <u>chosen latitude</u>: 54° North.

Astronavigation from Square One

DECLINATION (0°- 14°) **SAME** NAME AS LATITUDE

LAT 54°

LHA	7° Hc d Z	8° Hc d Z	9° Hc d Z	10° Hc d Z	11° Hc d Z	12° Hc d Z	13° Hc d Z	14° Hc d Z	LHA
0	43 00 +60 180	44 00 +60 180			47 00 +60 180	48 00	49 00 +60 180	50 00 +60 180	360
1	43 00 60 179	44 00 60 179		80	47 00 60 179	48 00	49 00 60 179	50 00 60 178	359
2	42 58 60 177	43 58 60 177	WHOLE	77	46 58 60 177	47			358
3	42 56 60 176	43 56 60 176	DEGREES OF	76	46 56 60 176	47	CALCULATED ALTITUDE,		357
4	42 53 60 174	43 53 60 174	LHA: 0°-69°	74	46 53 60 174	47	DECLINATION DIFFERENCE		356
5	42 50 +59 173	43 49 +60		73	46 49 +60 173	47	AND AZIMUTH		355
6	42 45 60 172	43 45	72	4	46 44 60 171	47			354
7	42 40 59 171		60 170	44	46 38 59 170	47			353
8	42 33 60	43 33 60 169	44 33 59 169	45 32 60 169	46 32 59 169	47 31 60 168	48 31 59 168	49 30 60 168	352
9	42 26 60 168	43 26 59 168	44 25 60 168	45 25 59 167	46 24 59 167	47 24 59 167	48 23 60 167	49 23 59 167	351
10	42 19 +59 167	43 18 +59 166	44 17 +60 166	45 17 +59 166	46 16 +59 166	47 15 +59 166	48 15 +59 165	49 14 +59 165	350
11	42 10 59 165	43 09 60 165	44 09 59 165	45 08 59 165	46 07 59 164	47 06 59 164	48 05 59 164	49 04 59 164	349
12	42 01 59 164	43 00 59 164	43 59 59 163	44 58 59 163	45 57 59 163	46 56 59 163	47 55 59 162	48 54 59 162	348
13	41 50 59 163	42 49 59 162	43 48 59 162	44 47 59 162	45 46 59 162	46 45 59 161	47 44 59 161	48 43 59 161	347
14	41 39 59 161	42 38 59 161	43 37 59 161	44 36 59 160	45 35 59 160	46 33 59 160	47 32 59 160	48 31 59 159	346
15	41 28 +58 160	42 26 +59 160	43 25 +59 159	44 24 +58 159	45 22 +59 159	46 21 +58 158			345
16			43 12 58 158	44 11 58 158	45 09 58 157	46 08 58 157		58	344
17			42 59 58 157	43 57 58 156	44 55 58 156	45 53 58 156	WHOLE	55	343
18			42 44 58 155	43 43 58 155	44 41 58 154	45 39 58 154	DEGREES OF	54	342
19			42 29 58 154	43 27 58 154	44 25 58 153	45 23 58 153	LHA: 291-360°	52	341
20		+58 153	42 14 +57 153	43 11 +58 152	44 09 +58 152	45 07 +57 152	46		340
21		57 152	41 57 58 152	42 55 57 151	43 52 58 151	44 50 57 150	45	150	339
22		57 151	41 40 57 150	42 37 58 150	43 35 57 149	44 32 57 149	45 29 57 149	46 26 57 148	338
23		57 149	41 22 57 149	42 19 57 149	43 16 57 148	44 13 57 148	45 10 57 147	46 07 56 147	337
24		57 148	41 04 57 148	42 01 57 147	42 58 56 147	43 54 57 146	44 51 56 146	45 47 56 146	336
25		+57 147	40 45 +56 147	41 41 +57 146	42 38 +56 146	43 34 +57 145	44 31 +56 145	45 27 +56 144	335
26		56 146	40 25 56 145	41 21 57 145	42 18 56 144	43 14 56 144	44 10 56 143	45 06 56 143	334
27		57 145	40 05 56 144	41 01 56 144	41 57 56 143	42 53 56 143	43 49 55 142	44 44 56 142	333
28	37 51 58 143	39 44 56 143	40 40 55 142	41 35 56 142	42 31 56 141	43 27 55 141	44 22 55 140	332
29	37 30 56	38 26 55 142	39 22 56 142	40 18 55 141	41 13 56 141	42 09 55 140	43 04 55 140	43 59 55 139	331
30	37 09 +55 141	38 04 +56 141	39 00 +56 140	39 56 +55 140	40 51 +55 140	41 46 +55 139	42 41 +55 138	43 36 +55 138	330
31	36 46 56 1	37 42 55 140	38 37 56 139	39 33 55 139	40 28 55 138	41 23 55 138	42 18 54 137	43 12 55 137	329
32	36 24 55 1	37 19 55 139	38 14 55 138	39 09 55 138	40 04 55 137	40 59 54 137	41 53 55 136	42 48 54 136	328
33	36 00 55 13	36 55 55 138	37 50 55 137	38 45 55 137	39 40 54 136	40 34 55 135	41 29 54 135	42 23 54 134	327
34	35 37 54 137	36 31 55 136	37 26 55 136	38 21 54 135	39 15 54 135	40 09 54 134	41 03 54 134	41 57 54 133	326
35	35 12 +55 136	36 07 +54 135	37 01 +55 135	37 56 +54 134	38 50 +54 134	39 44 +54 133	40 38 +53 133	41 31 +54 132	325
36	34 47 55 135	35 42 54 134	36 36 54 134	37 30 54 133	38 24 54 133	39 18 54 132	40 11 54 131	41 05 53 131	324
37	34 22 54 134	35 16 54 133	36 10 54 133	37 04 54 132	37 58 53 131	38 51 54 131	39 45 53 130	40 38 53 130	323
38	33 56 54 133	34 50 54 132	35 44 53 131	36 38 53 131	37 31 54 130	38 25 53 130	39 18 53 129	40 11 52 129	322
39	33 30 54 131	34 24 53 131	35 17 54 130	36 11 53 130	37 04 53 129	37 57 53 129	38 50 53 128	39 43 52 127	321
40	33 04 +53 130	33 57 +53 130	34 50 +54 129	35 44 +53 129	36 37 +53 128	37 29 +53 128	38 22 +53 127	39 15 +52 126	320
41	32 37 53 129	33 30 53 129	34 23 53 128	35 16 53		27	37 54 52 126	38 46 52 125	319
42	32 09 53 128	33 02 53 128	33 55 53 127	34 48 52	EXAMPLE 27	25	37 25 52 125	38 17 52 124	318
43	31 41 53 127	32 34 53 127	33 27 52 126	34 19 53	AFTERNOON SIGHT	24	36 56 52 124	37 48 51 123	317
44	31 13 53 126	32 06 52 126	32 58 52 125	33 50 53	LHA 54°	23	36 26 52 123	37 18 51 122	316
45	30 44 +53 125	31 37 +52 125	32 29 +52 124	33 21 52	DECLINATION 8°	22	35 57 +51 122	36 48 +51 121	315
46	30 15 52 124	31 08 52 124	32 00 52 123	32 52 51	> Hc 27° 03´	21	35 26 52 121	36 18 50 120	314
47	29 46 52 123	30 38 52 123	31 30 52 122	32 22 51	DIFF. +50	20	34 55 51 119	35 47 51 119	313
48	29 16 52 122	30 08 52 122	31 00 51 121	31 51	Z 116°	19	34 25 51 119	35 16 50 118	312
49	28 46 52 121	29 38 52 121	30 30 51 120	31		18	33 54 51 118	34 45 50 117	311
50	28 16 +52 120	29 08 +51 120	29 59 +51 119			17	33 23 +50 117	34 13 +50 116	310
51	27 45 52 119	28 37 51 119	29 28 51		51	16	32 51 50 116	33 41 50 115	309
52	27 15 51 118	28 06 51 118	28 57		29 48 50	15	32 19 50 115	33 09 50 114	308
53	26 43 51 117	27 34 51 117	28	51 116	29 16 50 116	30 06 51 115	30 57 50 114	31 47 50 114	32 37 49 113	307
54	26 12 51 116	27 03 50 116	27 53 51 115	28 44 50 115	29 34 50 114	30 24 50 113	31 14 50 113	32 04 50 112	306
55	25 40 +51 116	26 31 +50 115	27 21 +51 114	28 12 +50 114	29 02 +50 113	29 52 +50 112	30 42 +49 112	31 31 +50 111	305
56	25 08 51 115	25 59 50 114	26 49 50 113	27 39 50 113	28 29 50 112	29 19 49 112	30 09 49 111	30 58 50 110	304
57	24 36 51 114	25 27 50 113	26 17 50 113	27 07 50 112	27 57 49 111	28 46 50 111	29 36 49 110	30 25 49 109	303
58	24 04 50 113	24 54 50 112	25 44 50 112	26 34 50 111	27 24 49 110	28 13 49 110	29 03 49 109	29 52 49 108	302
59	23 31 50 112	24 21 50 111	25 11 50 111	26 01 50 110	26 51 49 109	27 40 49 109	28 29 49 108	29 18 49 107	301
60	22 58 +50 111	23 48 +50 110	24 38 +50 110	25 28 +49 109	26 17 +49 109	27 06 +49 108	27 55 +49 107	28 45 +48 107	300
61	22 25 50 110	23 15 50 109	24 05 49 109	24 54 50 108	25 43 49 107		49 106	28 11 48 106	299
62	21 52 50 109	22 42 49 109	23 31 50 108	24 21 49 107	25 10 49 106		49 105	27 37 48 105	298
63	21 19 49 108	22 08 50 108	22 58 49 107	23 47 49 106	24 35 49 106	WHOLE DEGREES	49 105	27 03 48 104	297
64	20 45 50 107	21 35 49 107	22 24 49 106	23 13 49 105	24 02 49 105	OF DECLINATIION	49 104	26 28 49 103	296
65	20 11 +50 107	21 01 +49 106	21 50 +49 105	22 39 +49 105	23 28 +49 104	24 17 +49 103	25 05 +49 103	25 54 +48 102	295
66	19 38 49 106	20 27 49 105	21 16 49 104	22 05 49 104	22 54 48 103	23 42 49 103	24 31 48 102	25 19 48 101	294
67	19 04 49 105	19 53 49 104	20 42 49 104	21 31 49 103	22 19 48 102	23 08 48 102	23 56 48 101	24 45 48 100	293
68	18 29 49 104	19 18 49 103	20 07 49 103	20 56 48 102	21 44 48 102	22 33 48 101	23 22 48 100	24 10 48 100	292
69	17 55 49 103	18 44 48 103	19 33 49 102	20 22 48 101	21 10 49 101	21 59 48 100	22 47 48 99	23 35 48 99	291
	7°	8°	9°	10°	11°	12°	13°	14°	

EXAMPLE 27 MORNING SIGHT
LHA 324°
DECLINATION 8°
> Hc 35° 42´
DIFF. +54
Z 134°

Table 19. NP 303 Rapid Sight Reduction Table Lat 54°

– navigating with the Sun

From **Table 19**:

Chosen latitude 54°	LHA 324° / Dec 8°	LHA 54° / Dec 8°
	Morning sight	**Afternoon sight**
Calculated altitude:	35° 42′	27° 03′
Difference (d)	+ 54	+ 50
Azimuth (Z)	134	116

Using **Table 20** to account for the minutes (<u>rounded</u>) of the Sun's declination:

Entering with Dec mins 15 and difference d = +54   Dec mins 20 / d = +50

		Morning sight		**Afternoon sight**
Diff (for mins of Dec)	+54	14′	+50	17′
Calculated Alt.		<u>35° 42′</u> +		<u>27° 03′</u> +
Calculated altitude		35° 56′		27° 20′

The azimuth (Z_n) must then be found from the azimuth angle (Z) according to the rules printed on each and every page of the main tables: in northern latitudes LHA > 180: Zn=Z and LHA < 180: Zn = 360–Z. The relevant LHA values being morning: 324° and afternoon: 54°

Z	(LHA > 180: Zn=Z)	134	(LHA < 180: Zn = 360–Z)	116	(360 – 116 = 244)
Z_n		134			244

Intercept = observed altitude – calculated altitude = $H_o - H_c$

	Morning sight	**Afternoon sight**
	35° 41.2′	27° 28.5′
	<u>35° 56.0′</u>	<u>27° 20.0′</u>
Intercept =	–14.8′ away	+ 8.5′ towards

8. Calculate the **latitude** of the noon sight:

Using **Equation 1** (p.154) as observer and Sun in same hemisphere and observer's latitude North is greater than Sun's declination:

                      latitude = zenith distance + declination   89° 60.0′
substituting: **zenith distance = 90° – observed altitude**   <u>44° 18.1′</u> –
                      latitude = (90° – observed altitude) + declination   45° 41.9′
                                                                                                           <u>8° 17.2′</u> +
                                                                                                                                     53° 59.1′

Astronavigation from Square One

TABLE 5 — Correction to Tabulated Altitude for Minutes of Declination

28 29 30	31 32 33	34 35 36	37 38 39	40 41 42	43 44 45	46 47 48	49 50 51	52 53 54	55 56 57	58 59 60	d
′ ′ ′	′ ′ ′	′ ′ ′	′ ′ ′	′ ′ ′	′ ′ ′	′ ′ ′	′ ′ ′	′ ′ ′	′ ′ ′	′ ′ ′	′
0 0 0	0 0 0	0 0 0	0 0 0	0 0 0	0 0 0	0 0 0	0 0 0	0 0 0	0 0 0	0 0 0	0
0 0 0	1 1 1	1 1 1	1 1 1	1 1 1	1 1 1	1 1 1	1 1 1	1 1 1	1 1 1	1 1 1	1
				1 1 1	1 1 1	1 1 2	2 2 2	2 2 2	2 2 2	2 2 2	2
d DIFFERENCE				2 2 2	2 2 2	2 2 2	2 2 3	3 3 3	3 3 3	3 3 3	3
FOR MINUTES OF				2 3 3	3 3 3	3 3 3	3 3 3	4 4 4	4 4 4	4 4 4	4
DECLINATION				3 3 3	3 3 4	4 4 4	4 4 4	5 5 5	5 5 5	5 5 5	5
3 3 4	4 4 4	4 4 4		4 4 4	4 4 4	4 4 5	5 5 5	5 5 6	6 6 6	6 6 6	6
4 4 4	4 4 4	5 5 5	5 5 5	**EXAMPLE 27 p.172**		5 6 6	6 6 6	6 6 7	7 7 7	7 7 7	7
4 4 4	4 5 5	5 5 5	5 6 6	**MORNING SIGHT**		6 6 7	7 7 7	7 7 8	8 8 8	8 8 8	8
				DEC MINS: 14.9′		7 7 7	7 8 8	8 8 8	8 9 9	9 9 9	9
5 5 5	5 5 6	6 6 6	6 6 6	**DEC DIFF +54 > 14′**		8 8 8	9 9 9	9 9 10	10 10 10	10 10 10	10
5 5 6	6 6 6	6 6 7	7 7 7			9 9 9	9 10 10	10 11 11	11 11 11	11 11 11	11
6 6 6	6 6 7	7 7 7	7 8 8	8 8 8	9 9 9	9 10 10	10 10 11	11 11 12	12 12 12	12 12 12	12
6 6 6	7 7 7	7 8 8	8 8 8	9 9 9	9 10 10	10 10 10	11 11 12	12 12 13	13 13 13	13 13 13	13
7 7 7	7 7 8	8 8 8	9 9 9	9 10 10	10 10 10	11 11 11	11 12 12	12 13 13	13 13 13	14 14 14	14
7 7 8	8 8 8	8 9 9	9 10 10	10 10 10	11 11 11	12 12 12	12 13 13	13 13 14	14 14 14	14 15 15	15
7 8 8	8 9 9	9 9 10	10 10 10	11 11 11	11 12 12	12 13 13	13 13 14	14 14 14	15 15 15	15 16 16	16
8 8 9	9 9 9	10 10 10	10 11 11	11 12 12	12 12 13	13 13 14	14 14 15	15 15 15	16 16 16	16 17 17	17
8 9 9	9 10 10	10 10 11	11 11 12	12 12 13	13 13 14	14 14 14	15 15 15	16 16 16	16 17 17	17 18 18	18
9 9 10	10 10 10	11 11 11	12 12 12	12 13 13	13 14 14	14 15 15	15 16 16	16 17 17	17 18 18	18 19 19	19
9 10 10	10 11 11	11 12 12	12 13 13	13 14 14	14 15 15	15 16 16	16 17 17	17 18 18	18 19 19	19 20 20	20
10 10 10	11 11 12	12 12 12	13 13 14	14 14 15	15 15 16	16 17 17	17 18 18	18 19 19	19 20 20	20 21 21	21
10 11 11	11 12 12	12 13 13	14 14 14	15 15 15	16 16 16	17 18	18 18 19	19 19 20	20 21 21	21 22 22	22
11 11 12	12 12 13	13 13 14	14 15 15	15 16 16	16 17	18 18 18	19 19 20	20 20 21	21 21 22	22 23 23	23
11 12 12	12 13 13	14 14 14	15 15	**EXAMPLE 27 p.172**		18 18 19	19 20 20	20 21 21	21 22 23	23 24 24	24
12 12 12	13 13 14	14 15 15	15 16	**AFTERNOON SIGHT**		20 20 20	20 21 21	22 22 22	23 23 24	24 25 25	25
12 13 13	13 14 14	15 15 16	16 16	**DEC MINS: 20.4′**		20 21	21 22 22	23 23 23	24 24 25	25 26 26	26
13 13 14	14 14 15	15 16 16	17 17	**DEC DIFF +50 > 17′**		21 22	22 22 23	23 24 24	25 25 26	26 27 27	27
13 14 14	14 15 15	16 16 17	17 18			22 23	23 23 24	24 25 25	26 26 27	27 28 28	28
14 14 15	15 15 16	16 17 17	18 18			23 23	23 24 25	25 26 26	26 27 27	28 29 29	29
14 14 15	16 16 16	17 18 18	18 19 20	20 20 21	22 22 22	23 24 24	24 25 26	26 26 27	28 28 28	29 30 30	30
14 15 16	16 17 17	18 18 19	19 20 20	21 21 22	22 23 23	24 24 25	25 26 26	27 27 28	28 29 29	30 30 31	31
15 15 16	17 17 18	18 19 19	20 20 21	21 22 22	23 23 24	25 25 26	26 27 27	28 28 29	29 30 30	31 31 32	32
15 16 16	17 17 18	19 19 19	20 20 20	21 21 22	23 24 24	25 26 26	27 28 28	29 29 30	30 31 31	32 32 33	33
16 16 17	18 18 19	19 20 20	21 21 21	22 22 23	23 24 25	26 26 27	28 28 29	29 30 31	31 32 32	33 33 34	34
16 17 18	18 19 19	20 20 21	22 22 23	23 24 24	25 26 26	27 27 28	29 29 30	30 31 32	32 33 33	34 34 35	35
17 17 18	19 19 20	20 21 22	22 23 23	24 25 25	26 26 27	28 28 29	29 30 31	31 32 32	33 34 34	35 35 36	36
17 18 18	19 20 20	21 22 22	23 23 24	24 25 25	26 27 27	28 29 29	30 30 31	32 32 33	34 34 35	36 36 37	37
18 18 19	20 20 21	22 22 23	23 24 25	25 26 27	27 28 28	29 30 30	31 32 32	33 34 34	35 35 36	37 37 38	38
18 19 20	20 21 21	22 23 23	24 25 25	26 27 27	28 29 29	30 31 31	32 32 33	34 34 35	36 36 37	38 38 39	39
19 19 20	21 21 22	23 23 24	25 25 26	27 27 28	29 29 30	31 31 32	33 33 34	35 35 36	37 37 38	39 39 40	40
19 20 20	21 22 22	23 24 25	25 26 27	27 28 29	29 30 31	31 32 32	33 34 35	36 36 37	38 38 39	40 40 41	41
20 20 21	22 22 23	24 24 25	26 27 27	28 29 29	30 31 32	32 33 34	34 35 36	36 37 38	38 39 40	41 42	42
20 21 22	22 23 24	24 25 26	27 27 28	29 29 30	31 32 32	33 34 34	35 36 37	37 38 39	39 40 41	42 43	43
21 21 22	23 23 24	25 26 26	27 28 29	29 30 31	32 32 33	34 34 35	36 37 37	38 39 40	40 41 42	43 43 44	44
21 22 22	23 24 25	26 26 27	28 28 29	30 31 32	32 33 33	34 35 36	37 38 38	39 40 40	41 42	44 44 45	45
21 22 23	24 25 25	26 27 28	28 29 30	31 31 32	33 34 34	35 36 37	38 38 39	40 41 41	42 44	44 45 46	46
22 23 24	24 25 26	27 27 28	29 30 31	31 32 33	34 34 35	36 37 38	38 39 40	41 42 42	44 45	45 46 47	47
22 23 24	25 26 26	27 28 29	30 30 31	32 33 34	34 35 36	37 38 38	39 40 41	42 42	45 46	46 47 48	48
23 24 24	25 26 27	28 29 29	30 31 32	33 33 34	35 36 37	38 38 39	40 41 42	42	45 46 47	47 48 49	49
23 24 25	26 27 28	28 29 30	31 32 32	33 34 35	36 37 38	38 39	41 42 42		46 47 48	48 49 50	50
24 25 26	26 27 28	29 30 31	31 32 33	34 35 36	37 37 38	39 40	**WHOLE**		47 48 48	49 50 51	51
24 25 26	27 28 29	29 30 31	32 33 34	35 36 36	37 38 39	40 41	**MINUTES OF**		48 49 49	50 51 52	52
25 26 26	28 28 29	30 31 32	33 34 34	35 36 37	38 39 40	40 41	**DECLINATION**		49 49 50	51 52 53	53
25 26 27	28 29 29	31 32 32	33 34 35	36 37 38	39 40 40	41 42			50 50 51	52 53 54	54
26 27 28	28 29 30	31 32 33	34 35 36	37 38 38	39 40 41	42 43 44	45 46 47	48 49 50	50 51 52	53 54 55	55
26 27 28	29 30 31	32 33 34	35 35 36	37 38 39	40 41 42	43 44 45	46 47 48	49 49 50	51 52 53	54 55 56	56
27 28 28	29 30 31	32 33 34	35 36 37	38 39 40	41 42 43	44 45 46	47 48 48	50 51 52	53 54 55	56 57 58	57
27 28 29	30 31 32	33 34 35	36 37 38	39 40 41	42 43 44	44 45 46	47 48 49	50 51 52	53 54 55	56 57 58	58
28 29 30	30 31 32	33 34 35	36 37 38	39 40 41	42 43 44	45 46 47	48 49 50	51 52 53	54 55 56	57 58 59	59

Table 20. NP 303 (vol. 3) Interpolation table for minutes of declination

– navigating with the Sun

$$\text{latitude} = (90° - 44° 18.1') + 8° 17.2 = 45° 41.9 + 8° 17.2'$$

latitude = 53° 59.1′ North

8. Plot the **latitude line, azimuths, intercepts** and **position lines**:
The chosen positions, azimuths and intercepts are plotted in **Figure 100**.

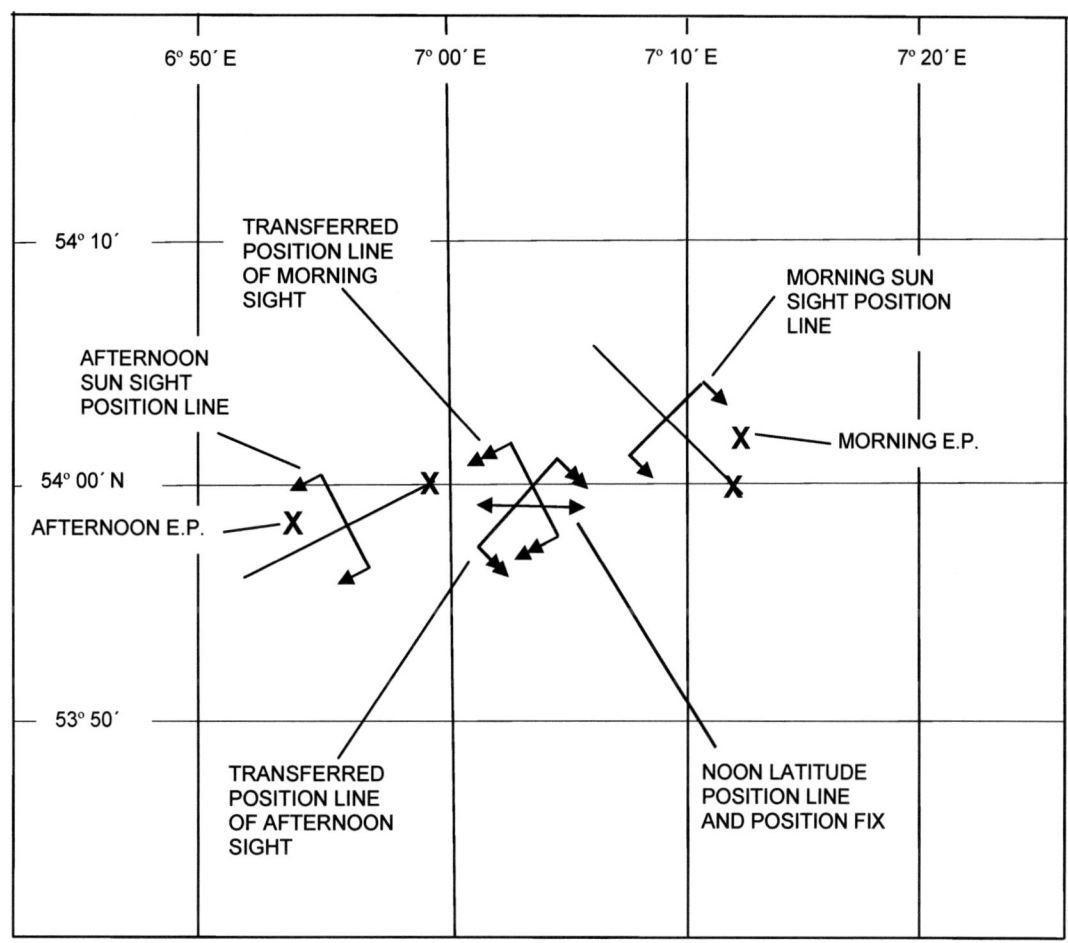

Figure 100. Plotting the Sun run Sun position lines

The plotting of the morning and latitude sights (**Figure 101**) is carried out in the following manner:-

The morning EP is plotted. The chosen position is plotted, the calculated azimuth is drawn in, together with the intercept and position line for the morning sight. The morning EP is then adjusted, by moving it to the position line, parallel to the azimuth. This gives the best estimate of position at the morning sight from the data collected. The course run (250°) is then drawn in from the adjusted morning EP. The EP at the meridian sight is plotted, based of the timed run between morning and meridian sights. The latitude achieved from the meridian passage is drawn in. The position line from the morning site is then transferred to the Meridian EP. The point at which this line crosses the latitude line is the best estimate of position at that moment from the data available.

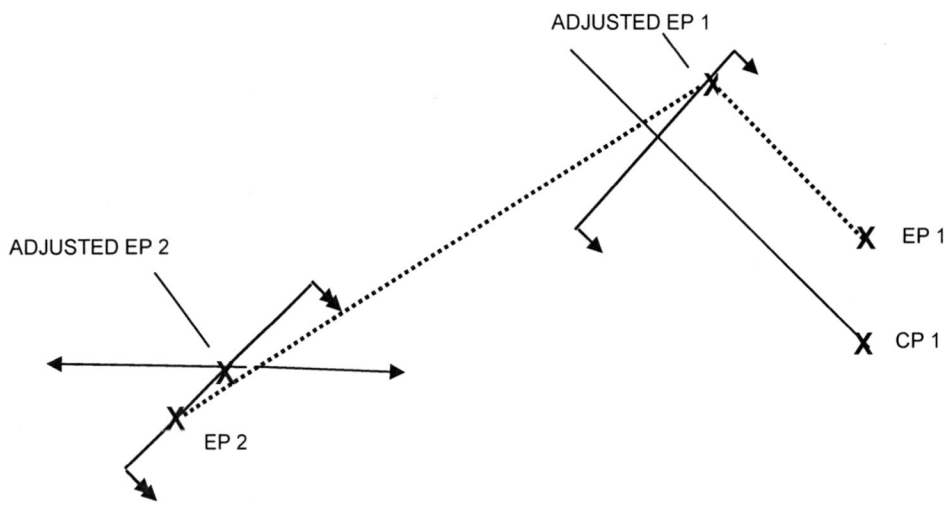

Figure 101 Detail of plotting of morning and meridian passage sights

The afternoon run and sights can then be added to the plot, drawn from the adjusted meridian position to give an afternoon EP. The afternoon chosen position, azimuth and position line is plotted and EP adjusted to give the best available EP for the given data. The afternoon position line can then be transferred back to the meridian EP to give a typical cocked hat fix of three position lines, giving the best Estimated Position at the meridian passage. When plotting position considerable distances from land, small scale charts prevent plotting with sufficient accuracy. Plotting sheets are then needed, explained in the Appendix (p.294).

Section 7: Navigating with the Moon

The Moon tends to be relatively neglected as a navigationally useful body, despite its frequent presence in the sky in the day when there is an horizon. It is true that its proximity to Earth introduces some complications to sight reduction but these can be overcome by <u>understanding</u> their cause and effect. The essential key to the Moon is to ask the question as to <u>why</u> these astro-navigational peculiarities arise.

By the end of Section 7, the following KEY FACTS should be understood:-

- **The proximity of the Moon to the Earth and its relatively rapidly-changing elliptical orbit requires that detailed adjustments for semi-diameter, horizontal parallax and augmentation are necessary in order to attain an observed altitude.**

- **The Moon is involved in cyclical eclipses, of both the Sun and itself when viewed from the Earth.**

- **The Moon is present in the daytime sky for a considerable period of its monthly course and offers good opportunities for daylight sights, especially leading up to first quarter and from last quarter.**

- **The Greenwich Hour Angle of the Moon varies continuously and this must be corrected for on an hourly basis as the tabled values are derived using a fixed hourly average GHA rate change of 14° 19´. The *v* value for each hour is the actual GHA change over the subsequent hour and is used to find the additional increment from the minutes and seconds *v* correction table, to adjust for deviations from the average value.**

- **The Declination of the Moon, measured North or South of the celestial equator, varies cyclically over the year. The Moon's declination value for each whole hour needs to be adjusted by a *d* value whose size corresponds to the rate of declination change occurring during each subsequent hour.**

- **Augmentation is the name given to the change in apparent diameter of the Moon with changing observed altitude from a given position, caused by the significant diameter of the Earth when compared to the relatively small Earth-Moon distance. Augmentation is corrected for in the Moon's altitude correction table.**

- **The Moon's proximity and varied orbital path produce a rapidly changing horizontal parallax (H.P.) and its hourly value is used to find an additional adjustment to the apparent altitude (H_a) to achieve an observed altitude (H_o).**

7.1 The Moon in general – what do we know about her?

The Moon, Earth's only natural satellite lies at a distance of 380,000km. The origin of its name is unclear, but is probably a distillation of the Old English: *mona*, the German: *mond*, Latin: *mensis* and Greek: *men*. The Earth and Moon share a very similar geological composition indicating that the Moon was probably formed as a result of a huge impact explosion between Earth and a very sizeable meteor possibly as long ago as 4.5 billion years[7.1]. It has a diameter of 3,500km, a quarter that of Earth and it has a relative mass of just $1/80^{th}$.

The close proximity of the Moon means that it has a significant diameter and of any of the heavenly bodies, it is the most varied both in position and appearance. Rotating around the Earth as the Earth orbits the Sun (**Figure 101**), the Moon has particularly complex phases and cycles, its orbit being a form of precession.

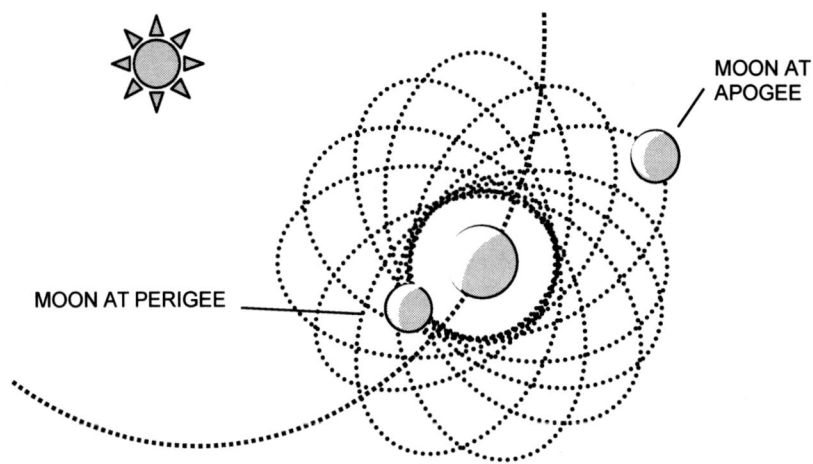

Figure 101. The complex geometry of the Moon's rotation round Earth and Sun

7.2 The Moon in navigation – what makes it peculiar?

The Moon is a useful navigation tool, often present in the daytime as well as at dawn and dusk. <u>Prior</u> to the invention of a stable, reliable, sea-going chronometer it was the vital link in estimating longitude. Tables were drawn up, called lunars, of predicted angular distances between the Moon, certain stars and the Sun, linked to Greenwich Time[7.2]. By measurement of the observed angular distances at local time and comparing this with known values in tables, it was possible to deduce the Greenwich Time of the observation.

As introduced in **Section 1.17** and expanded in **Section 5**; if the navigator observes a body at a particular altitude and knows the GMT/UT at which that altitude should be found, then that time can be converted to longitude, locating the navigator's particular meridian on the Earth's sphere. John Harrison and his chronometers, almost single-handedly, usurped the Moon's paramount importance as the arbiter of longitude. Lunars were calculated and used for navigation in the BBC filmed aspect of the reconstructed voyage of Cook in the ship Endeavour.

The Moon's most important peculiarity is that it does not keep a 24 hour day, but a rather variable one lasting 24 hours and 50 minutes or so, crossing any given meridian approximately 50 minutes <u>later</u> each subsequent day, appearing 14° or so to the East of any given longitude as each 24 hour period passes. It would be a rather variable time-piece if we did not have the Sun. The time between the Moon's daily meridian passages varies cyclically between 24 hours and 39 minutes to 24 hours and 69 minutes, as a consequence of its elliptical orbit. The local time of its crossing at the Greenwich meridian cannot therefore be relied on to be the local time of crossing all other meridians, unlike the stalwart Sun, which can.

Whereas the Sun crosses 15° of longitude per hour, the Moon crosses an average of only 14.5° per hour, varying from 14.3° to 14.6°. The apparent Sun traverses 360° in 24 hours, the Moon only around 347°, varying from 343-350°. To appreciate the usefulness and the particular complications of using the Moon for navigation, it is necessary to understand something of the complexity of its movement relative both to the Earth and Sun.

7.3 The Moon we see – why do we only seem to see one face?

The Moon does not now rotate independently on its own axis, unlike the Sun, Earth and planets which do. Consequent upon this, the same part of the Moon's globe constantly faces the Earth. In the very distant past the Moon did rotate, but it slowly came to a relative rest, showing the one face we know well with its dusty *maria* or "seas". This strange arrangement of the Moon is termed its **synchronous rotation** with Earth and is due to complex gravitational effects between the Earth and its close satellite that lays just 30 Earth-diameters distant [7.3]. The combined gravitational centre of the Earth-Moon system lies <u>inside</u> the Earth itself.

The eccentricity of the Moon's elliptical orbit results in an apparent, slight and alternate turning of the Moon's face, called a **libration** and due to this almost 60% of the surface can be seen from Earth at varying times[7.4]. The Moon's side we see, termed the near side, demonstrates the results of old volcanic activity and meteoric impaction. Contrastingly, the contrary side is relatively featureless, remaining as it does, facing away from the Earth, bathed in light during early and late Moon phases. There is, perhaps sadly for romantics, no permanently dark side of the Moon. It is also worthy of note that the Moon is not a very good reflector; it has a low **albedo** of 0.07 or so; meaning that it only reflects 7% of the light that is incident upon it [7.5].

7. 4 The Moon's varying orbit – how long does it take?

The Moon moves in an elliptical orbit, with an **eccentricity** value of 0.5. This equates to the major or long axis of the orbit being 13% greater in length than the minor or shorter axis and this is one of the <u>main</u> causes of the Moon's relative navigational complexities. The plane of the orbit makes an angle of 5° inclination with that of the Earth around our Sun. It is less helpful to relate the plane of the Moon's orbit to that of Earth's equator, as the former varies from 16-28° over the lunar cycle, shown in **Figure 102**, caused by Earth's fixed axial inclination.

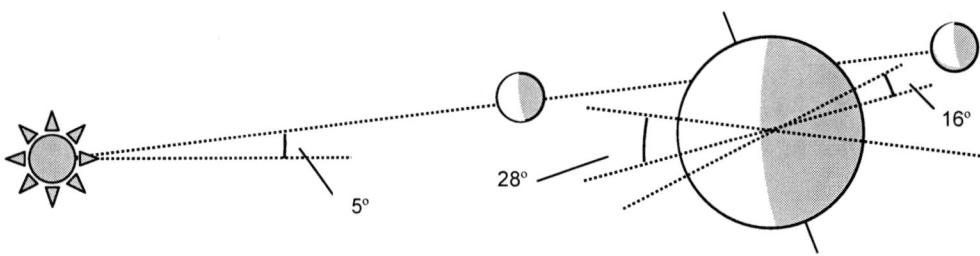

Figure 102. The complex geometry relating the orbital planes of the Moon and Earth

The Moon completes one orbit of the Earth in 27.3 days and this is known as its **sidereal month** (Latin *sidus*: a star). The Moon begins and ends this orbit at the same point in the sky when viewed from Earth, with identical celestial coordinates of GHA and Declination; in short, at the same place in the background of stars. But if observed closely from Earth, the Moon would be seen <u>not</u> have completed its detailed visual cycle.

This 27.3 day period does <u>not</u> therefore correspond to the phases of the Moon that we see from Earth as these are dependent upon the <u>relative</u> positions of the Moon, Earth <u>and</u> Sun, which change continually. During the 27.3 day period, the Earth will have progressed along its own orbital path by an arc approximating 30°, changing the positional relationship of the Earth and Moon with respect to the Sun (**Figure 103**).

It is the illumination and reflection of the Sun's light <u>as seen from Earth</u> that produces the Moon's phases and this cyclical change repeats itself in a period called a **lunation** or **synodic month** (Greek *synodus*: a meeting or conjunction). The Moon reaches the point where it has the <u>same angular relationship</u> to the Earth and Sun after 29.3 days (**Figure 104**). From Earth at the start of its cycle, the Moon is not visible, lying as it does between Earth and the Sun and none of its reflected light is directed to Earth. This is the period of the true birth of the new Moon.

– navigating with the Moon

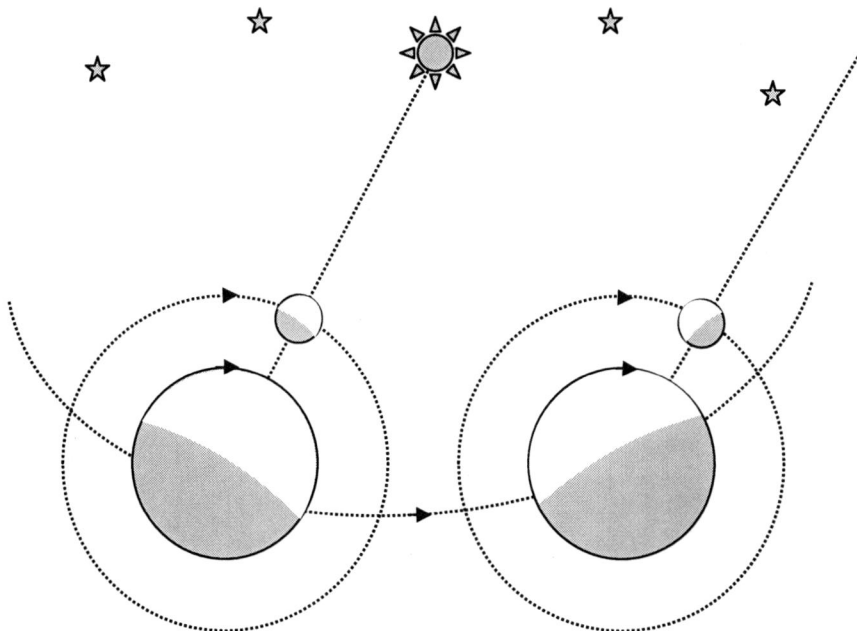

Figure 103. The positions of Sun, Earth and Moon after 1 sidereal month – 27.3 days

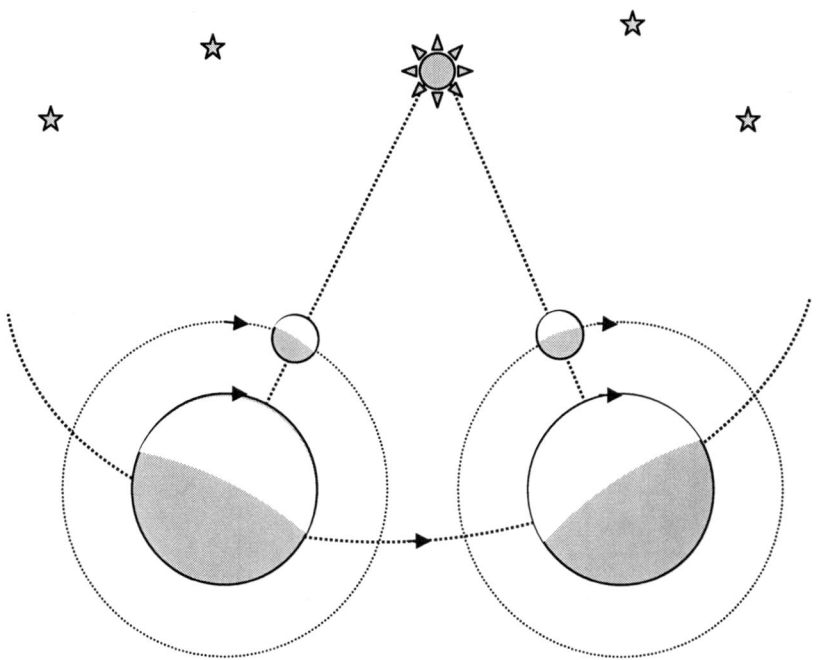

Figure 104. Positions of Sun, Earth and Moon after one synodic month – 29.3 days

Astronavigation from Square One

7.5 The phases of the moon – why do we see the Moon as different shapes?

Throughout the 29.3 days of the synodic month, the Moon is, but for the rare exception of a total eclipse of the Moon, constantly bathed in the Sun's light. <u>One</u> <u>half</u> of its surface; that which faces the Sun, is <u>continually</u> <u>illuminated</u> and to a <u>distant</u> <u>observer</u> out in space, the appearance of the Moon would be similar to that in **Figure 105**, where all phases throughout the synodic month would be similar in appearance.

The phases we see from our viewpoint on Earth are dependent upon the <u>degree</u> to which that illuminated face is <u>visible</u> <u>on</u> <u>Earth</u>. These appearances are also to some extent, influenced by the observer's latitude, the time of year and the declination of the Moon. Details of the phases are usually recorded in astro-nautical almanacs and appear in the bottom right corner of the NP 314 Sun and Moon pages.

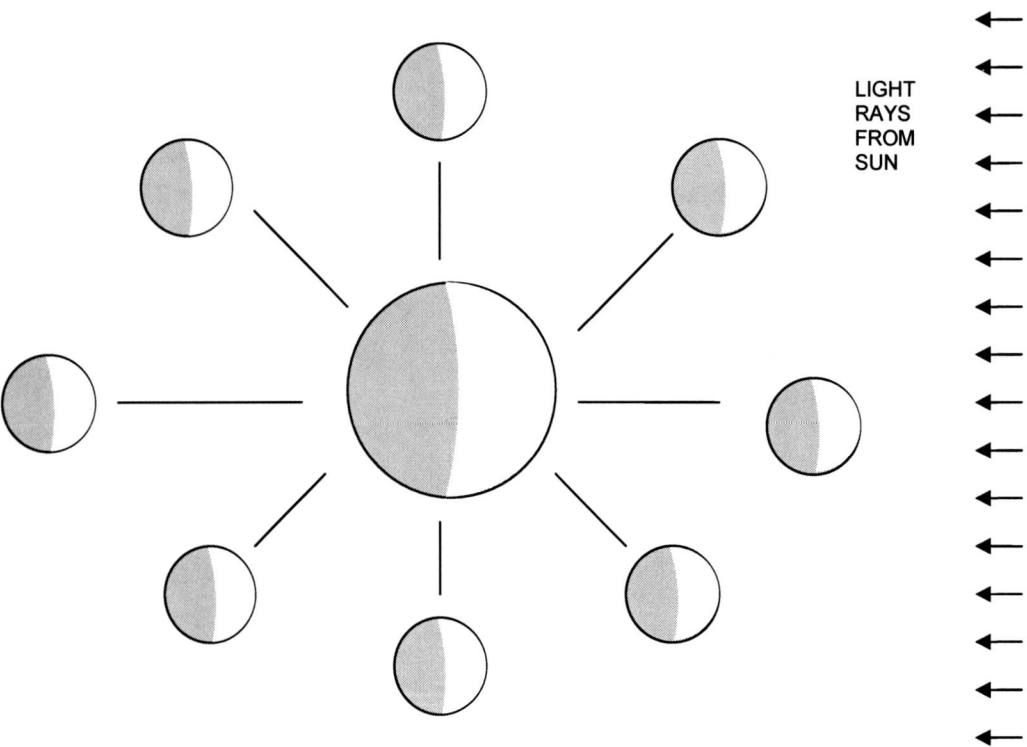

Figure 105. A polar view of the Earth and Moon over a synodic month

When the Moon on its orbital path lies nearest the Sun, its illuminated face is <u>invisible</u> on Earth as no reflected light can reach us (position A in **Figure 106**). This is the true birth of the new Moon. The crescent that we call the new Moon subsequently becomes visible on Earth as the relative angle between the Earth, Moon and Sun just begins to permit a small amount of the reflected light to be seen on Earth. At this point the Moon has moved a short way round its orbit from the true new Moon position and just the edge of its illuminated hemisphere is visible on Earth (position B). The visible crescent continues to grow as a waxing crescent (German *wachsen*: to increase), resulting in positions C and then D.

Following this, the half-Moon position is reached at the point called the first quarter (point E), where the Moon is a quarter way round its monthly journey. At point E, the angle between the observer, the Moon and the Sun is 90° or so and half the illuminated Sun-facing surface of the Moon is visible from Earth.

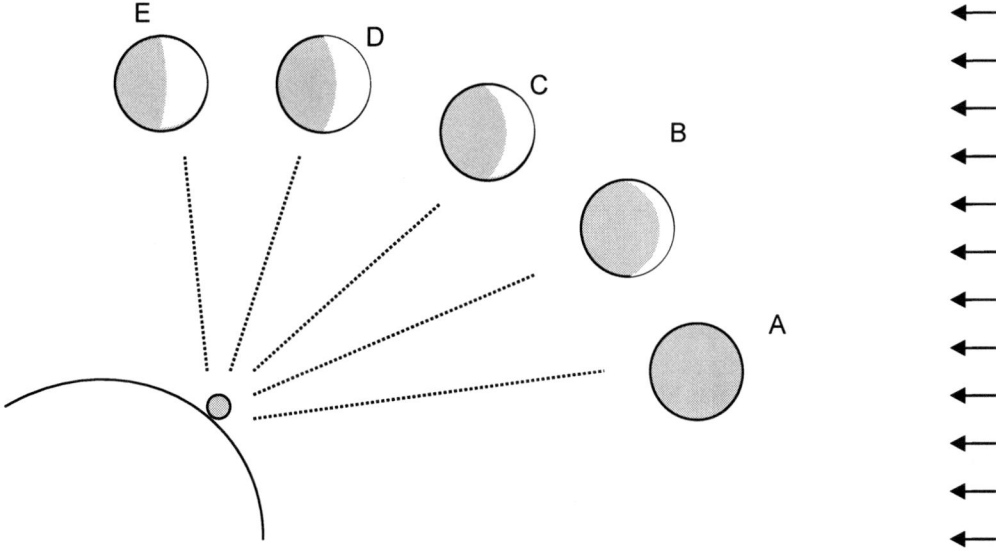

Figure 106. An earth-bound observer's view from new Moon to first quarter

The size of the visible face then continues to increase reaching the waxing gibbous (humped) phase (E-H **Figure 107**) and finally reaching full Moon (point I), where the whole of the illuminated hemisphere is visible from Earth as the Sun, observer and Moon are approximately in line or conjunction.

Full Moon is followed by the waning gibbous phase, leading to the last quarter, when the Moon again lies at a relative angle of 90° to the Sun and when viewed from Earth half the illuminated hemisphere is visible. The final phase passes through the waning crescent and on towards the next new Moon, when the Earth, Moon and Sun are once again in line, a condition termed **syzygy** –pronounced zi-zi-gee (Greek: *syzgia*: union). What a great name for a yacht!

Astronavigation from Square One

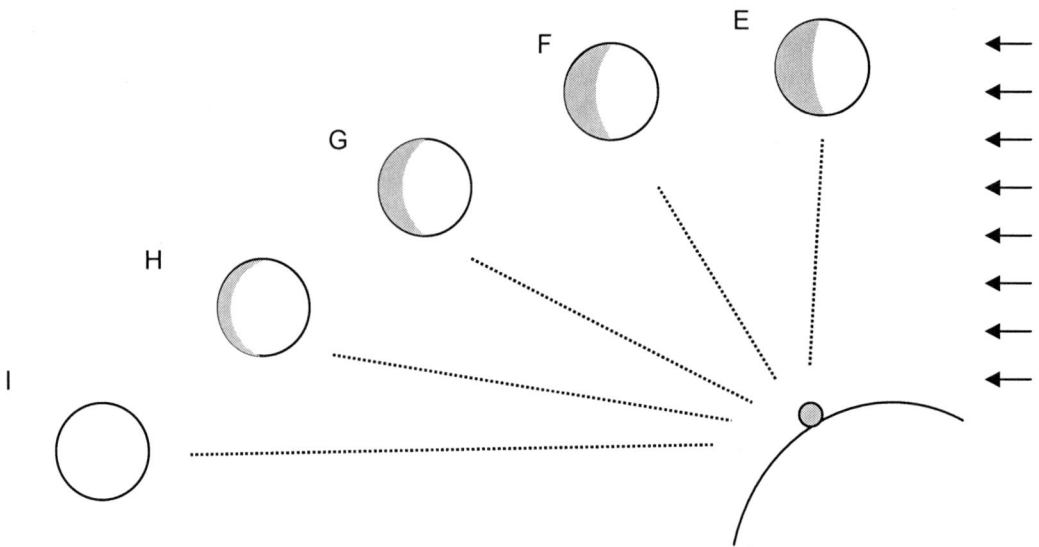

Figure 107. An Earth-bound observer's view from first quarter to full Moon

7.6 The time of day and the Moon's phase – is there a regular relationship?

We are used to the phases of the Moon and their progressive stages through the monthly cycle but not often aware of the cycles in relation to day and night. They are of particular importance to the navigator, especially at dawn and dusk twilight, the first and last opportunities in the day to get a sight.

At the time of the new Moon, the Moon's position in its orbit is near **conjunction** with the Sun. In simple terms, the Earth, Moon and Sun lie in a reasonably straight line in space; they are in transit, to use a term from coastal pilotage. If the Moon were visible on Earth at this time, rather than being obscured by the Sun's glare, it would be seen to rise just before the Sun, move across our sky from East to West still near the Sun and then set just before the Sun.

Between new Moon and first quarter and lagging the Sun's movements by 50 or so minutes per day, it escapes the glare and becomes visible in the western sky at sunset. By the first quarter, the Moon, having attained its half-moon appearance, tends to rise around midday and set around midnight. By full Moon, the time of its rise has slipped back to dusk in the evening and it sets by dawn, while towards its last quarter, again appearing as a half-moon, it will rise around midnight and set towards midday.

For good sights, the Moon needs to be relatively high to avoid excessive parallax and refractive errors because of its proximity. A new Moon, being so very close to the direction to the Sun, is all but invisible and therefore of little help to the navigator at dawn and dusk. By the first quarter however, rising at midday and setting at midnight, it is of use for afternoon

– navigating with the Moon

sights and at dusk. By full-moon, rising at dusk and setting at dawn, it is low in the sky when the horizon is visible and of dubious navigational use. It is present all night, but the navigator is then without a visible horizon. The "horizon" visible in good moonlight is not the true horizon. By last quarter however, rising at midnight and setting at midday, the Moon is again useful, for sights at dawn and in the forenoon. There are therefore distinct times when the Moon and Sun can be sighted simultaneously and used to produce an immediate fix from two-position lines.

Summarizing the most useful times; new Moon towards first quarter gives opportunities for afternoon and dusk twilight sights. Last quarter towards new Moon gives opportunities for dawn twilight and morning sights.

7.7 Apparent Moon size, its orbit and Augmentation – Why does its size change?

The Moon appears to change in size when viewed from Earth and this size is relevant to the navigator taking Moon sights as the semi-diameter correction will vary and therefore needs to be corrected for. The apparent size varies for two reasons. The first relates to the elliptical orbit which results in the Moon being closest to the Earth at one point and furthest away at a point half an orbit later.

The point at which it is closest to Earth is called the **perigee** and the point furthest away is called the **apogee** and the times of their occurrence are not rigidly fixed but vary cyclically. The cycle, called the **full moon cycle** begins when **perigee** occurs at full Moon and it lasts for 14 cycles or **lunations** until perigee and full Moon again coincide. This rather complicated process is caused by a form of **precession**, a gravitational effect, which for simplicity can be likened to the wobble of the axis of a spinning child's top. A similar but much more sedate wobbling movement happens to the Moon as it orbits the Earth producing a complex pattern.

The complex elliptical orbital shape of the Moon is caused by the relative gravitational attraction of the Moon by the Earth and Sun, which also change cyclically as the Moon orbits the Earth and as the Earth orbits the Sun. In some positions of the gravitational pull of Earth and Sun are additive and at other times less so and their resultant force dictates the Moon's relative position. When the forces are at an added maximum, perigee coincides with full Moon and the Sun, Moon and Earth are in line, the Moon reaching its closest point to Earth.

The second reason for the change in apparent size of the Moon is called **Augmentation**[7.6] and is consequent upon the ratio between radius of the Earth and the distance to the Moon being significant, due to the Moon's nearness. This causes a small decrease in the observer-Moon distance which slightly increases the Moon's apparent diameter, as the observed altitude increases. An observer viewing the Moon overhead is an Earth's radius (~4000 miles or 1.6%) nearer to it than when viewing it on the horizon (**Figure 108**). A correction for this factor needs to be made and is <u>included</u> with the correction for refraction which also varies with altitude in the Moon altitude table of the astro-nautical almanac (**Table 23** p.197).

187

Astronavigation from Square One

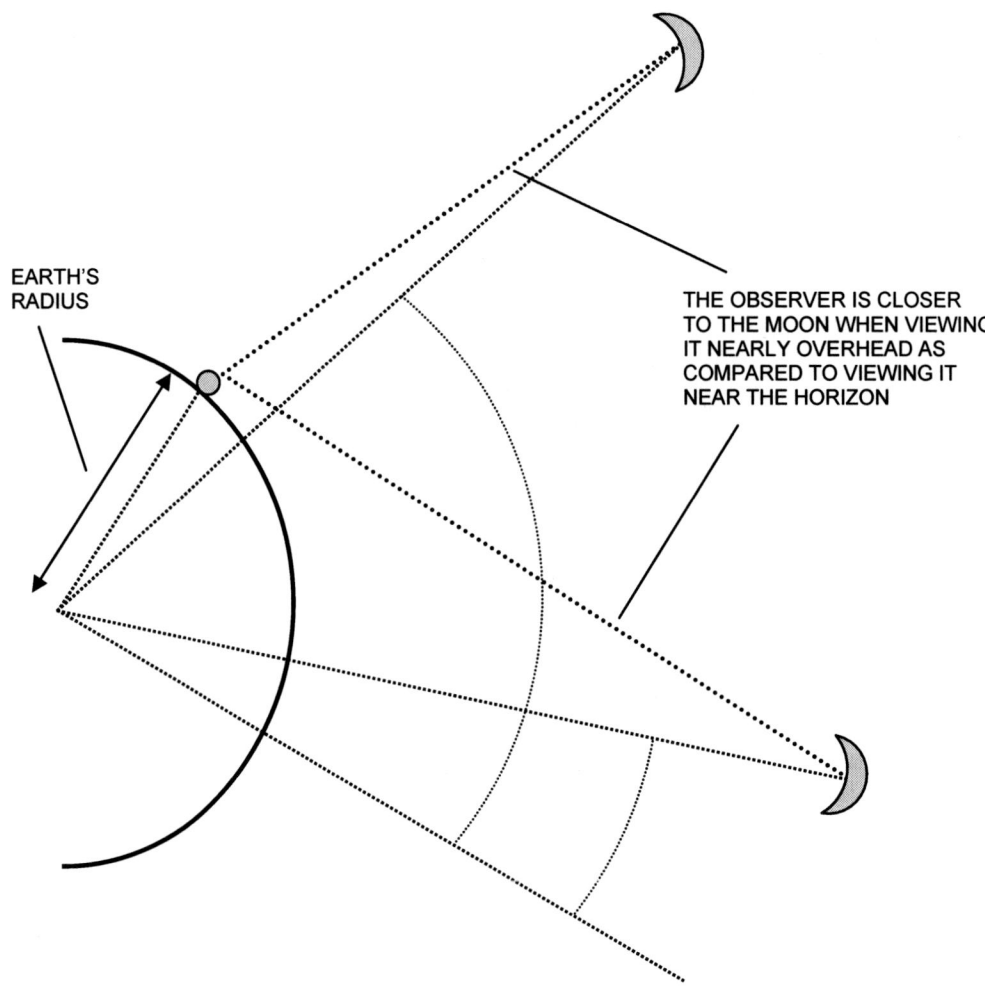

Figure 108. The effect of augmentation on apparent Moon diameter

7.8 The Moon's parallax - complicating sight reduction

The generally close proximity of the Moon also means that parallax is of great importance, as introduced in **Section 2.3**(p.40). Parallax in the astro-nautical context is the apparent change in the position of an object when viewing it from the surface of the Earth as compared to the theoretical view from the Earth's centre. This factor is very much larger when sighting a close object than one more distant. This is a particular problem when viewing the Moon relatively horizontally compared to when viewing it more vertically, as shown in **Figure 109**.

– navigating with the Moon

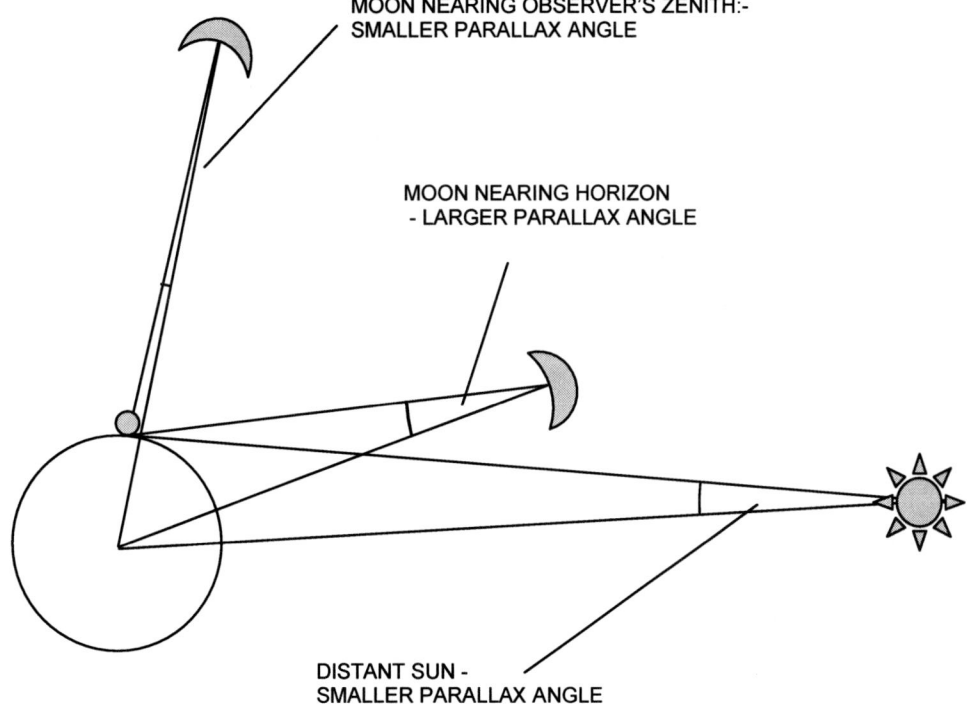

Figure 109. The relative parallax angles of high/low and near/far objects

The parallax changes over the Moon's cycle are of such importance in altitude measurements that detailed figures are listed for the Moon's horizontal parallax which needs to be accounted for when reducing Moon sights. **NP 314** in its tri-daily pages includes an hourly figure for parallax changes and this will be further discussed in **Section 7.13** p.196 on Moon sight reduction.

7.9 The Moon and eclipses – what influences their occurrence?

An eclipse occurs when a heavenly body obscures the view from Earth of another body. An eclipse of the Sun occurs at or near a new Moon, when the Moon, by reason of its orbit, is interpositioned between the Sun and Earth. A shadow is cast by the Moon as it obscures light rays from the Sun that would otherwise have fallen on the Earth as in **Figure 110**.

The Sun may be completely obscured: a total eclipse, or less than completely obscured: a partial eclipse. Eclipse data usually appears in an astro-nautical almanac; NP 314 has useful charts detailing the extent of eclipse visibility and details are widely available elsewhere[7.7].

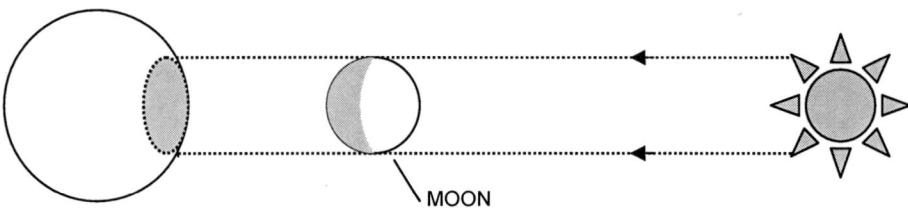

Figure 110. Positional conditions for a total eclipse of the Sun

There is a partial eclipse of the Sun by the Moon on January 4th 2011, visible from Europe, North Africa, the Middle East, western Asia and north-west India. There are further partial eclipses on June 1st (visible in eastern Asia, northern Alaska and Canada and northern Scandinavia, Greenland and Iceland), July 1st (visible only in the Southern Ocean) and November 25th (visible in southern South Africa, Tasmania, New Zealand and Antarctica).

An eclipse of the Moon is caused by the Earth interposing itself between the Moon and the Sun. The shadow of the Earth is cast onto the Moon as it obscures light rays from the Sun that would otherwise have fallen on the Moon (**Figure 111**). Eclipses of the Moon occur around a full Moon, when the Earth is positioned between the Sun and the Moon.

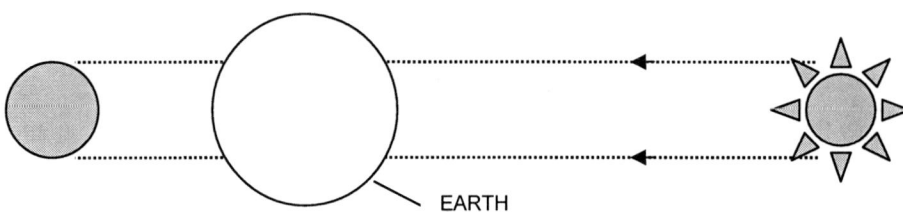

Figure 111. Positional conditions for a total eclipse of the Moon

In 2011 there will be a total eclipse of the Moon on June 15th, seen in Australasia, Japan, Asia, Africa, Europe and South America. There is a further total eclipse of the Moon on December 10th which will be visible in western and central North America, Australasia, Asia, Eastern Africa, Eastern Europe and Iceland.

Given the continuous orbiting of the Moon round Earth and Earth round the Sun, then why are eclipses not more frequent? The answer is that the inclination of the Moon's orbit (5°) to the plane of the path of the Earth round the Sun means that the three bodies are seldom in a straight enough line (a **syzygy**). If the Moon crosses the point at which the orbital planes intersect then an eclipse will occur and if near to that point, an eclipse may occur.

Given the predictable movement of the heavenly bodies, it is not surprising that eclipses of the Sun and Moon occur on a cyclical basis. What seems amazing is that this was observed and described by the ancient Babylonians, whilst so-called "modern man" just stares at "soaps". The cyclical occurrence was subsequently named the Saros cycle by Edmund Halley (of Comet fame). It is based on a time period of 18 years 11 days and 8 hours and involves yet another measure of a lunar month, the draconic month that is perhaps thankfully, not of immediate interest to navigators.

7.10 Moonrise and Moonset – how does this influence sight taking?

If you plan to use the Moon for a sight, the process is slightly more complex than that for the Sun as it is much nearer to Earth and needs more correction to a sextant altitude to achieve an observed altitude. But, having absorbed the previous paragraphs detailing the peculiarities of the Moon's movements, then the complexity should be more transparent. The navigator first needs to decide if the Moon is going to be visible in the sky above him at the time of his proposed sight. Referring to the phase of the Moon and its broad relationship to the time of day, covered in **Section 7.5** p.184 will give some indication.

The astro-nautical almanac daily page will give details the times of moonrise and moonset based on Greenwich. These times can be broadly used for local times as precision is not commonly needed. Should precision be needed, tables are given in **NP 314** for detailed interpolation for varying longitude (**Section 8.11** p.221). Details are also given of the age and phase of the Moon and the percentage of its face that will be illuminated.

NP 314 has a table on its Sun and Moon tri-daily page for moonrise and moonset, from latitude 72° North to 60° South for the three days of the relevant page and the subsequent day (**Table 21**). Inspection of this table shows that above latitude 66° North, it is relatively common for the Moon not to appear at all, or be present above the horizon for the whole of a 24 hour period. Like the Sun, these periods are related to the Moon's declination.

If the Moon's declination is in a northern period then the Moon might be above the horizon for 24 hours, depending on the latitude of observation and the extent of its northern path over the celestial sphere. The reverse is true when it has a more southerly declination, at some northern latitudes, the Moon may be entirely absent for a period. **Table 21** indicates with the dark bars, that the Moon at latitude 68° North stays below the horizon from the 23rd-25th March 2011.

2011 MARCH 23, 24, 25 (WED., THURS., FRI.)

UT	SUN GHA	SUN Dec	MOON GHA	MOON v	MOON Dec	MOON d	MOON HP	Lat.	Twilight Naut.	Twilight Civil	Sunrise	Moonrise 23	Moonrise 24	Moonrise 25	Moonrise 26
d h	° ′	° ′	° ′		° ′			°	h m	h m	h m	h m	h m	h m	h m
23 00	178 17.7	N 0 48.0	315 54.6	5.8	S20 11.1	7.9	59.9	N 72	02 56	04 31	05 40	■■■■	■■■■	■■■■	■■■■
01	193 17.9	49.0	330 19.4	5.8	20 19.0	7.8	59.9	N 70	03 24	04 40	05 41	■■■■	■■■■	■■■■	■■■■
02	208 18.1	50.0	344 44.2	5.8	20 26.8	7.6	59.8	68	03 36	04 45	05 45	01 32			
03	223 18.3	51.0	359 09.0	5.7	20 34.4	7.5	59.8	66	03 50	04 55	05 47	00 46	02 50	04 32	04 50
04	238 18.5	52.0	13 33.7	5.7	20 41.9	7.3	59.8	64	04 01	05 00	05 50	00 16	01 59	03 16	03 58
05	253 18.7	52.9	27 58.4	5.8	20 49.2	7.2	59.7	62	04 11	05 05	05 52	00 02	01 27	02 40	03 26
06	268 18.9	N 0 53.9	42 23.2	5.7	S20 56.4	7.0	59.7	60	04 18	05 09	05 54	■ 04	01 02	02 14	03 02
W 07	283 19.1	54.9	56 47.9	5.7	21 03.4	6.9	59.7	N 58	04 25	05 13	05 56		00 42	01 53	02 43
E 08	298 19.2	55.9	71 12.6	5.7	21 10.3	6.8	59.6	56	04 31	05 16	05 58		00 27	01 36	02 27
D 09	313 19.4	56.9	85 37.3	5.7	21 17.1	6.6	59.6	54	04 36	05 19	05 59		00 13	01 22	02 14
N 10	328 19.6	57.9	100 02.0	5.7	21 23.7	6.4	59.6	52	04 41	05 22	06 00		00 01	01 09	02 02
S 11	343 19.8	58.9	114 26.7	5.7	21 30.1	6.3	59.5	50	04 45	05 25	06 02	23 52	24 58	00 58	01 51
D 12	358 20.0	N 0 59.8	128 51.4	5.6	S21 36.4	6.1	59.5	45	04 53	05 30	06 04	23 35	24 35	00 35	01 29
A 13	13 20.2	1 00.8	143 16.0	5.7	21 42.5	6.0	59.4	N 40	04 59	05 34	06 06	23 21	24 17	00 17	01 11
Y 14	28 20.4	01.8	157 40.7	5.7	21 48.5	5.9	59.4	35	05 04	05 37	06 08	23 09	24 01	00 01	00 56
15	43 20.6	02.8	172 05.4	5.7	21 54.4	5.6	59.4	30	05 08	05 40	06 09	22 58	24 43	00 43	
16	58 20.8	03.8	186 30.1	5.7	22 00.0	5.6	59.3	20	05 14	05 44	06 12	22 40	23 25	24 20	00 20
Y 17	73 20.9	04.8	200 54.8	5.7	22 05.6	5.3	59.3	N 10	05 17	05 47	06 14	22 25	23 05	24 01	00 01
18	88 21.1	N 1 05.8	215 19.5	5.7	S22 10.9	5.3	59.2	0	05 19	05 49	06 15	22 06	23 05	23 43	24 36
19	103 21.3	06.7	229 44.2	5.7	22 16.2	5.0	59.2	S 10	05 18	05 49	06 17	21 48	22 47	23 43	24 36
20	118 21.5	07.7	244 08.9	5.7	22 21.2	5.0	59.2	20	05 15	05 48	06 19	21 31	22 28	23 25	24 20
21	133 21.7	08.7	258 33.6	5.7	22 26.2	4.7	59.1	30	05 11	05 46	06 20	21 12	22 08	23 05	24 02
22	148 21.9	09.7	272 58.3	5.7	22 30.9	4.6	59.1	35	05 08	05 45	06 21	20 51	21 46	22 43	23 41
23	163 22.1	10.7	287 23.0	5.7	22 35.5	4.5	59.0	40	05 04	05 43	06 22	20 39	21 32	22 30	23 29
								45	05 03	05 38	06 07	20 24	21 17	22 15	23 16
								S 50	04 57	05 35	06 07	20 08	20 59	21 57	22 59
24 00	178 22.3	N 1 11.7	301 47.7	5.8	S22 40.0	4.3	59.0	S 50	04 57	05 35	06 07	19 47	20 36	21 34	22 39
01	193 22.4	12.7	316 12.5	5.8	22 44.3	4.2	59.0	52	04 54	05 34	06 08	19 37	20 25	21 24	22 29
02	208 22.6	13.6	330 37.3	5.7	22 48.5	4.0	58.9	54	04 51	05 33	06 08	19 26	20 13	21 12	22 18
03	223 22.8	14.6	345 02.0	5.8	22 52.5	3.8	58.9	56	04 47	05 31	06 08	19 13	19 59	20 58	22 06
04	238 23.0	15.6	359 26.8	5.9	22 56.3	3.7	58.8	58	04 43	05 29	06 08	18 59	19 43	20 41	21 52
05	253 23.2	16.6	13 51.7	5.8	23 00.0	3.5	58.8	S 60	04 38	05 27	06 09	18 42	19 23	20 22	21 35
06	268 23.4	N 1 17.6	28 16.5	5.9	S23 03.5	3.4	58.8	Lat.	Sunset	Twilight Civil	Twilight Naut.	Moonset 23	Moonset 24	Moonset 25	Moonset 26
07	283 23.6	18.6	42 41.4	5.8	23 06.9	3.3	58.7								
T 08	298 23.8	19.5	57 06.2	6.0	23 10.2	3.0	58.7	°	h m	h m	h m	h m	h m	h m	h m
H 09	313 24.0	20.5	71 31.2	5.9	23 13.2	3.0	58.6	N 72				■■■■	■■■■	■■■■	■■■■
U 10	328 24.1	21.5	85 56.1	5.9	23 16.2	2.7	58.6	N 70				■■■■	■■■■	■■■■	■■■■
R 11	343 24.3	22.5	100 21.0	6.0	23 18.9	2.6	58.6	68							
S 12	358 24.5	N 1 23.5	114 46.0	6.0	S23 21.5	2.5	58.5	66				04 23			
D 13	13 24.7	24.5	129 11.0	6.1	23 24.0	2.3	58.5	64				05 10	05 09	05 28	07 07
A 14	28 24.9	25.5	143 36.1	6.1	23 26.3	2.2	58.4	62				05 41	06 01	06 44	07 58
Y 15	43 25.1	26.4	158 01.2	6.1	23 28.5	2.0	58.4	60				06 04	06 33	07 21	08 30
16	58 25.3	27.4	172 26.3	6.1	23 30.5	1.9	58.4	N 58				06 23	06 57	07 47	08 54
17	73 25.5	28.4	186 51.4	6.2	23 32.4	1.7	58.3								
18	88 25.7	N 1 29.4	201 16.6	6.3	S23 34.1	1.5	58.3	N 58	18 22	19 02	19 49	06 39	07 16	08 07	09 13
19	103 25.8	30.4	215 41.9	6.2	23 35.6	1.4	58.2	56	18 21	18 59	19 43	06 52	07 32	08 24	09 28
20	118 26.0	31.4	230 07.1	6.3	23 37.0	1.3	58.2	54	18 20	18 56	19 38	07 04	07 45	08 39	09 42
21	133 26.2	32.3	244 32.4	6.4	23 38.3	1.1	58.2	52	18 19	18 53	19 34	07 14	07 57	08 51	09 54
22	148 26.4	33.3	258 57.8	6.5	23 39.4	1.0	58.1	50	18 19	18 51	19 29	07 23	08 08	09 02	10 04
23			273 23.1	6.5	23 40.4	0.8	58.1	45	18 17	18 46	19 21	07 42	08 30	09 25	10 26
25 00			287 48.6	6.5	S23 41.2	0.6	58.0	N 40	18 16	18 43	19 15	07 58	08 48	09 44	10 43
			302 14.1	6.5	23 41.8	0.5	58.0	35	18 15	18 40	19 09	08 12	09 03	09 59	10 58
			316 39.6	6.6	23 42.4	0.3	58.0	30	18 14	18 38	19 05	08 23	09 16	10 13	11 11
			331 05.2	6.6	23 42.7	0.3	57.9	20	18 12	18 34	19 00	08 44	09 39	10 36	11 33
			345 30.8	6.7	23 43.0	0.0	57.9	N 10	18 11	18 32	18 56	09 01	09 58	10 55	11 51
			359 56.5	6.7	23 43.0	0.0	57.8	0	18 10	18 30	18 54	09 18	10 16	11 14	12 09
			14 22.2	6.8	S23 43.0	0.2	57.8	S 10	18 09	18 30	18 54	09 34	10 34	11 32	12 26
07	283 28.1	44.2	28 48.0	6.8	23 42.8	0.4	57.8	20	18 08	18 30	18 55	09 52	10 54	11 52	12 45
08	298 28.3	45.2	43 13.8	6.9	23 42.4	0.5	57.7	30	18 07	18 31	18 58	10 12	11 16	12 13	13 06
F 09	313 28.5	46.1	57 39.7	6.9	23 41.9	0.7	57.7	35	18 06	18 31	19 01	10 24	11 29	12 28	13 19
R 10	328 28.7	47.1	72 05.6	7.0	23 41.3	0.8	57.6	40	18 06	18 33	19 04	10 37	11 44	12 43	13 33
I 11	343 28.9	48.1	86 31.6	7.1	23 40.5	0.9	57.6	45	18 05	18 34	19 08	10 54	12 02	13 01	13 50
D 12	358 29.1	N 1 47.1	100 57.7	7.1	S23 39.6	1.1	57.6	S 50	18 04	18 37	19 14	11 14	12 25	13 24	14 11
A 13	13 29.2	48.1	115 23.8	7.2	23 38.5	1.2	57.5	52	18 04	18 38	19 17	11 23	12 36	13 35	14 21
Y 14	28 29.4	49.1	129 50.0	7.2	23 37.3	1.3	57.5	54	18 04	18 39	19 20	11 34	12 48	13 47	14 32
15	43 29.6	50.0	144 16.2	7.3	23 36.0	1.5	57.5	56	18 03	18 40	19 24	11 46	13 02	14 01	14 45
16	58 29.8	51.0	158 42.5	7.4	23 34.5	1.6	57.4	58	18 03	18 42	19 28	12 00	13 18	14 17	14 59
17	73 30.0	52.0	173 08.9	7.4	23 32.9	1.8	57.4	S 60	18 03	18 44	19 33	12 17	13 37	14 37	15 17
18	88 30.2	N 1 53.0	187 35.3	7.5	S23 31.1	1.9	57.3		SUN			MOON			
19	103 30.4	54.0	202 01.8	7.6	23 29.2	2.0	57.3	Day	Eqn. of Time		Mer. Pass.	Mer. Pass. Upper	Mer. Pass. Lower	Age	Phase
20	118 30.6	55.0	216 28.4	7.6	23 27.2	2.1	57.3		00ʰ	12ʰ					
21	133 30.8	55.9	230 55.0	7.7	23 25.1	2.3	57.2	d	m s	m s	h m	h m	h m	d	%
22	148 30.9	56.9	245 21.7	7.7	23 22.8	2.5	57.2	23	06 49	06 40	12 07	03 04	15 33	19	81
23	163 31.1	57.9	259 48.4	7.9	S23 20.3	2.5	57.1	24	06 31	06 22	12 06	04 02	16 31	20	71
	SD 16.1	d 1.0	SD 16.2		15.9		15.7	25	06 13	06 04	12 06	05 00	17 28	21	61

Table 21. NP 314-11 March 23rd-25th Sun & Moon GHA, Dec & rising/setting data

7.11 The Moon's meridian passage

NP 314 on its daily page gives the time of an upper and lower meridian passage, two passages per day! It is not of great importance to the navigator but worthy of explanation. It is a consequence of the fixed tilt of the Earth's axis. Navigationally, this situation is not seen unless the observer is in high latitudes where the upper passage becomes apparent.

The Moon by reason of its position, which changes slowly and continually, reflects the Sun's light and is therefore visible over the Earth's sphere on which lays the Greenwich meridian. In the top part of **Figure 112**, the Moon is at meridian passage over Greenwich, but because of the fixed axial obliquity, the higher latitudes are not exposed to the Moon's reflected light. This is the <u>lower</u> meridian passage of the Moon.

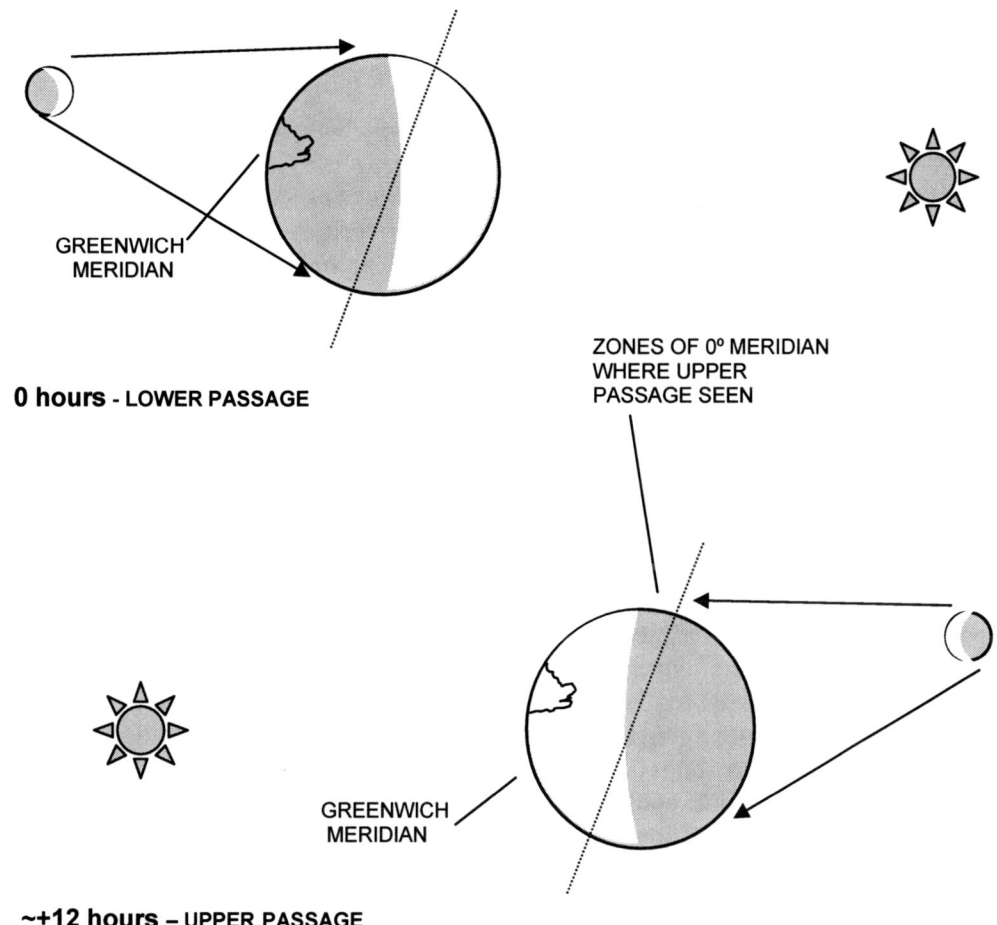

Figure 112. Showing the two meridian passages of the Moon per 24 hours

Only later, as the Earth rotates, is the Moon's relative position such that it becomes visible just at those high latitudes on the Greenwich meridian where a further meridian pass occurs, as shown in the lower part of **Figure 112**.

These periods are a little less than 12 hours apart as the Moon is also moving slowly on its monthly course. **Table 21** p.192 in the lower right-hand corner, shows the upper and lower meridian passages for the Moon on March 23rd–25th.

The Moon's meridian passage can be used to determine longitude using an interpolation table in **NP 314**, in the same table used for interpolating the Local Time for accurate assessment of moonrise and moonset. The difference in time on a given day for the local meridian passage and that at Greenwich is used enter the table from which is extracted a longitude value. The time of local passage is of course, before Greenwich in easterly longitudes and after it in westerly ones. The table is shown in **Section 8.11** p.221.

7.12 The *Moon's GHA and Declination* – complex or merely logical?

The movement of the Moon around the Earth is not at a constant rate due largely to the elliptical shape of its orbit and the affect this has on orbital speed. As a result of this, the rate of change of the Moon's Greenwich Hour Angle (celestial longitude) varies quite significantly within each hour of every day and has to be accounted for when reducing Moon sights.

NP 314 gives the Greenwich Hour Angle for every hour based on the lowest hourly rate of change of 14° 19′ (remembering that the Sun changes by 15° or so every hour). Any increase in the rate above this base figure is adjusted by adding a specific value, called the *v* correction to the GHA value for hours, minutes and seconds. The NP 314 page for the 23rd-25th March 2011 (**Table 21**) clearly shows the hourly GHA and *v* value. The Sun's GHA values do not have a *v* value assigned as the rate of change of GHA for the Sun is negligible. The size of each *v* value is arrived at by subtracting 14° 19′ from the listed GHA for each hour.

The Moon GHA *v* value found adjacent to the whole hour is used in the column adjacent to those for minutes and seconds of the appropriate minute page. The page showing the increments for 48 & 49 minutes is depicted in **Table 22**. The *v* value can vary from 0-16 or so, with its corresponding correction ranging up to almost 18′. The correction is always added. Note that the table also contains corrections for the Sun, Moon, planets and stars (represented by Aries). The whole of the table applies to 48 & 49 minute increments with each second accounted for in the left-hand column.

Declination adjustments for the Moon change more quickly than those for the Sun, but as for the Sun, only the whole hour figure is used together with the *d* value found adjacent to the whole hour figure. No separate value is needed for the minutes and seconds of the hour of observation, as these are wholly accounted for by the *d* correction, found on the

– *navigating with the Moon*

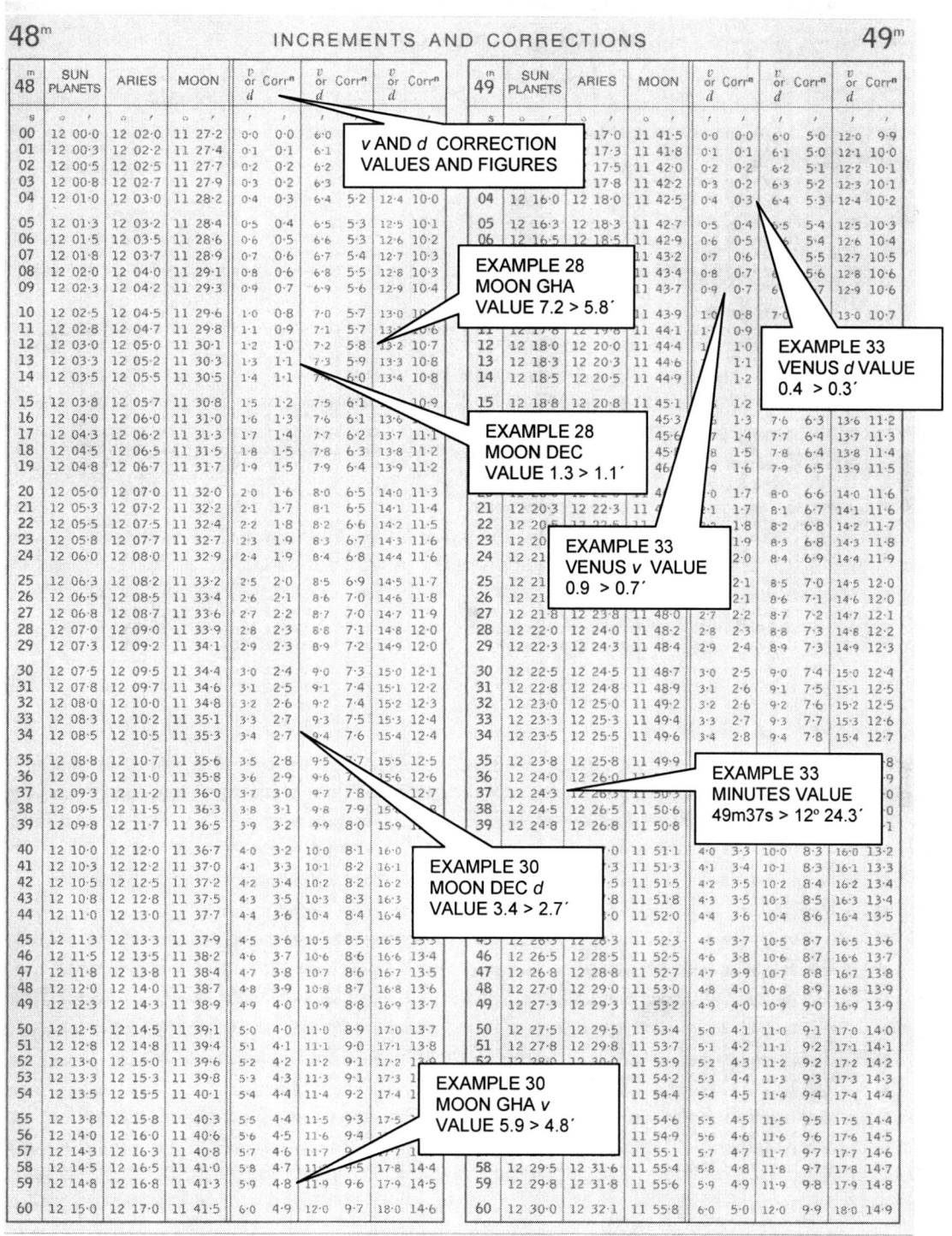

Table 22. NP 314-11 48 & 49 minute increment and *v* and *d* correction data

page corresponding to the minutes number of the observation..

Example 28. Find the GHA and Declination of the Sun <u>and</u> Moon at 1449hrs 54s GMT/UT on Friday25th March 2011 using **Tables 21** and **22**.

		Sun:	Moon:
GHA whole hour - 1400 hrs	=	28° 29.4'	129° 50.0'
increm. for 48m 54s	=	12° 13.5'+	11° 40.1'
v value for rate change 7.2	=	only Moon	5.8' +
GHA	=	**40° 42.9'**	**141° 35.9'**
Dec: whole hour - 1400 hrs.	=	1° 49.1' N	23° 37.3' S
d value Sun – foot of page = 1.0 (inc.)		0.8' + (48m page)	
d value Moon – adjacent = 1.3 (dec.)			1.1' –
Declination	=	**1° 49.9' N**	**23° 36.2' S**

A rapid and coarse check can be made on the calculations by making sure that the GHA and Dec figures arrived at do appropriately match those listed for the next whole hour, depending whether declination is increasing or decreasing. If they do not tally, then check your figures very carefully as you have made an error.

7.13 *Sighting the Moon* – how does it differ from Sun-sights?

Like the Sun, the Moon's sextant altitude reading (H_s) must be corrected to arrive at an observed altitude (H_o); that value which would have been attained had the observation been made from the Earth's centre. Having achieved an apparent altitude (H_a), by correcting for sextant error and dip, the sight is then corrected for semi-diameter, parallax and refraction as usual BUT IN ADDITION and <u>only</u> for the Moon, there is <u>included</u> a further adjustment for augmentation.

NP 314 simplifies these four latter adjustments into two values found in the Moon altitude correction table that is split horizontally into two parts (**Table 23**) entered with apparent altitude (H_a). The upper half provides a correction for the two variables affected by the altitude of the sight; namely refraction and augmentation. The lower half corrects for the two variables affected by the relative proximity of the Moon; namely horizontal parallax and semi-diameter.

The table looks a little fearsome with bands of 5° increments of apparent altitude across the top and with their respective minutes down the left side. Once the correct column is identified, the main correction (for refraction and augmentation) can be read off and noted down. The same column is then followed to the lower part of the table where the rows have horizontal parallax (HP) corrections listed down the left and right hand sides. The HP value for the whole hour of the observation is taken from the tri-daily Sun and Moon table (see **Table 22** p.192) adjacent to the Moon's declination figure and used to identify the correct row, which is traced across to the

ALTITUDE CORRECTION TABLES 0°–35°— MOON

App. Alt.	0°–4° Corrn	5°–9° Corrn	10°–14° Corrn	15°–19° Corrn	20°–24° Corrn	25°–29° Corrn	30°–34° Corrn	App. Alt.
00	0° 34.5	5° 58.2	10° 62.1	15° 62.8	20° 62.2	25° 60.8	30° 58.9	00
10	36.5	58.5	62.2	62.8	62.2	60.8	58.8	10
20	38.3	58.7	62.2	62.8	62.1	60.7	58.8	20
30	40.0	58.9	62.3	62.8	62.1	60.7	58.7	30
40	41.5	59.1	62.3	62.8	62.0	60.6	58.6	40
50	42.9	59.3	62.4	62.7	62.0	60.6	58.5	50
00	1° 44.2	6° 59.5	11° 62.4	16° 62.7	21° 62.0	26° 60.5	31° 58.4	00
10	45.4	59.7	62.4	62.7	61.9	60.4	58.3	10
20	46.5	59.9	62.5	62.7	61.9	60.4	58.2	20
30	47.5	60.0	62.5	62.7	61.9	60.3	58.2	30
40	48.4	60.2	62.5	62.7	61.8	60.3	58.2	40
50	49.3	60.3	62.6	62.7	61.8	60.2	58.1	50
00	2° 50.1	7° 60.5	12° 62.6	17° 62.7	22° 61.7	27° 60.1	32° 58.0	00
10	50.8	60.6	62.6	62.6	61.7	60.1	57.9	10
20	51.5	60.7	62.6	62.6	61.6	60.0	57.8	20
30	52.2	60.9	62.7	62.6	61.6	59.9	57.8	30
40	52.8	61.0	62.7	62.6	61.6	59.9	57.7	40
50	53.4	61.1	62.7	62.6	61.5	59.8	57.6	50
00	3° 53.9	8° 61.2	13° 62.7	18° 62.5	23° 61.5	28° 59.7	33° 57.5	00
10	54.4	61.3	62.7	62.5	61.4	59.7	57.4	10
20	54.9	61.4	62.7	62.5	61.4	59.6	57.4	20
30	55.3	61.5	62.8	62.5	61.3	59.5	57.3	30
40	55.7	61.6	62.8	62.4	61.3	59.5	57.2	40
50	56.1	61.6	62.8	62.4	61.2	59.4	57.1	50
00	4° 56.4	9° 61.7	14° 62.8	19° 62.4	24° 61.2	29° 59.3	34° 57.0	00
10	56.8	61.8	62.8	62.4	61.1	59.3	56.9	10
20	57.1	61.9	62.8	62.3	61.1	59.2	56.9	20
30	57.4	61.9	62.8	62.3	61.0	59.1	56.8	30
40	57.7	62.0	62.8	62.3	61.0	59.1	56.7	40
50	58.0	62.1	62.8	62.2	60.9	59.0	56.6	50

HP	L U	L U	L U	L U	L U	L U	L U	HP
54.0	0.3 0.9	0.3 0.9	0.4 1.0	0.5 1.1	0.6 1.2	0.7 1.3	0.9 1.5	54.0
54.3	0.7 1.1	0.7 1.2	0.8 1.2	0.8 1.3	0.9 1.4	1.1 1.5	1.2 1.7	54.3
54.6	1.1 1.4	1.1 1.4	1.1 1.4	1.2 1.5	1.3 1.6	1.5 1.8	1.5 1.8	54.6
54.9	1.4 1.6	1.5 1.6	1.5 1.6	1.6 1.7	1.6 1.8	1.8 1.9	1.9 2.0	54.9
55.2	1.8 1.8	1.8 1.8	1.9 1.8	1.9 1.9	2.0 2.0	2.1 2.1	2.2 2.2	55.2
55.5	2.2 2.0	2.2 2.0	2.3 2.1	2.3 2.1	2.4 2.2	2.4 2.3	2.5 2.4	55.5
55.8			2.3	2.7 2.3	2.7 2.4	2.8 2.4	2.9 2.5	55.8
56.1			2.5	3.0 2.5	3.1 2.6	3.1 2.6	3.2 2.7	56.1
56.4			2.7	3.4 2.7	3.4 2.8	3.5 2.8	3.5 2.9	56.4
56.7			2.9	3.8 2.9	3.8 3.0	3.8 3.0	3.9 3.0	56.7
57.0	3.1		3.1	4.1 3.1	4.2 3.2	4.2 3.2	4.2 3.2	57.0
57.3	3.3		3.3	4.5 3.3	4.5 3.3	4.5 3.4	4.6 3.4	57.3
57.6	3.5		3.5	4.9 3.5	4.9 3.5	4.9 3.5	5.2 3.6	57.6
57.9	5.3 3.8	5.3	5.2 3.8	5.2 3.7	5.2 3.7	5.2 3.7	5.2 3.7	57.9
58.2	5.6 4.0	5.6 4.0	5.6 4.0	5.6 4.0	5.6 3.9	5.6 3.9	5.6 3.9	58.2
58.5	6.0 4.2	6.0 4.2	6.0	6.0 4.2	6.0 4.1	5.9 4.1	5.9 4.1	58.5
58.8	6.4 4.4	6.4 4.4	6.4 4.4	6.3 4.4	6.3 4.3	6.3 4.3	6.2 4.2	58.8
59.1	6.8 4.6	6.8 4.6	6.7 4.6	6.7 4.6	6.7 4.5	6.6 4.5	6.6 4.4	59.1
59.4	7.2 4.8	7.1 4.8	7.1 4.8	7.1 4.8	7.0 4.7	7.0 4.7	6.9 4.6	59.4
59.7	7.5 5.1	7.5 5.0	7.5		7.3 4.8	7.3 4.8	7.2 4.8	59.7
60.0	7.9 5.3	7.9 5.3	7.9		7.7 5.0	7.6 4.9	7.6 4.9	60.0
60.3	8.3 5.5	8.3 5.5	8.2		8.0 5.2	7.9 5.1	7.9 5.1	60.3
60.6	8.7 5.7	8.7 5.7	8.6		8.4 5.4	8.2 5.3	8.2 5.3	60.6
60.9	9.1 5.9	9.0 5.9	9.0		8.7 5.6	8.6 5.4	8.6 5.4	60.9
61.2	9.5 6.2	9.4 6.1	9.4		9.1 5.8	8.9 5.6	8.9 5.6	61.2
61.5	9.8 6.4	9.8 6.3	9.7		9.4 5.9	9.2 5.8	9.2 5.8	61.5

DIP

Ht. of Eye	Corrn	Ht. of Eye	Ht. of Eye	Corrn	Ht. of Eye
m		m	ft.		ft.
2.4	−2.8	8.0	9.5	−5.5	31.5
			9.9	−5.6	32.7
			10.3	−5.7	33.9
			10.6	−5.8	35.1
			11.0	−5.9	36.3
			11.4	−6.0	37.6
			11.8	−6.1	38.9
3.8	−3.4	12.6	12.2	−6.2	40.1
4.0	−3.5	13.3	12.6	−6.3	41.5
4.3	−3.6	14.1	13.0	−6.4	42.8
4.5	−3.7	14.9	13.4	−6.5	44.2
4.7	−3.8	15.7	13.8	−6.6	45.5
5.0	−3.9	16.5	14.2	−6.7	46.9
5.2	−4.0	17.4	14.7	−6.8	48.4
5.5	−4.1	18.3	15.1	−6.9	49.8
	−4.2				
	−4.9	24.9	18.4	−7.6	60.5
7.9	−5.0	26.0	18.8	−7.7	62.1
8.2	−5.1	27.1	19.3	−7.8	63.8
8.5	−5.2	28.1	19.8	−7.9	65.4
8.8	−5.3	29.2	20.4	−8.0	67.1
9.2	−5.4	30.4	20.9	−8.1	68.8
9.5		31.5	21.4		70.5

MOON CORRECTION TABLE

The correction is in two parts: the first correction is taken from the upper part of the table with argument apparent altitude, and the second from the lower part, with argument HP, in the same column as that from which the first correction was taken. Separate corrections are given in the lower part for lower (L) and upper (U) limbs. All corrections are to be added to apparent altitude, but 30' is to be subtracted from the altitude of the upper limb.

For corrections for pressure and temperature see page A4.

For bubble sextant observations ignore dip, take the mean of upper and lower limb corrections and subtract 15' from the altitude.

App. Alt. = Apparent altitude = Sextant altitude corrected for index error and dip.

Callout boxes:
- EXAMPLE 30 MOON ALTITUDE MAIN CORRECTION VALUE: 16° 39.7′ > 62.7′
- EXAMPLE 29 MOON ALTITUDE MAIN CORRECTION VALUE: 34° 14.1′ > 56.9′
- EXAMPLE 30 HORIZONTAL PARALLAX ADJUSTMENT FIGURE: 6.3′
- EXAMPLE 29 HORIZONTAL PARALLAX ADJUSTMENT FIGURE: 6.2′
- EXAMPLES 29 & 30 MOON HORIZONTAL PARALLAX ADJUSTMENT 58.8′

Table 23. Part of the NP 314 Moon altitude correction table

correct column of lower and upper limb corrections. The values from this table are always added to the apparent altitude figure.

There is one further important step if the upper limb of the Moon has been used for the sight; that a mean Moon diameter of 30′ must be subtracted from the final result to give the observed altitude (H_o). Note that the mean diameter of the Moon is used and not the actual diameter that day as the 30′ mean diameter is used in the construction of the tables.

Two examples follow; the first comparing altitude calculations for the Sun and Moon and the second is a fully worked example of a Moon sight with a 3 dimensional diagram (**Figure 112**).

Example 29. A sextant sight of the lower limb of the Moon was taken on Thursday March 24th at 0630. Later in the day a sun sight was taken. Both sights were identical. Correct both sights to observed altitudes. Eye height was 4.2m and sextant index error 2.1′ off the arc. Use **Table 23** for the Moon and **Table 5** (p.94) for the Sun

		Sun	Moon
Sextant sight (H_s)	=	34° 15.6′	34° 15.6′
Index error 2.1′ off		2.1′ +	2.1′ +
		34° 17.7′	34° 17.7′
Eye height 4.2m		3.6′ −	3.6′ −
Apparent altitude (H_a)	=	34° 14.1′	34° 14.1′
Main correction		14.8′ +	56.9′
Moon HP (adjacent *d*) 58.8′			6.2′ +
Observed altitude (H_o)	=	**34° 28.9′**	**35° 17.2′**

Example 30. A sight was taken of the lower limb of the Moon on Thursday March 24th 2011 timed at 0648 hours and 20 seconds GMT/UT (deck watch was 3s slow) from an estimated position of 45° 32′ North and 15° 45′ West. The height of eye was 6m. The sextant altitude was 16° 46.2′ with an "on the arc" index error of 2.2′.

Using information from the NP 314 GHA, Declination and Altitude in **Tables 21, 22 & 23**, reduce the sight and derive the calculated altitude, intercept, azimuth and position line

In this example, for clarity, the same 1-8 stages as used for the Sun sight reduction in **Section 6** will be used. On this occasion, the calculated altitude will be computed using the modified cosine formula for calculator and the ABC formula for determining the azimuth.

1. **Chosen position**: Latitude 45° 32.0′ North. Longitude 15° 45.0′ West

– *navigating with the Moon*

2. Observed altitude (H_o)

$H_s \pm IE - Dip = H_a +$ main correction $+$ **horiz. parallax corrn.** $= H_o$

Sextant altitude (Hs) =	16° 46.2′
Sextant Index error	2.2′ − (on the arc, so taken off)
	16° 44.0′
Dip correction 6m	4.3′ − (**Table 5** p.94 - always subtracted)
Apparent alt. (H_a)	16° 39.7′
Main corrn.	62.7′ + (**Table 23** - added for lower limb)
	17° 42.4′
Hor. paral. Corrn. 58.8′	6.3′ + (**Table 23** - lower half)
Observed alt.(H_o)	**17° 48.7′**

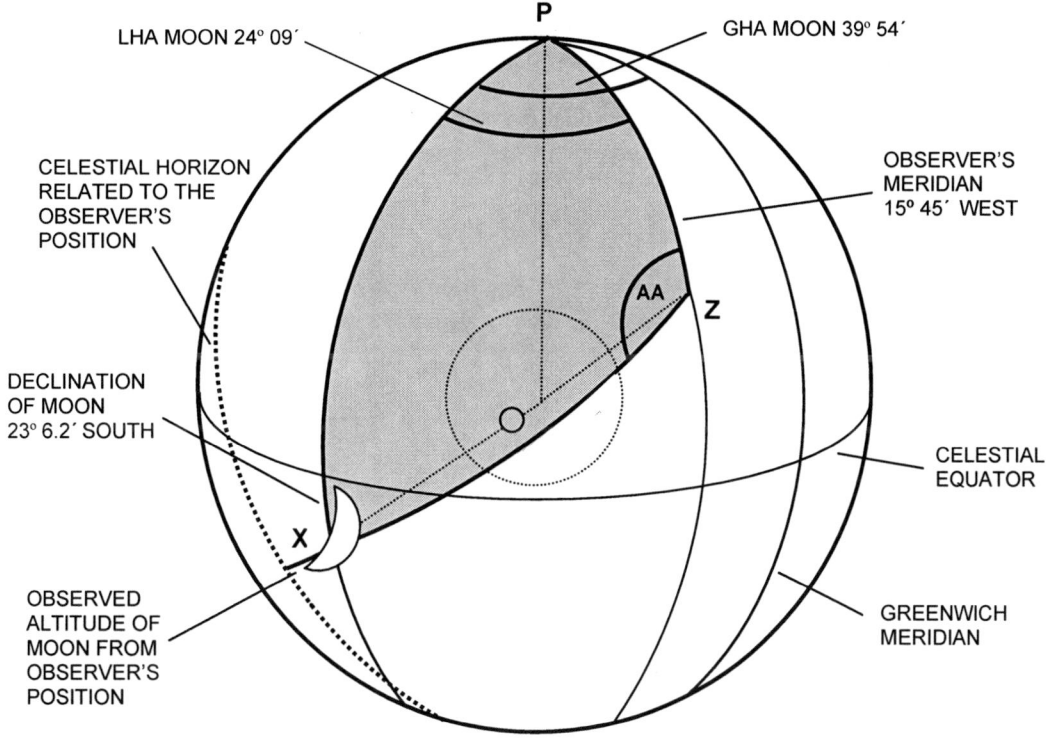

Figure 112. Representation in 3-D of Example 30

Astronavigation from Square One

3. **Greenwich Hour Angle** of Moon: daily page of almanac:

   Time of sight 0648 hrs 20s + 3s watch correction = 0648 23s GMT/UT

GHA Moon at 0600 hrs	28° 16.5′	(**Table 21**)
Increment for 48 mins 23s	11° 32.7′ +	(**Table 22**)
v value for rate change (5.9)	4.8′ +	(**Table 22**)

   GHA of Moon at 0648 23s   39° 54.0′

   Longitude of observation   15° 45.0′ − (subtract for West)

   Local Hour Angle of Moon   24° 09.0′ (= 9′/60+24° =24.15°)

4. **Declination** of Moon: daily page of almanac:

   Declination at 0600 hrs: 23° 03.5′ South
   d value 3.4 corrn. 48 min page: 2.7′ + (increasing)

   Declination of Moon 0648 23s   23° 06.2′ South (= 6.2/60+23=23.1033)

5. **Calculated altitude**: using the modified cosine formula (**Section 6.15** p.165) and calculator to work out the calculated altitude from the given observer's latitude and Moon's Local Hour Angle and Declination:

   sin Calc. Alt. = (*cos* Lat x *cos* Dec x *cos* LHA) ± (*sin* Lat x *sin* Dec)

   In this case the two products are subtracted as Lat and Dec are different names.

   sin Cal. Alt. = (*cos* Lat x *cos* Dec x *cos* LHA) − (*sin* Lat x *sin* Dec)

   Substituting :   Lat = 45° 32′ North = 32/60′ + 45° = 45.5333°
   Dec = 23° 06.2′ South = 23.1033° South
   LHA = 24° 09.0′ = 24.1500°

   sin Calc. Alt. = (*cos* 45.5333 x *cos* 23.1033 x *cos* 24.15°) − (*sin* 45.5333 x *sin* 23.1033)

sin Calc. Alt.	=	0.5879	−	0.2800
sin Calc. Alt.	=		0.3079	
Calc. Alt.	=		arcsin 0.3079	
Calculated altitude	=	17.9327°	= 17° + (0.9327 x 60′) =	**17° 56.0** ′

6. **Intercept**: observed altitude (H_o) − calculated altitude (H_c) = intercept

   H_o = 17° 48.7′
   H_c = 17° 56.0′ −
   − 7.3′

   Intercept =   − 7.3 ′

– navigating with the Moon

The intercept is negative as H_o is less than H_c and therefore lies <u>away</u> from the geographical position of the Moon.

7. **Azimuth:** Using the ABC formula **Section 4.14** p.110:

The azimuth found is in quadrantal form and needs converting to a true bearing.

$$A = tan\ Lat \div tan\ LHA$$

$$B = tan\ Dec \div sin\ LHA$$

$$C = A \pm B$$

$$tan\ Az = (1 \div C \div cos\ Lat)$$

$$A = tan\ 45°\ 32' \div tan\ 24°\ 09\ ' = 2.2722 \quad A = \mathbf{2.2722}$$

A is given the name North/South according to the LHA and latitude. If LHA is either 0-90° or 270-360°, A is given the name opposite to the name of latitude. If the LHA 90-270°, A is given the same name as the name of latitude.

In this case the LHA is 24° 09´ and is therefore assigned as **South**

$$B = tan\ 23°\ 06.2' \div sin\ 24°\ 09' = 1.0427 \quad B = \mathbf{1.0427}$$

(Note that figures for A, B and C are absolute and any negative is ignored).

B is given the same name as declination of the observed body: North or South.

In this case as declination is South then B is assigned **South**

$C = A \pm B$: added if names of A and B are the same i.e. both North <u>or</u> South

: subtracted if names of A and B, i.e. North and South, are different.

In this case, as **A** is **South** and **B** is **South**, both in the same hemisphere, then $C = A + B$

$$C = 2.2722 + 1.0427 = \quad C = \mathbf{3.3149}$$

If C is positive, it is assigned the prefix South in North latitudes and North in South latitudes. If C is negative, it is assigned the prefix North in North latitudes and South in South latitudes.

In this case C is positive and therefore prefixed South as latitude is North: **C** is **South**.

$$tan\ Az = (1 \div C \div cos\ Lat)$$

$tan\ Az = (1 \div 3.3149 \div cos\ 45° 32') = 0.3017 \div 0.7005 = 0.4307$

$arctan\ 0.4307 = 23.3015° = (0.3015 \times 60) + 23 = 23° 18.1'$

Azimuth = 23° 18.1' (quadrantal)

Rules for converting quadrantal azimuth to true azimuth:

Azimuth is prefixed with name of C. In this case, Azimuth is **South**.

Azimuth is suffixed East or West according to the LHA. Azimuth is East if LHA > 180° and West if LHA < 180°.

In this case LHA = 24° and therefore Azimuth is suffixed **West**.

Azimuth = South 23° 18.1' East

Converting to a true bearing: Azimuth = 180° + 23° 18.1' = 203° 18.1'

Azimuth = 203° (rounded down)

8. Plot the azimuth and intercept: the chosen position, azimuth, intercept and position line are plotted in **Figure 113**.

– navigating with the Moon

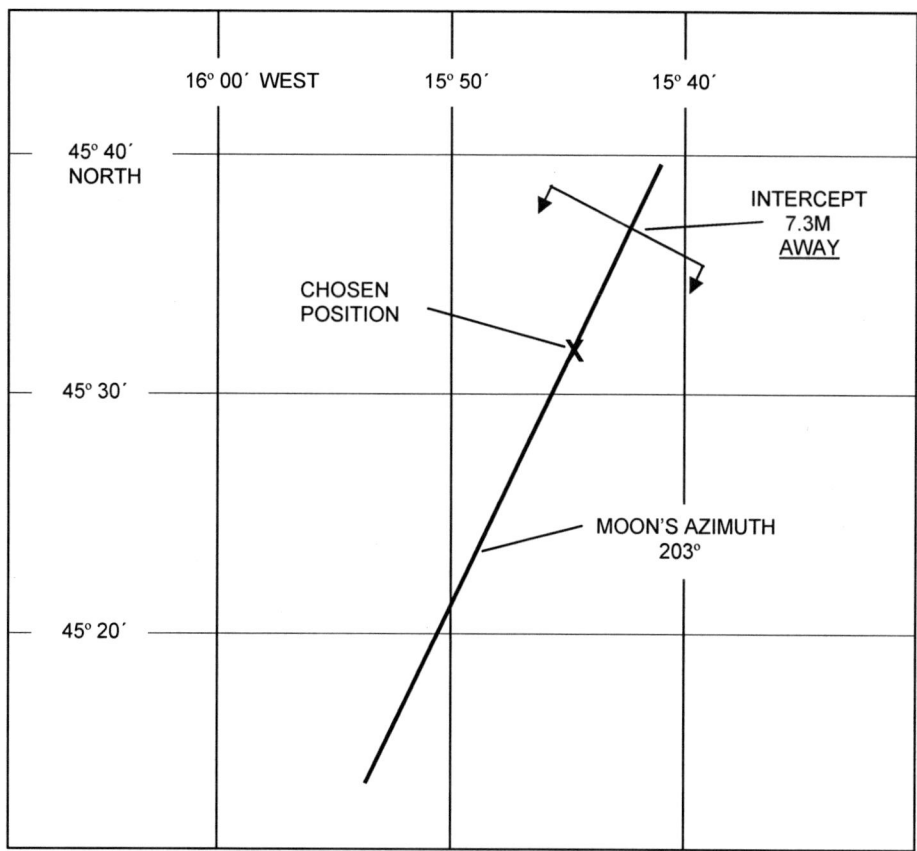

Figure 113. Plotting of the intercept on the azimuth for Example 30

As has been detailed in **Section 6.11** (p.150), it is possible to use a sight reduction proforma and an example is shown in **Figure 114**.

Astronavigation from Square One

Moon Sight

Date									
						−30′			
Estimated position	Latitude	°	. ′	N / S	Longitude		°	. ′	W / E
						+30′			

	Day	Hour	Min		GHA Hours		°	. ′
Zone time					Min Sec		°	. ′
Zone					v value __ : corrn. +			. ′
GMT					**GHA**		°	. ′
GMT / UT					Chosen longitude −W+E		°	. ′

	Day	Hour	Min	Sec		Declination Hrs __	°	. ′
Watch time						d __ +/− Min page __	°	. ′
Watch error						**Declination: N / S**	°	. ′
GMT / UT								

Horizontal Parallax		

Chosen Lat. N / S
Declination N / S
SAME / CONTRARY

Sextant Altitude (Hs)	°	. ′
Index Error on −/ off +		. ′
	°	. ′
Height of Eye m dip −		. ′
Apparent Altitude (Ha)	°	. ′
Main Correction +		. ′
HP correction +		. ′
UL correction −30′	°	. ′
Observed Altitude Ho	°	. ′

Chosen Lat	° N / S

Azimuth Z	°
Zn	°

Hc	°	. ′
d +/−	. ′	+
Hc	°	. ′

Ho	°	. ′
Hc	°	. ′
Intercept		. ′

Chosen position N / S	° W / E	°	. ′
Intercept	towards / away		
Azimuth	° T		

Figure 114. Example of a Moon Sight Proforma

Section 8: Navigating with the Planets

Planets are creatures of habit, adhering to well worn paths around our solar system. Not always easy to see, they may be there in a background of stars or obscured by the glare of the Sun, but at other times a planet can be the brightest thing in the night sky. When sights are possible at twilight they are straight forward to reduce with few of the minor complications faced with the Sun and Moon. The key is to know where to look for them.

By the end of Section 8, the following KEY FACTS should be understood:-

- **The planets orbit the Sun like our Earth, the orbits of Venus and Mercury inside that of Earth, Mars, Jupiter and Saturn outside.**

- **The navigational planets are close enough and large enough to reflect sufficient light to be seen with the naked eye and so are useful for sights.**

- **The planets are relatively close to Earth when compared to the Stars.**

- **The brightness of planets varies with their orbital proximity to Earth**

- **Planets are not always brighter than the brightest stars.**

- **The closer planets move relatively quickly across the celestial sphere when compared to the outer planets, but very much more slowly than the Moon.**

- **The orbit of Venus, within that of Earth causes it to appear after sunset in the evening <u>or</u> before dawn in the morning.**

- **Mercury, the innermost planet is not useful for navigation but can on occasions be confused with other planets.**

- **The annual planet diagram is a useful way of describing the planetary positions in relation to the Sun, each other and the Stars.**

- **Because of their proximity and elliptical orbital shapes, the GHA of planets has a three day *v* value used to derive a correction factor applied to the GHA figures for hours and minutes.**

- **Planet declination, like that for the Sun, results in them appearing relatively low in the sky. They have a three day *d* value used to derive a correction factor applied to the hourly declination figure.**

- **The relative proximity of Venus and Mars requires a small correction for parallax and a tiny correction for the semi-diameter of Venus.**

8.1 The Planets – what are they and where are they?

Within our heliocentric solar system, in addition to our Earth, there are other bodies in orbit around our Sun called planets. Recently, sailing on a Moonless night in the Mediterranean, Jupiter rose majestically in the early hours, bright enough to produce a moonbeam-like path of light to our boat, a great privilege we felt. Planets, like our Moon, do not produce their own light, but passively reflect our Sun's light.

For a planet to be useful for astro-navigation on Earth there are several prerequisites: the planet must be big <u>enough</u> and <u>close enough</u> to both the Sun and to Earth, to reflect a significant amount of light such that it is <u>visible</u> on Earth through a sextant (**Figure 115**). It is now generally agreed by the International Astronomical Union that there are 8 true planets and several lesser bodies currently called planetoids. In order of distance from the Sun the true planets are: Mercury, Venus, Earth, Mars, Jupiter, Saturn, Uranus and Neptune.

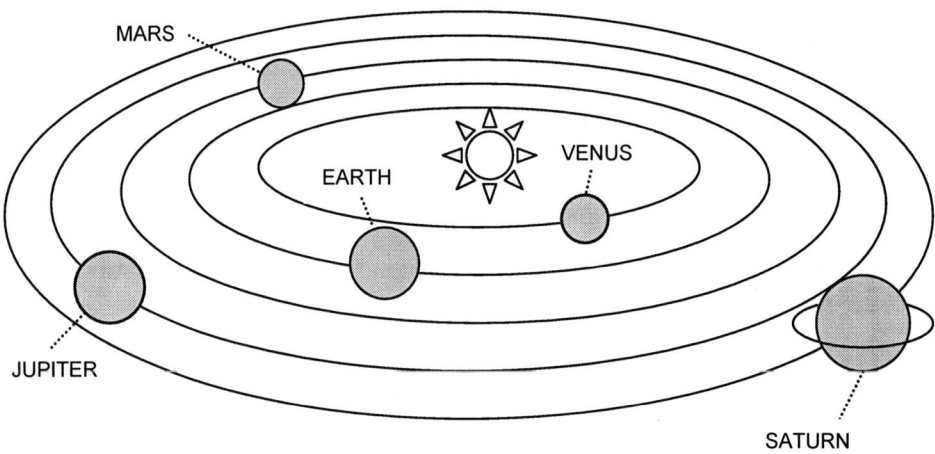

Figure 115. Earth and the navigational useful planets

Pluto together with others such as Ceres and Eris in the outer reaches of our solar system, are currently described as dwarf planets[8.1]. True planets are one of two types; relatively small, high density, largely rock versions called "terrestrial" planets such as Mercury, Venus, Earth and Mars and the low-density "gas giants" such as Jupiter, Saturn, Uranus and Neptune. The naming of the planets stems from the Greek Pantheon of Gods almost certainly rooted in Sumerian and Babylonian mythology and subsequently translated by the Romans to name their own Gods. The Roman names, with the Greek in brackets are: Mercury (Hermes), Venus (Aphrodite), Mars (Ares), Jupiter (Zeus) and Saturn (Kronos)[8.2].

This list did, in ancient times, include both our Sun and Moon but not Earth, given the geocentric view prevailing at the time. Earth was only generally accepted to be a planet in the 17th century and therefore has no similarly Graeco-Roman name. When further planets were discovered in the 18th and 19th centuries, the original tradition was maintained with Uranus (Ouranos) and Neptune (Poseidon).

8.2 *The Navigational Planets* – what features are common to the useful planets?

Only four planets, Venus, Mars, Jupiter and Saturn fulfil the navigationally-useful criteria of visibility, size, and relative proximity to the Sun and Earth to be useful for navigational purposes. Their orbital periods differ from that of Earth and they move across the celestial sphere (**Figure 115**). They are therefore visible in our skies at varying times and periods, the key factor for the navigator being their possible visibility for sights at dawn and dusk twilight.

Mercury, the innermost planet in our solar system and not navigationally useful, is not easily seen as its greatest angular separation from the Sun is only 28° so it is readily obscured by the glare. **NP 314** gives details of the periods between which Mercury can be seen, either low in the West after sunset or low in the East before sunrise during the period of civil twilight. This is of relevance if Mercury happens, by reason of its obit, to be in the vicinity of a regularly used planet when it might produce confusion.

Venus, our near neighbour, has a mass and diameter similar to the Earth, but rotates around its axis once every 243 days and around the Sun in approximately 226 days with a comparative orbital radius 0.7 times that of the Earth. Mars, which appears to have acquired a reputation as a vast planet, has a surprising diameter half that of Earth and a mass of just a tenth. Mars rotates on its own axis in about 1 Earth day and around the sun in 1.8 years, at an orbital radius 1.5 times that of Earth.

Jupiter and Saturn are truly vast, the former with a diameter 11 times and a mass 300 times that of Earth, the latter, 9 times Earth's diameter and with a mass of almost 100 times. Both Jupiter and Saturn rotate completely once every 10 hours but take respectively 12 and 29 years to orbit the Sun, at radii of 5 and 10 times that of the Earth. What they lack in reflective solidity, they make up in huge diameter.

8.3 Planetary eccentricities and inclinations – defining character and orbital features?

The planets have relatively circular solar orbits governed by Kepler's laws, outlined in **Section 5.6**. Their orbits are quite flat, as depicted in **Figure 115**, kept so by the Sun's gravitational force. They have low figures of **eccentricity**, the useful way to describe the elongation factor of their elliptical orbits as described in **Section 5.7** (p.118). The Earth's eccentricity value is 0.017, equivalent to a difference between the length of the long and short axes of its orbital ellipse of only 0.01%. The eccentricity figures for the other navigational planets are: Venus 0.007, Mars 0.093, Jupiter 0.048 and Saturn 0.054, Mars having the most elliptical orbit of the group.

The planets also have low **inclinations**, a figure which describes the angle which their orbital paths make with the plane of the Sun's equator: Venus 3.8°, Earth 7.2°, Mars 5.6°, Jupiter 6.1°, Saturn 5.5°. Importantly, this means that they always appear relatively low in the sky. All the planets have an axial tilt to some degree, which will produce their own particular and peculiar seasons.

8.4 The brightness of planets – are they always brighter than stars?

Planets can only be seen if they are in a position to reflect some light from the Sun that is visible on Earth. In broad terms, the nearer the planet to Earth then the brighter it will appear to be, but if a planet lies on or near a line between Earth and the Sun, it continues to reflect light, but none in the direction of Earth similar to the new Moon in **Section 7.5** p.184.

As well as the usual details on GHA and Declination (see below), the daily pages in NP 314 give other useful data. The brightness of each navigational planet is given on each tri-daily page (**Table 24**), adjacent the planet name. Consequent upon the changing relationship between the distance of a planet from both the Sun and Earth, their brightness, properly called **apparent magnitude** (the brightness that a planet appears to have when viewed from Earth), varies significantly over the year and planets in general are not always brighter than our brightest star (Sirius –1.5).

Magnitude is a logarithmic brightness scale and its origins are explained in detail in **Section 9.4** (p.235). Rather perversely, the brighter the object, the lower its value, the Sun having an apparent magnitude of –26.7. An increase of 1 unit on the scale indicates a halving of apparent brightness, a decrease of 1 unit to a doubling.

Interpreting the figures published in NP 314-2011; Venus is at her brightest at the beginning of the year (– 4.1), then dimming slightly over the spring months to a value of – 3.9 at which she stays for the remainder of the year. Venus is the brightest object in our sky, excepting the Moon and Sun, its brightness being of the order of 11 times that of Sirius (–1.5) our brightest star.

Mars begins the year only twice as bright as *Polaris* at +1.2, dimming marginally to +1.4 in the summer months, then tripling in brightness to a maximum of + 0.2 in January 2012. Jupiter begins the year brightly at – 2.3, second only to Venus, then fading a little in the spring (– 2.1) before almost doubling in brightness by November (– 2.9), then fading slightly again by January (– 2.6). Saturn begins the year half as bright as Sirius at + 0.8, brightening to + 0.4 in April before fading again to + 0.9 in July to September, dulling further to +0.8 at the year's end.

– *navigating with the planets*

2011 NOVEMBER 15, 16, 17 (TUES., WED., THURS.)

UT	ARIES GHA	VENUS −3.8 GHA / Dec	MARS +0.9 GHA / Dec	JUPITER −2.9 GHA / Dec	SATURN +0.8 GHA / Dec	STARS (name / SHA / Dec)
15 00	53 44.3	159 22.3 S23 31.1	259 09.1 N12 32.2	22 22.5 N11 10.7	210 40.0 S 7 11.4	
01	68 46.8	174 21.4 .. 31.5	274 10.3 .. 31.8	37 25.3 .. 10.6	225 42.2 .. 11.5	
02	83 49.3	189 20.5 .. 31.9	289 11.6 .. 31.4	52 28.1 .. 10.5	240 44.4 .. 11.6	
03	98 51.7	204 19.6 .. 32.4	304 12.9 .. 31.0	67 30.8 .. 10.4	255 46.6 .. 11.7	
04	113 54.2	219 18.7 .. 32.8	319 14.2 .. 30.6	82 33.6 .. 10.3	270 48.8 .. 11.8	
05	128 56.6	234 17.8 .. 33.2	334 15.4 .. 30.2	97 36.3 .. 10.2	285 51.0 .. 11.9	
06	143 59.1	249 16.9 S23 33.6	349 16.7 N12 29.8	112 39.1 N11 10.2	300 53.2 S 7 12.0	Alioth 166 22.1 N55 53.5
07	159 01.6	264 16.0 .. 34.1	4 18.0 .. 29.4	127 41.8 .. 10.1	315 55.4 .. 12.1	153 00.2 N49 15.1
T 08	174 04.0	279 15.2 .. 34.5	19 19.2 .. 29.0	142 44.6 .. 10.0	330 57.6 .. 12.2	27 45.1 S46 54.3
U 09	189 06.5	294 14.3 .. 34.9	34 20.5 .. 28.6	157 47.3 .. 09.9	345 59.8 .. 12.3	275 47.2 S 1 11.7
E 10	204 09.0	309 13.4 .. 35.3	49 21.8 .. 28.2	172 50.1 .. 09.8	1 02.0 .. 12.4	217 57.2 S 8 42.6
S 11	219 11.4	324 12.5 .. 35.8	64 23.1 .. 27.8	187 52.8 .. 09.7	16 04.2 .. 12.5	
D 12	234 13.9	339 11.6 S23 36.2	79 24.3 N12 27.4	202 55.6 N11 09.6	31 06.4 S 7 12.6	126 12.3 N26 40.6
A 13	249 16.4	354 10.7 .. 36.6	94 25.6 .. 27.0	217 58.4 .. 09.5	46 08.6 .. 12.7	357 44.5 N29 09.7
Y 14	264 18.8	9 09.8 .. 37.0	109 26.9 .. 26.6	233 01.1 .. 09.4	61 10.8 .. 12.8	Altair 62 09.5 N 8 54.2
15	279 21.3	24 08.9 .. 37.4	124 28.2 .. 26.2	248 03.9 .. 09.3	76 13.0 .. 12.9	Ankaa 353 16.5 S42 14.5
16	294 23.7	39 08.0 .. 37.8	139 29.4 .. 25.8	263 06.6 .. 09.2	91 15.2 .. 12.9	Antares 112 28.0 S26 27.4
17	309 26.2	54 07.1 .. 38.3	154 30.7 .. 25.4	278 09.4 .. 09.1	106 17.4 .. 13.0	
18	324 28.7	69 06.2 S23 38.7	169 32.0 N12 25.0	293 12.1 N11 09.0	121 19.6 S 7 13.1	Arcturus 145 57.1 N19 07.2
19	339 31.1	84 05.3 .. 39.1	184 33.3 .. 24.6	308 14.9 .. 08.9	136 21.8 .. 13.2	Atria 107 31.4 S69 02.9
20	354 33.6	99 04.4 .. 39.5	199 34.5 .. 24.2	323 17.7 .. 08.8	151 24.0 .. 13.3	Avior 234 18.3 S59 32.7
21	9 36.1	114 03.5 .. 39.9	214 35.8 .. 23.8	338 20.4 .. 08.7	166 26.2 .. 13.4	Bellatrix 278 32.9 N 6 21.6
22	24 38.5	129 02.6 .. 40.3	229 37.1 .. 23.4	353 23.2 .. 08.7	181 28.4 .. 13.5	Betelgeuse 271 02.2 N 7 24.5
23	39 41.0	144 01.7 .. 40.7	244 38.4 .. 23.0	8 25.9 .. 08.6	196 30.6 .. 13.6	
16 00	54 43.5	159 00.8 S23 41.1	259 39.7 N12 22.6	23 28.6 N11 08.5	211 32.8 S 7 13.7	Canopus 263 56.2 S52 42.0
01	69 45.9	173 59.9 .. 41.5	274 40.9 .. 22.2	38 31.4 .. 08.4	226 35.0 .. 13.8	Capella 280 35.7 N46 00.5
02	84 48.4	188 59.0 .. 41.9	289 42.2 .. 21.8	53 34.1 .. 08.3	241 37.2 .. 13.9	Deneb 49 32.4 N45 19.8
03	99 50.9	203 58.1 .. 42.3	304 43.5 .. 21.4	68 36.9 .. 08.2	256 39.4 .. 14.0	Denebola 182 35.0 N14 30.2
04	114 53.3	218 57.2 .. 42.7	319 44.8 .. 21.0	83 39.6 .. 08.1	271 41.6 .. 14.1	Diphda 348 56.8 S17 55.2
05	129 55.8	233 56.3 .. 43.1	334 46.1 .. 20.6	98 42.4 .. 08.0	286 43.8 .. 14.2	
06	144 58.2	248 55.4 S23 43.5	349 47.3 N12 20.2	113 45.1 N11 07.9	301 46.0 S 7 14.3	Dubhe 193 53.3 N61 40.9
W 07	160 00.7	263 54.5 .. 43.9	4 48.6 .. 19.8	128 47.9 .. 07.8	316 48.2 .. 14.4	Elnath 278 13.7 N28 37.0
E 08	175 03.2	278 53.6 .. 44.3	19 49.9 .. 19.4	143 50.7 .. 07.7	331 50.4 .. 14.5	Eltanin 90 47.1 N51 29.5
D 09	190 05.6	293 52.7 .. 44.7	34 51.2 .. 19.0	158 53.4 .. 07.6	346 52.6 .. 14.6	Enif 33 48.2 N 9 56.0
N 10	205 08.1	308 51.8 .. 45.1	49 52.5 .. 18.6	173 56.2 .. 07.5	1 54.8 .. 14.7	Fomalhaut 15 25.1 S29 33.5
E 11	220 10.6	323 50.9 .. 45.5	64 53.7 .. 18.3	188 58.9 .. 07.4	16 57.0 .. 14.8	
S 12	235 13.0	338 50.0 S23 45.9	79 55.0 N12 17.9	204 01.6 N11 07.3	31 59.2 S 7 14.9	Gacrux 172 02.6 S57 10.6
D 13	250 15.5	353 49.1 .. 46.3	94 56.3 .. 17.5	219 04.4 .. 07.2	47 01.4 .. 15.0	Gienah 175 53.7 S17 36.4
A 14	265 18.0	8 48.2 .. 46.7	109 57.6 .. 17.1	234 07.1 .. 07.2	62 03.6 .. 15.1	Hadar 148 50.2 S60 25.7
Y 15	280 20.4	23 47.3 .. 47.1	124 58.9 .. 16.7	249 09.9 .. 07.1	77 05.8 .. 15.2	Hamal 328 01.7 N23 31.3
16	295 22.9	38 46.4 .. 47.4	140 00.2 .. 16.3	264 12.6 .. 07.0	92 08.0 .. 15.3	Kaus Aust. 83 45.6 S34 22.7
17	310 25.4	53 45.5 .. 47.8	155 01.4 .. 15.9	279 15.4 .. 06.9	107 10.2 .. 15.4	
18	325 27.8	68 44.6 S23 48.2	170 02.7 N12 15.5	294 18.1 N11 06.8	122 12.4 S 7 15.5	Kochab 137 21.1 N74 06.3
19	340 30.3	83 43.7 .. 48.6	185 04.0 .. 15.1	309 20.9 .. 06.7	137 14.6 .. 15.6	Markab 13 39.3 N15 16.4
20	355 32.7	98 42.8 .. 49.0	200 05.3 .. 14.7	324 23.6 .. 06.6	152 16.9 .. 15.7	Menkar 314 15.9 N 4 08.3
21	10 35.2	113 41.9 .. 49.3	215 06.6 .. 14.3	339 26.4 .. 06.5	167 19.1 .. 15.8	Menkent 148 09.3 S36 25.6
22	25 37.7	128 41.0 .. 49.7	230 07.9 .. 13.9	354 29.1 .. 06.4	182 21.3 .. 15.9	Miaplacidus 221 39.8 S69 45.7
23	40 40.1	143 40.1 .. 50.1	245 09.2 .. 13.5	9 31.9 .. 06.3	197 23.5 .. 16.0	
17 00	55 42.6	158 39.2 S23 50.5	260 10.5 N12 13.1	24 34.6 N11 06.2	212 25.7 S 7 16.1	Mirfak 308 41.5 N49 54.3
01	70 45.1	173 38.3 .. 50.9	275 11.7 .. 12.7	39 37.4 .. 06.1	227 27.9 .. 16.2	Nunki 76 00.0 S26 16.8
02	85 47.5	188 37.4 .. 51.2	290 13.0 .. 12.3	54 40.1 .. 06.1	242 30.1 .. 16.3	Peacock 53 21.2 S56 41.9
03	100 50.0	203 36.5 .. 51.6	305 14.3 .. 11.9	69 42.9 .. 06.0	257 32.3 .. 16.4	Pollux 243 28.9 N27 59.7
04	115 52.5	218 35.5 .. 52.0	320 15.6 .. 11.5	84 45.6 .. 05.9	272 34.5 .. 16.5	Procyon 245 00.7 N 5 11.6
05	130 54.9	233 34.6 .. 52.3	335 16.9 .. 11.1	99 48.3 .. 05.8	287 36.7 .. 16.6	
06	145 57.4	248 33.7 S23 52.7	350 18.2 N12 10.7	114 51.1 N11 05.7	302 38.9 S 7 16.7	Rasalhague 96 07.8 N12 33.3
07	160 59.9	263 32.8 .. 53.1	5 19.5 .. 10.3	129 53.8 .. 05.6	317 41.1 .. 16.8	Regulus 207 44.7 N11 54.4
T 08	176 02.3	278 31.9 .. 53.4	20 20.8 .. 09.9	144 56.6 .. 05.5	332 43.3 .. 16.9	Rigel 281 12.8 S 8 11.2
H 09	191 04.8	293 31.0 .. 53.8	35 22.1 .. 09.6	159 59.3 .. 05.4	347 45.5 .. 16.9	Rigil Kent. 139 54.0 S60 52.9
U 10	206 07.2	308 30.1 .. 54.2	50 23.3 .. 09.2	175 02.1 .. 05.3	2 47.7 .. 17.0	Sabik 102 14.2 S15 44.3
R 11	221 09.7	323 29.2 .. 54.5	65 24.6 .. 08.8	190 04.8 .. 05.2	17 49.9 .. 17.1	
S 12	236 12.2	338 28.3 S23 54.9	80 25.9 N12 08.4	205 07.6 N11 05.1	32 52.1 S 7 17.2	Schedar 349 40.9 N56 36.5
D 13	251 14.6	353 27.4 .. 55.3	95 27.2 .. 08.0	220 10.3 .. 05.0	47 54.3 .. 17.3	Shaula 96 21.9 S37 06.7
A 14	266 17.1	8 26.5 .. 55.6	110 28.5 .. 07.6	235 13.1 .. 04.9	62 56.5 .. 17.4	Sirius 258 33.7 S16 43.9
Y 15	281 19.6	23 25.6 .. 56.0	125 29.8 .. 07.2	250 15.8 .. 04.8	77 58.7 .. 17.5	Spica 158 31.8 S11 13.3
16	296 22.0	38 24.7 .. 56.3	140 31.1 .. 06.8	265 18.6 .. 04.7	93 00.9 .. 17.6	Suhail 222 52.7 S43 28.7
17	311 24.5	53 23.8 .. 56.7	155 32.4 .. 06.4	280 21.3 .. 04.6	108 03.1 .. 17.7	
18	326 27.0	68 22.9 S23 57.0	170 33.7 N12 06.0	295 24.1 N11 04.5	123 05.3 S 7 17.8	Vega 80 40.0 N38 48.0
19	341 29.4	83 22.0 .. 57.4	185 35.0 .. 05.6	310 26.8 .. 04.4	138 07.5 .. 17.9	Zuben'ubi 137 07.0 S16 05.4
20	356 31.9	98 21.1 .. 57.7	200 36.3 .. 05.2	325 29.5 .. 04.3	153 09.7 .. 18.0	SHA / Mer. Pass.
21	11 34.3	113 20.2 .. 58.1	215 37.6 .. 04.8	340 32.3 .. 04.3	168 11.9 .. 18.1	h m
22	26 36.8	128 19.2 .. 58.4	230 38.9 .. 04.4	355 35.0 .. 04.2	183 14.1 .. 18.2	Venus 104 17.4 13 25
23	41 39.3	143 18.3 .. 58.8	245 40.2 .. 04.0	10 37.7 .. 04.1	198 16.3 .. 18.3	Mars 204 56.2 6 41
Mer. Pass. 20 17.8	h m	v −0.9 d 0.4	v 1.3 d 0.4	v 2.7 d 0.1	v 2.2 d 0.1	Jupiter 328 45.2 22 22 / Saturn 156 49.4 9 52

Table 24. NP 314 Aries, planet and star data table for Nov 15th-17th 2011

8.5 The Planets and the celestial sphere – how do they move around it?

Astro-nautical almanacs usually give some general data regarding the likely periods of the year when particular planets will be potentially visible and useful to the navigator as well as advice on when to be wary. To help the navigator locate the navigational planets, the **Sidereal Hour Angle** (SHA) for each is given in small print for every third day in the lower right corner in NP 314 daily page (**Table 24**). Sidereal Hour Angle was introduced in **Section 1.7** (p.11). Note that the word *sidereal* usually relates to stars (Latin *sidus*: a star), but it is used here to denote a planet's position in relation to the stars.

Plainly the planets are always somewhere in our heavens. By relating them to their background of stars as seen from Earth, the likely positions of the planets can be identified. As 2011 elapses, Venus will range over the whole arc of 360° of the celestial sphere, but consequent upon its orbit inside Earth's and apparent proximity to the Sun, it can only be seen by an observer over a relatively narrow arc after sunset in the West or before sunrise in the East. Mars, also being close, ranges over an arc of 231° over the year. Contrastingly, the larger more slowly orbiting planets Jupiter and Saturn, will range over smaller arcs of 41° and 15° respectively, being consistently present in relatively narrow areas of the background of stars on the celestial sphere over the year. The planet position arcs for 2011 are illustrated in **Figure 116**.

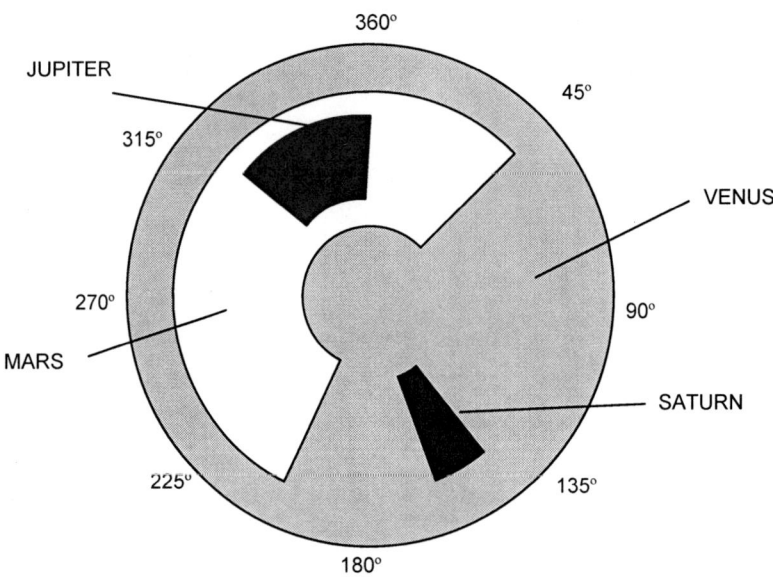

Figure 116. The planets relative Sidereal Hour Angle arcs in 2011

8.6 The special orbit of Venus – a short-cut to finding her?

Venus races Earth round our Sun on the inside track in 226 days. She is visible rising in the south-eastern sky three or so hours before sunrise in January 2011, disappearing as the Sun makes its presence felt at dawn. Her time of rising becomes later as the year progresses until early July, when her rising only just precedes the Sun. From mid-July until the end of September, she is <u>invisible</u> to an Earth-bound observer as her orbit carries her behind the Sun from our viewpoint. In October she escapes the Sun's glare to appear in the evening sky, just visible at sunset on the western horizon and then becomes visible for an increasing time as the late-autumn and winter evenings progress, as she travels West, setting after the Sun.

The declination of Venus lies in the southern hemisphere in the winter months of early 2011, reaching 21° South in early February. Her declination then progressively decreases until she crosses into the northern hemisphere in late April reaching a peak of 23° North in early July. By mid-September she again lies in southern celestial hemisphere, reaching a maximum of almost 25° South in late November.

Why does she appear to do this? It is a function of the facts that the planes of her orbit and Earth's are at angle of some 3.4° and that the Earth's axis is inclined at 23° with respect to the Sun. From our viewpoint on Earth, just as the axial inclination makes the Sun appear to move over the year from the southern celestial hemisphere to the northern and then back again, so the same process appears to us on Earth, to happen to the planets.

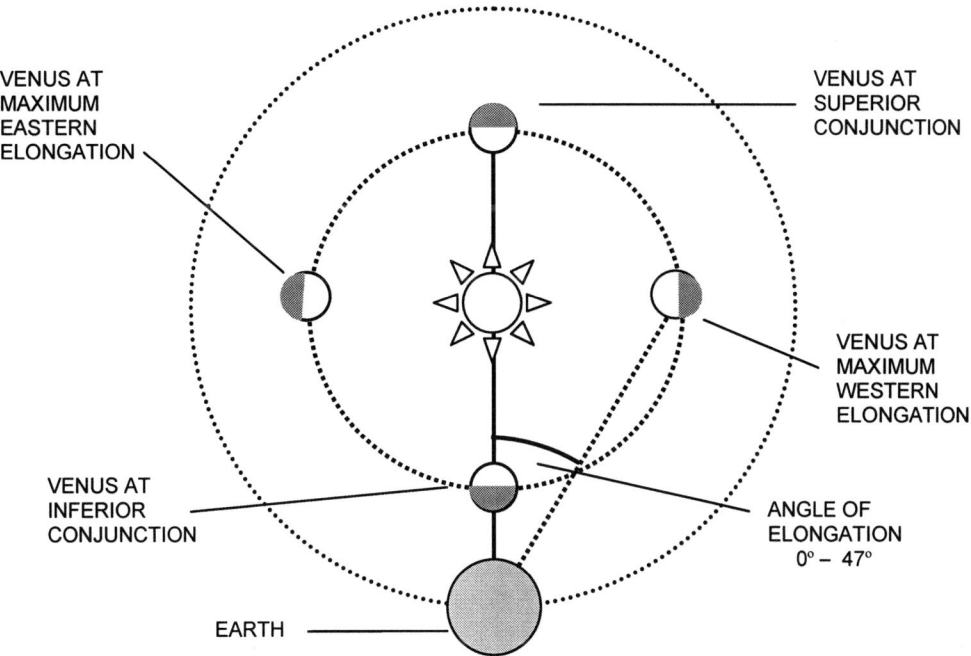

Figure 117. The special features of the orbit of Venus

Identifying Venus is relatively straightforward; she is bright, usually very bright and hovers in the vicinity of the Sun and is the only planet bright enough to be seen with any ease in daylight, when not overwhelmed by solar glare. Because of her unique orbital position she either rises in the East before the Sun at dawn twilight or sets in the West after the Sun in fading evening twilight, or she is not visible at all, hidden by glare. The reasons for this are revealed in the orbital peculiarities between the Earth, Venus and the Sun.

The orbit of Venus (**Figure 117**) lies nearer to the Sun than that of Earth. A line has been drawn from the Sun, to the Earth and then to Venus, creating an angle between the Sun and the planet as viewed from Earth. This angle is called the **elongation of Venus** with respect to the Sun. As Venus moves around her orbit, the elongation varies from 0-47°, the maximum angle varying slightly over time as the orbit is a little elliptical. Importantly, this relative movement also changes the degree to which her illuminated face is visible on Earth which to us on Earth will be seen as varying brightness which is not just a function of distance. Venus exhibits phases like our Moon and its apparent brightness is similarly dependent upon the relative position of the Earth and Sun. Their effect needs to be allowed for in sight reduction.

Conjunction is a particular position of elongation when the angle between the Sun and Venus observed from the Earth is 0°, at which point their Greenwich Hour Angles (celestial longitude) are identical. Both bodies are then lying on the same celestial meridian at which point Venus is said to be in conjunction with the Sun, a superior conjunction, when Venus lies in the part of its orbit beyond the Sun with respect to Earth and an inferior conjunction when she lies between the Sun and Earth.

Leading up to, during and after these particular periods, Venus would not be visible on Earth, but hidden by the Sun's glare. Very occasionally the relative positions of Earth and Venus result in a perfectly aligned inferior conjunction. This results in a visible transit of Venus across the face of the Sun. Such an event occurred in 2004, the next being in 2012. Transits occur in pairs 8 years apart every 120 years or so[8.3]. If you missed the last one make sure you see it in 2012.

When Venus is at its maximum angle, she is said to be at her greatest elongation, either eastwards, or westwards. Leading up to, during and after these periods, when Venus lies clear of the Sun's glare when viewed from Earth, she is seen in all her majesty. Picture the observer watching from the Earth as it turns on its axis in an easterly direction, with Venus at or near its greatest elongation westwards of the Sun: she would be seen to rise in the sky prior to dawn and be visible for a dawn twilight sight before the Sun rises and obscures Venus with its glare. Equally, when Venus lies at or near its greatest eastern elongation, she would become visible in the sky as the Sun sets and be visible for a dusk twilight sight, prior to setting further to the West, following the Sun. Both of these situations are plainly useful to the navigator at either dawn or dusk twilight..

Venus can also reach a conjunction with the other visible planets, appearing very close to its neighbour when viewed from Earth. Venus is in conjunction with Jupiter on May 11th and with Mars on May 22nd 2011.

8.7 A word about Mercury – keeping a wary eye?

NP 314-11 describes the instances when Mercury is visible either low in the West after sunset or in the East before sunrise. It is visible at the following times:

Morning	Evening
Jan 1 – Feb 13	Mar 7 – Apr 2
Apr 18 – Jun 5	Jun 20 – Aug 9
Aug 25 – Sep 19	Oct 12 – Nov 28
Dec 10 – Dec 31	

In the mornings, Mercury is brighter at the end of each period and in the evenings it is brighter at the beginning of the listed period. In all its brightness varies from a dim +3.1 (half as bright as *Polaris* (+2.0) to a considerable −1.4, almost as bright as *Sirius* (−1.5) and approximately 60 times as bright as in its dull phase.

During 2011, because of the particular arrangement of their relative orbits, Mercury can be confused with Jupiter in the early evenings in mid-March and in the early mornings of the first half of May. Jupiter is always the brightest of the pair. In the second half of April, Mars can be confused with Mercury in the mornings, the former being brighter. Again in mid-May in the mornings, Mars lies near Mercury and the latter is the brighter. Venus can be confused with Mercury in the early mornings of late April to late May and from mid-October to mid-November in the evenings.

8.8 Planets with orbits larger than Earth's – a different pattern of movement?

The planets with orbital radii greater than that of the Earth follow a different pattern to that of Venus. Look at the rotation of Mars in **Figure 118**. From Earth, Mars appears to move more independently of the Sun when compared to Venus. Mars has maximum elongation greater than that of Venus in its orbit outside that of Earth. At certain points in their relative orbits, Earth and Mars come into **conjunction** with the Sun. In the vicinity of this point, Mars is furthest from the Earth and appears dim as compared to that when much closer.

Mars and Earth then continue in their orbits and at some point will again be in alignment with the Earth and Sun but in this arrangement the Earth is positioned between Sun and Mars. This is called **opposition**. At this point, being relatively closer to Earth it will appear increasingly bright. When, I wonder, will man reach Mars and be able to witness a transit of Earth across the face of the Sun?

It follows that with the more slowly orbiting planets Jupiter and Saturn, conjunction and opposition with the Earth are relatively rare occurrences but instances in which two planets are in near alignment when seen from Earth are relatively frequent.

Astronavigation from Square One

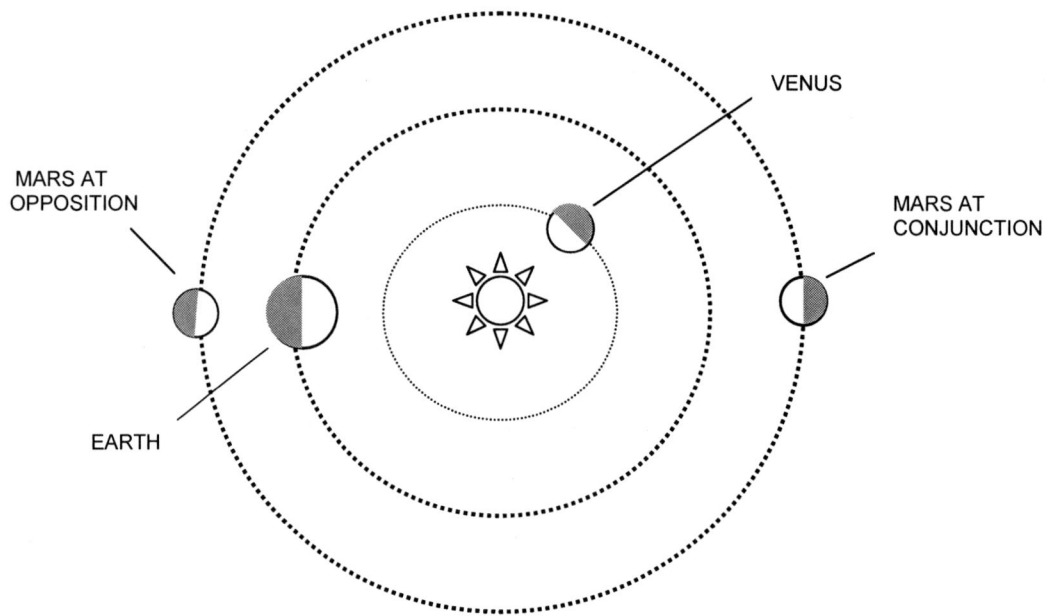

Figure 118. The features of the orbit of Mars – typical of an outer planet

At the beginning of the year, the orbit of Mars results in the planet lying beyond the Sun and invisible to us as observers until mid-April in the morning sky with an SHA of approximately 350° and a declination of 3° North, below the star *Alpheratz*. NP 314 indicates that Its westward elongation increases over the year, passing North of *Aldebaran* in early July, South of *Pollux* in September and North of *Regulus* in November. Mars is in conjunction with Mercury on April 19th and May 20th, with Jupiter on May 1st and with Venus on May 22nd.

Jupiter begins the year at an SHA of 2°, visible in the evenings, southwest of *Alpheratz* with a declination of 2° South. From late March it passes close to the Sun and becomes invisible, obscured by the Sun's glare. It appears again in the morning sky in late April with an SHA of 340° and a declination of 7° North southeast of *Alpheratz*, reaching opposition in late October when it will be visible throughout the night, rising as the Sun sets and setting as the Sun rises. Jupiter lies in conjunction with Mercury on March 16th and May 10th and with Mars on May 1st and with Venus on May 11th.

Saturn the slowest of planetary movers rises just after midnight at the beginning of the year with an SHA 163° and a declination of 4° South, and remains in this vicinity for the year. Saturn is in opposition on April 4th, being visible throughout the night. By late September, being visible only in the evening sky, it then passes behind the Sun and is obscured by glare. It appears again in late October but in the morning sky for the remainder of the year and passing North of *Spica* at the end of October.

– navigating with the planets

Why do planets disappear from evening skies then reappear in the morning? It is a fact that as all move around their orbits, when seen from Earth, at some stage they will pass behind the Sun – being obscured before, during and after this event by the Sun's glare if not its physical bulk. When their movements appear to the observer on Earth to get progressively nearer the Sun, they will be seen just after the Sun sets before setting themselves over the horizon. When they progress to the other side of the Sun, as viewed from Earth, they will rise before the Sun at dawn, becoming invisible as the Sun rises and obscures their reflected light with its glare. The slower a planet's orbit then the longer it will be obscured by the Sun.

8.9 *The NP 314 Planet diagram summary chart* – a helping hand?

NP 314 has a superb but initially daunting summary diagram, reproduced in **Figure 119**. Its aim is to give a general and relatively instant guide to the likely availability of planets and stars throughout each month over the year. It does this by using a single indicator, that of the approximate time of meridian passage; that moment when a body reaches its maximum height in the sky as it crosses the navigator's own meridian.

The local mean time of meridian passage is arrayed on the y-axis and months of the year on the x-axis. Lines representing sidereal hour angles cross the table obliquely from top-left to bottom-right. The Sun's path including a minor variation for the equation of time, runs centrally from left to right with a band either side, 45 time-minutes in width, corresponding to that zone where proximity to the Sun would obscure a body by glare. The Sun sits in the central band of the diagram, its meridian passage plainly centred on 1200 hours on the y-axis. The paths of the 4 navigational planets and Mercury are plotted through the stars, which are represented by the lines of sidereal hour angle, 0-360°.

The table is instructive in the following broad ways:

It indicates where a planet lies in the sky, in terms of its SHA and at what time in the 24 hour period it is likely to be visible. For example, Jupiter, its meridian passage following the Sun by 5 hours in early January, means it will be visible in the southern sky at evening twilight, setting in the West, in the very late evening. Saturn begins the year with its meridian passage running 6 hours ahead of the Sun, rising around midnight. By April its meridian passage has slipped back to midnight, indicating that it will be visible in the sky for most of the dark hours, rising in the East in the evening.

The planet diagram indicates if a planet or a star band (of SHA) is likely to be too close to the Sun to provide a sight at evening or morning twilight. For example, Mars is obscured by the Sun and its glare in the early months of the year, only escaping to be seen in the morning sky in late April, before sunrise. Venus, its meridian passage running three hours before the Sun in January will rise in the East in the early hours until it too is obscured by the glare of sunrise. The time of its meridian passage slowly increases until July when it rises only just before the Sun and then spends three months obscured by glare, reappearing at sunset in October.

215

Astronavigation from Square One

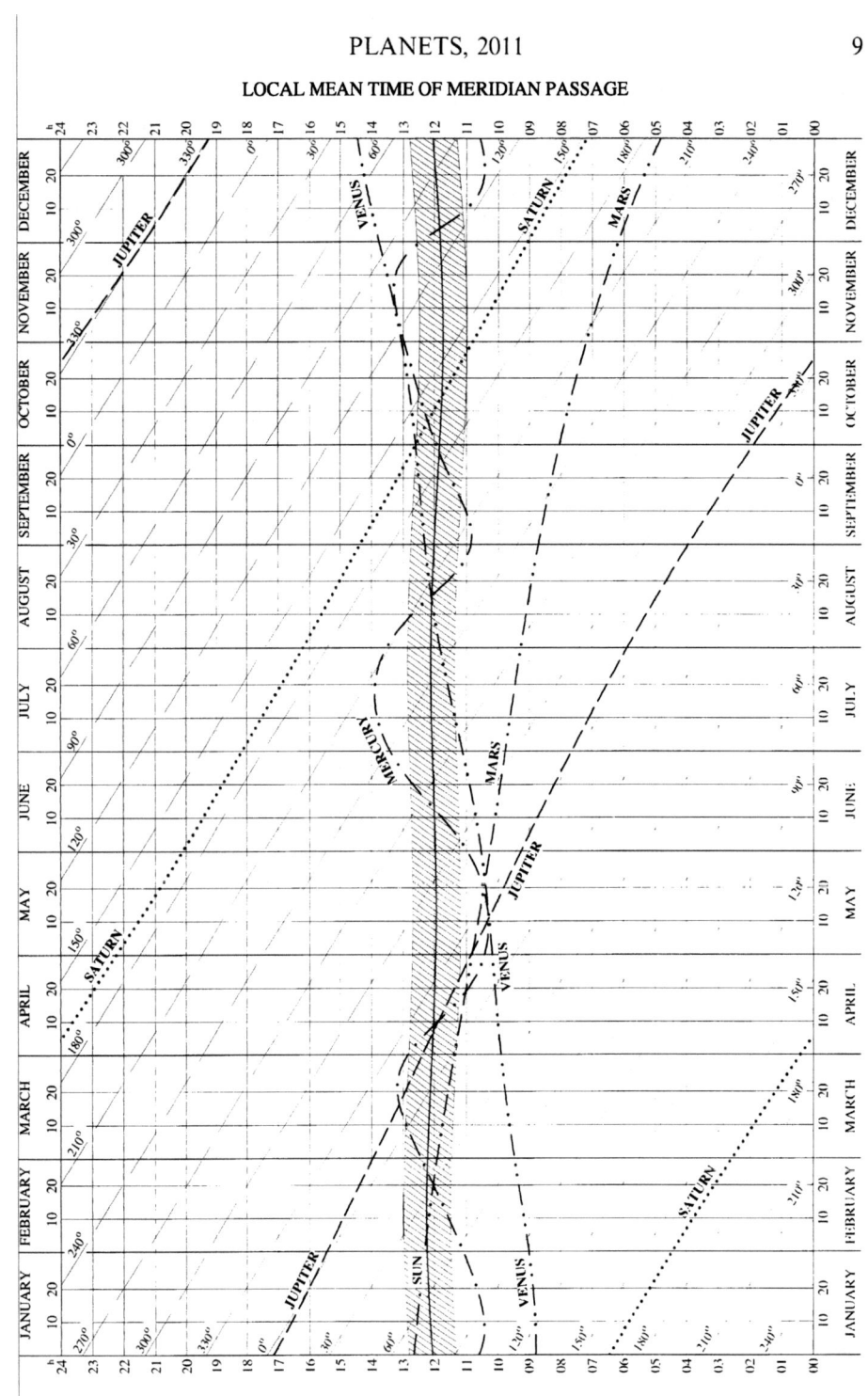

Figure 119. The NP 314-11 Planet diagram

The chart gives a clear indication when planets are in the proximity of one another and might be confused. In the year 2011 it happens that during the month of May, Mercury, Venus, Mars and Jupiter will appear relatively close to one another to us on Earth. The likely times of possible confusion have been indicated in the previous planet specific paragraphs.

The bands of Sidereal Hour Angle crossing the chart indicate which stars will be seen in the sky at what times of the year. In January for instance, Sirius our brightest star, with an SHA of 258° (in the 240-260° band on the chart), has its meridian passage around midnight. This indicates that stars in this band will rise in the early evening and be present through the night sky, setting towards dawn.

The corollary of this is that the chart also tells the navigator what star SHA range will not be visible to an observer. In January for example, stars in SHA bands 0-90°, whose meridian passage values range from 1200-1800hrs or so will rise after the Sun and be present in the sky during daylight hours, overhead but invisible to the navigator.

8.10 *Planetary Greenwich Hour Angle and Declination*

Retuning to the celestial sphere and viewed from our centrally placed Earth; over a period the planets can be seen to move across the static stars and like the Sun and Moon, their positions can be mathematically computed for any moment on any day. By a comparison between their calculated altitude and a corrected sextant altitude at a given time, a position line on an azimuth can be obtained.

NP 314 gives an hourly GHA and Declination for the planets (**Table 24** p.209) and shares with the Sun, 60 individual tables for increments of minutes and seconds of GHA. The planetary GHA, like that for the Sun is based on a standard hourly rate of change of 15°. At the foot of the each planet column is listed a *v* correction value which is valid for the three days.

The *v* value is used to extract a correction on the relevant minute increment page. This corrects for the variation in GHA consequent upon each planet's individual orbital pattern. All corrections are additive except some of those for Venus and Mars where indicated.

It is worth emphasising that attaining a specific GHA for a planet is like that for the Moon. They both require a GHA value for the whole hour AND the increment for minutes and seconds AND a *v* value correction. Remember, the Sun does NOT have a *v* value correction as its GHA variation is negligible.

Planetary declination, like that of the Sun, changes at a much slower rate and does not need an incremental addition for minutes but does have a *d* value correction value to cope with lesser degrees of variation over each 3 day period, again found at the foot of the page. The corresponding correction is found on same minute increment page as that for the GHA *v* correction.

8.11 *Twilight at varying latitude and longitude* – Sun-up and Sun-down?

A planet sight needs to be made during twilight when there is a visible horizon. **NP 314** has on its tri-daily pages, the values for twilight from 72° North to 60° South, with their divisions into both civil and nautical twilight. Somewhat perversely, the former period is the useful one, with both bright stars or planets and the horizon being visible. The latter period of "nautical twilight" is of doubtful relevance to the use of a sextant as rightly suggested in the guidance notes of NP 314: the definition of the extremes of nautical twilight has the Sun's zenith distance as 102° (i.e. 12° below the horizon) when it is likely to be too dim to see the horizon in any event.

Most interestingly, the tables also give exotic information at the higher latitudes, in the region of the Arctic and Antarctic circles. Here, according to season and latitude, there may be twilight all night or the Sun may remain above the horizon throughout the night in the mid-summer months. In the mid-winter months, indications are given as to the latitudes where the Sun will be below the horizon all day. **Table 25** shows the Sun and Moon daily data table for November 15th -17th with sunrise and sunset details.

Most navigators will be in less illustrious positions than on the prime meridian at Greenwich, so to find the GMT of local sunrise, sunset and twilight, the published times will need to be adjusted for the observer's longitude. The observer's approximate latitude is used to enter the Latitude aspect of the table and the figure is extracted with eye interpolation as necessary. Should accuracy be required then NP 314 has a specific table for this which is reproduced below (**Table 26**); the sunrise/moonrise interpolation table.

To adjust for longitude needs a standard arc to time table, subtracting for <u>easterly</u> longitudes, as they are ahead of Greenwich and adding for <u>westerly</u> longitudes, which lag behind. Alternatively, the longitude in time adjustment ($\div$ 15) can be used (**Section 4.3** p.90).

Example 31. Between what approximate <u>local times</u> can a navigator expect to take morning twilight sights on Tuesday November 15th 2010 at latitude 30° 25′ South, 50° 35′ West?

Consulting the tri-daily table (**Table 25**) gives the times in GMT/UT of sunrise, sunset and twilight at a range of latitudes on the Greenwich meridian (72° North to 60° South):

	civil twilight	**sunrise**	
Greenwich (0°)	0429	0456	at lat. 30° South
	0418	0446	at lat. 35° South

As accuracy is not usually critical it would be reasonable to take the 30° South times and adjust them to the longitude of the observer.

Longitude 50° 35′ West converts to time using the Arc to Time table (**Table 1** p.28) or by dividing it by 15 "longitude in time". The latter will be used.

    50° 35′ = 35/60 + 50 = 50.5833° ÷ 15 = 3.3722hr   0300 00
    0.3722 hours = 0.3722 x 60m = 22.332m.   22
    0.332m = 0.332 x 60s = 19.92 = ~ 20s   20 +

    Long. 50° 35′ West lags Greenwich by: 0322 20 = 3hr 22m 20s

2011 NOVEMBER 15, 16, 17 (TUES., WED., THURS.)

UT	SUN		MOON					Lat.	Twilight		Sunrise	Moonrise			
	GHA	Dec	GHA	v	Dec	d	HP		Naut.	Civil		15	16	17	18
d h	° '	° '	° '	'	° '	'	'	°	h m	h m	h m	h m	h m	h m	h m
15 00	183 53.3	S18 20.0	314 28.6	10.2	N20 59.2	4.6	55.7	N 72	07 02	08 37	11 17	▓▓	18 19	20 43	22 51
01	198 53.2	20.6	328 57.8	10.3	20 54.6	4.8	55.7	N 70	06 51	08 14	09 54	16 37	19 03	21 04	23 00
02	213 53.1	21.3	343 27.1	10.3	20 49.8	4.9	55.7	68	06 43	07 56	09 16	17 39	19 32	21 21	23 08
03	228 53.0	.. 21.9	357 56.4	10.2	20 44.9	5.0	55.8	66	06 35	07 41	08 49	18 14	19 53	21 34	23 15
04	243 52.9	2						64	06 29	07 29	08 29	18 39	20 10	21 44	23 20
05	258 52.8	2	*EXAMPLE 32 p.220*					62	06 23	07 19	08 13	18 59	20 24	21 54	23 25
06	273 52.7	S18 2	*SUNRISE AT*					60	06 18	07 10	07 59	19 14	20 36	22 01	23 29
07	288 52.6	2	*GREENWICH MERIDIAN*					N 58	06 14	07 02	07 48	19 28	20 46	22 08	23 33
T 08	303 52.4	2	*60-62° NORTH*					56	06 09	06 55	07 38	19 39	20 55	22 14	23 36
U 09	318 52.3	.. 2	*0759 - 0813*					54	06 06	06 49	07 29	19 50	21 03	22 20	23 39
E 10	333 52.2	2						52	06 02	06 43	07 21	19 59	21 10	22 25	23 41
S 11	348 52.1	2						50	05 59	06 38	07 14	20 07	21 16	22 29	23 44
D 12	3 52.0	S18 27.7	128 20.4	10.4	N19 55.9	5.9	56.0	45	05 51	06 27	06 58	20 24	21 30	22 38	23 49
A 13	18 51.9	28.3	142 49.8	10.4	19 50.0	6.1	56.0	N 40	05 44	06 17	06 46	20 38	21 41	22 46	23 53
Y 14	33 51.8	29.0	157 19.2	10.5	19 43.9	6.2	56.0	35	05 37	06 08	06 35	20 49	21 50	22 53	23 57
15	48 51.7	.. 29.6	171 48.7	10.4	19 37.7	6.3	56.0	30	05 31	06 00	06 25	21	21 59	22 59	24 00
16	63 51.6	30.2	186 18.1	10.5	19 31.4	6.4	56.1	20	05 19	05 46	06 09	21	*EXAMPLE 38 p.259*		
17	78 51.5	30.9	200 47.6	10.5	19 25.0	6.5	56.1	N 10	05 07	05 33	05 55	21	*DAWN TWILIGHT*		
18	93 51.4	S18 31.5	215 17.1	10.6	N19 18.5	6.6	56.1	0	04 54	05 19	05 41	21	*AT GREENWICH*		
19	108 51.3	32.1	229 46.7	10.5	19 11.9	6.7	56.1	S 10	04 39	05 05	05 27	22	*MERIDIAN 35° NORTH*		
20	123 51.2	32.8	244 16.2	10.6	19 05.2	6.8	56.2	20	04 21	04 49	05 13	22	*0608 - 0635*		
21	138 51.1	.. 33.4	258 45.8	10.5	18 58.4	6.9	56.2	30	03 58	04 29	04 56	22			
22	153 51.0	34.0	273 15.3	10.6	18 51.5	7.1	56.2	35	03 44	04 18	04 46	22			
23	168 50.9	34.7	287 44.9	10.7	18 44.4	7.1	56.2	40	03 26	04 04	04 34	22			
								45	03 03	03 46	04 21	23			
16 00	183 50.8	S18 35.3	302 14.6	10.6	N18 37.3	7.2	56.3	S 50	02 33	03 25	04 04	23 26	23 56	24 22	00 22
01	198 50.7	35.9	316 44.2	10.7	18 30.1	7.3	56.3	52	02 17	03 14	03 56	23 34	24 02	00 02	00 26
02	213 50.6	36.6	331 13.9	10.6	18 22.8	7.4	56.3	54	01 57	03 02	03 47	23 42	24 09	00 09	00 31
03	228 50.4	.. 37.2	345 43.5	10.7	18 15.4	7.5	56.3	56	01 32	02 48	03 37	23 52	24 16	00 16	00 36
04	243 50.3	37.8	0 13.2	10.8	18 07.9	7.7	56.4	58	00 56	02 31	03 26	24 03	00 03	00 24	00 42
05	258 50.2	38.5	14 43.0	10.7	18 00.2	7.7	56.4	S 60	////	02 11	03 13	24 15	00 15	00 34	00 48
06	273 50.1	S18 39.1	29 12.7	10.7	N17 52.5	7.8	56.4		Sunset	Twilight		Moonset			
W 07	288 50.0	39.7	43 42.4	10.8	17 44.7	7.9	56.4	Lat.		Civil	Naut.	15	16	17	18
E 08	303 49.9	40.3	58 12.2	10.8	17 36.8	8.0	56.5								
D 09	318 49.8	.. 41.0	72 42.0	10.8	17 28.8	8.1	56.5	°	h m	h m	h m	h m	h m	h m	h m
N 10	333 49.7	41.6	87 11.8	10.9	17 20.7	8.2	56.5	N 72	12 12	14 51	16 26	▓▓	14 46	14 05	13 41
E 11	348 49.6	42.2	101 41.7	10.8	17 12.5	8.3	56.5	N 70	13 35	15 14	16 37	14 41	14 01	13 42	13 29
S 12	3 49.5	S18 42.9	116 11.5	10.9	N17 04.2	8.4	56.6	68	14 13	15 32	16 46	13 38	13 31	13 24	13 19
D 13	18 49.3	43.5	130 41.4	10.9	16 55.8	8.4	56.6	66	14 39	15 47	16 53	13 03	13 08	13 10	13 10
A 14	33 49.2	44.1	145 11.3	11.0	16 47.4	8.6	56.6	64	14 59	15 59	17 00	12 38	12 50	12 58	13 03
Y 15	48 49.1	.. 44.7	159 41.3	10.9	16 38.8	8.7	56.6	62	15 16	16 10	17 05	12 18	12 35	12 48	12 57
16	63 49.0	45.3	174 11.2	11.0	16 30.1	8.7	56.7	60	15 29	16 19	17 10	12 01	12 23	12 39	12 52
17	78 48.9	46.0	188 41.2	10.9	16 21.4	8.9	56.7	N 58	15 41	16 26	17 15	11 47	12 12	12 31	12 47
18	93 48.8	S18 46.6	203 11.1	11.0	N16 12.5	8.9	56.7	56	15 51	16 33	17 19	11 35	12 03	12 25	12 43
19	108 48.7	47.2	217 41.1	11.1	16 03.6	9.0	56.8	54	16 00	16 40	17 23	11 25	11 54	12 18	12 39
20	123 48.5	47.8	232 11.2	11.0	15 54.6	9.1	56.8	52	16 08	16 46	17 27	11 15	11 47	12 13	12 36
21	138 48.4	.. 48.5	246 41.2	11.0	15 45.5	9.2	56.8	50	16 15	16 51	17 30	11 07	11 40	12 08	12 33
22	153 48.3	49.1	261 11.2	11.1	15 36.3	9.3	56.8	45	16 31	17 02	17 38	10 49	11 25	11 57	12 26
23	168 48.2	49.7	275 41.3	11.1	15 27.0	9.4	56.9	N 40	16 43	17 12	17 45	10 34	11 13	11 48	12 20
17 00	183 48.1	S18 50.3	290 11.4	11.1	N15 17.6	9.5	56.9	35	16 54	17 21	17 52	10 22	11 03	11 40	12 15
01	198 48.0	50.9	304 41.5	11.2	15 08.1	9.5	56.9	30	17 04	17 29	17 58	10 11	10 54	11 33	12 11
02	213 47.8	51.5	319 11.7	11.1	14 58.6	9.6	56.9	20	17 20	17 43	18 10	09 52	10 38	11 21	12 03
03	228 47.7	.. 52.2	333 41.8	11.2	14 49.0	9.8	57.0	N 10	17 35	17 57	18 22	09 36	10 24	11 11	11 56
04	243 47.6	52.8	348 12.0	11.1	14 39.2	9.8	57.0	0	17 48	18 10	18 35	09 21	10 11	11 01	11 49
05	258 47.5	53.4	2 42.2	11.2	14 29.4	9.8	57.0	S 10	18 02	18 24	18 51	09 05	09 58	10 51	11 43
06	273 47.4	S18 54.0	17 12.4	11.2	N14 19.6	10.0	57.1	20	18 17	18 41	19 09	08 49	09 44	10 40	11 36
07	288 47.3	54.6	31 42.6	11.2	14 09.6	10.1	57.1	30	18 34	19 00	19 32	08 30	09 28	10 27	11 28
T 08	303 47.1	55.2	46 12.8	11.3	13 59.5	10.1	57.1	35	18 44	19 12	19 46	08 19	09 18	10 20	11 23
H 09	318 47.0	.. 55.9	60 43.1	11.3	13 49.4	10.2	57.1	40	18 56	19 27	20 04	08 06	09 08	10 12	11 18
U 10	333 46.9	56.5	75 13.3	11.3	13 39.2	10.3	57.2	45	19 10	19 44	20 27	07 50	08 55	10 02	11 12
R 11	348 46.8	57.1	89 43.6	11.3	13 28.9	10.3	57.2	S 50	19 26	20 06	20 58	07 32	08 39	09 50	11 04
S 12	3 46.7	S18 57.7	104 13.9	11.3	N13 18.6	10.5	57.3	52	19 34	20 17	21 15	07 23	08 32	09 45	11 01
D 13	18 46.5	58.3	118 44.2	11.3	13 08.1	10.5	57.3	54	19 43	20 29	21 34	07 13	08 24	09 39	10 57
A 14	33 46.4	58.9	133 14.5	11.4	12 57.6	10.6	57.3	56	19 53	20 43	22 00	07 02	08 15	09 32	10 53
Y 15	48 46.3	18 59.5	147 44.9	11.3	12 47.0	10.7	57.3	58	20 05	21 00	22 40	06 49	08 04	09 25	10 48
16	63 46.2	19 00.1	162 15.2	11.4	12 36.3	10.7	57.4	S 60	20 18	21 21	////	06 33	07 52	09 16	10 43
17	78 46.0	00.7	176 45.6	11.4	12 25.6	10.8	57.4								
18	93 45.9	S19 01.3	191 16.0	11.3	N12 14.8	10.9	57.4		SUN			MOON			
19	108 45.8	02.0	205 46.3	11.4	12 03.9	11.0	57.4	Day	Eqn. of Time		Mer.	Mer. Pass.		Age	Phase
20	123 45.7	02.6	220 16.7	11.5	11 52.9	11.0	57.5		00ʰ	12ʰ	Pass.	Upper	Lower		
21	138 45.6	.. 03.2	234 47.2	11.4	11 41.9	11.2	57.5	d	m s	m s	h m	h m	h m	d	%
22	153 45.4	03.8	249 17.6	11.4	11 30.7	11.1	57.5	15	15 33	15 28	11 45	03 09	15 34	20	81
23	168 45.3	04.4	263 48.0	11.4	N11 19.6	11.3	57.6	16	15 23	15 18	11 45	03 59	16 24	21	72
	SD 16.2	d 0.6	SD	15.2	15.4	15.6		17	15 13	15 07	11 45	04 49	17 13	22	62

Table 25. NP 314 -11 data table for November 15ᵗʰ -17ᵗʰ to show sunrise – sunset

Astronavigation from Square One

Many scientific calculators will have a hexadecimal key (base 60) for inputting either angles or time without the need to convert to decimals (base 10) - "D°M′S". This enables hours/minutes/seconds or degrees/minutes/seconds to be entered directly.

	civil twilight	sunrise
Greenwich GMT/UT	0429 00	0456 00
Adjusted for longitude	0332 20	0332 20 + (added for West long.)
At lat. 30° 25′ South:	0751 20	0818 20 GMT/UT
Zone time 50° West + 3	0300 00	0300 00 − (subtracted: GMT > LT)
Local Times:	**0451 00**	**0517 00** (approximate).

Increased accuracy may be required and this can be attained by using the latitude interpolation table (**Table 26**). But it is not the easiest of tables with which to grapple.

Example 32. The Admiral has decreed that the ensign shall be raised at exactly the moment of sunrise on Wednesday November 16th 2010. Find an accurate estimation of the GMT/UT of sunrise on board the flagship at position: latitude 61° 25′ North and longitude 34° 17.5′ West.

NP 314 data on sunrise to the nearest minute is for the middle of the 3 days per daily page at the Greenwich meridian. The figures are based on a Sun semi-diameter of 16′ and 34′ of horizontal refraction. The figures indicate the moment that the upper limb of the Sun should just appear on an otherwise clear horizon.

The figures given in the daily page are:

      Sunrise 60° North: 0759 hrs on the Greenwich meridian
      Sunrise 62° North: 0813   ″

Interpolation is required to find the Greenwich time of sunrise at latitude 61° 25′ and then to convert that figure to the observer's longitude.

Using NP 314: **Table 26** p.221 for latitude interpolation:

The upper table is entered at the left hand side with the value that the required latitude exceeds the nearest tabled latitude below the value required. The second entry value, used to enter the top line of the table, is the difference in times on the required day, between the tabled value below the required latitude and the tabled value at the latitude above the required latitude.

In this case, the tabled value below the required latitude is 60° North. This is 1° 25′ less than the required latitude 61° 25′.

– navigating with the planets

TABLES FOR INTERPOLATING SUNRISE, MOONRISE, ETC.
TABLE I—FOR LATITUDE

Tabular Interval			Difference between the times for consecutive latitudes													
10°	5°	2°	5ᵐ	10ᵐ	15ᵐ	20ᵐ	25ᵐ	30ᵐ	35ᵐ	40ᵐ	45ᵐ	50ᵐ	55ᵐ	60ᵐ	1ʰ05ᵐ	1ʰ10ᵐ	1ʰ15ᵐ	1ʰ20ᵐ

° ′	° ′	° ′	m	m	m	m	m	m	m	m	m	m	m	m	h m	h m	h m	h m
0 30	0 15	0 06	0	0	1	1	1	1	2	2	2	2	2	2	0 02	0 02	0 02	0 02
1 00	0 30	0 12	0	1	1	2	2								05	05	05	05
1 30	0 45	0 18	1	1	2	3	3								07	07	07	07
2 00	1 00	0 24	1	2	3	4	5								10	10	10	10
2 30	1 15	0 30	1	2	4	5	6								12	13	13	13
3 00	1 30	0 36	1	3	4	6	7								0 15	0 15	0 16	0 16
3 30	1 45	0 42	2	3	5	7	8								18	18	19	19
4 00	2 00	0 48	2	4	6	8	9	11	13	14	15	16	18	19	20	21	22	22
4 30	2 15	0 54	2	4	7	9	11	13	15	16	18	19	21	22	23	24	25	26
5 00	2 30	1 00	2	5	7	10	12	14	16	18	20	22	23	25	26	27	28	29
5 30	2 45	1 06	3	5	8	11	13								0 29	0 30	0 31	0 32
6 00	3 00	1 12	3	6	9	12	14								32	33	34	36
6 30	3 15	1 18	3	6	10	13									36	37	38	40
7 00	3 30	1 24	3	7	10	14	17	20	23	26	29	31	34	37	39	41	42	44
7 30	3 45	1 30	4	7	11	15	18	22	25	28	31	34	37	40	43	44	46	48
8 00	4 00	1 36	4	8	12										0 48	0 51	0 53
8 30	4 15	1 42	4	8	13	17									0 53	0 56	0 58
9 00	4 30	1 48	4	9	13	18	22								0 58	1 01	1 04
9 30	4 45	1 54	5	9	14	19	24								1 04	1 08	1 12
10 00	5 00	2 00	5	10	15	20	25	30	35	40	45	50	55	60	1 05	1 10	1 15	1 20

Table I is for interpolating the LMT of sunrise, twilight, moonrise, etc., for latitude. It is to be entered, in the appropriate column on the left, with the difference between true latitude and the nearest tabular latitude which is *less* than the true latitude; and with the argument at the top which is the nearest value of the difference between the times for the tabular latitude and the next higher one; the correction so obtained is applied to the time for the tabular latitude; the sign of the correction can be seen by inspection. It is to be noted that the interpolation is not linear, so that when using this table it is essential to take out the tabular phenomenon for the latitude *less* than the true latitude.

TABLE II—FOR LONGITUDE

Long. East or West	Difference between the times for given date and preceding date (for east longitude) or for given date and following date (for west longitude)																	
	10ᵐ	20ᵐ	30ᵐ	40ᵐ	50ᵐ	60ᵐ	1ʰ+ 10ᵐ	20ᵐ	30ᵐ	1ʰ+ 40ᵐ	50ᵐ	60ᵐ	2ʰ10ᵐ	2ʰ20ᵐ	2ʰ30ᵐ	2ʰ40ᵐ	2ʰ50ᵐ	3ʰ00ᵐ

°	m	m	m	m	m	m	m	m	m	m	m	m	h m	h m	h m	h m	h m	h m
0	0	0	0	0	0	0	0	0	0	0	0	0	0 00	0 00	0 00	0 00	0 00	0 00
10	0	1	1	1	1	2	2	2	2	3	3	3	04	04	04	04	05	05
20	1	1	2	2	3	3	4	4	5	6	6	7	07	08	08	09	09	10
30	1	2	2	3	4	5	6	7	7	8	9	10	11	12	12	13	14	15
40	1	2	3	4	6	7	8	9	10	11	12	13	14	16	17	18	19	20
50	1	3	4	6	7	8	10	11	12	14	15	17	0 18	0 19	0 21	0 22	0 24	0 25
60	2	3	5	7	8	10	12	13	15	17	18	20	22	23	25	27	28	30
70	2	4	6	8	10	12	14	16	17	19	21	23	25	27	29	31	33	35
80	2	4	7	9	11	13	16	18	20	22	24	27	29	31	33	36	38	40
90	2	5	7	10	12	15	17	20	22	25	27	30	32	35	37	40	42	45
100	3	6	8	11	14	17	19	22	25	28	31	33	0 36	0 39	0 42	0 44	0 47	0 50
110	3	6	9	12	15	18	21	24	27	31	34	37	40	43	46	49	0 52	0 55
120	3	7	10	13	17	20	23	27	30	33	37	40	43	47	50	53	0 57	1 00
130	4	7	11	14	18	22	25	29	32	36	40	43	47	51	54	0 58	1 01	1 05
140	4	8	12	16	19	23	27	31	35	39	43	47	51	54	0 58	1 02	1 06	1 10
150	4	8	13	17	21	25	29	33	38	42	46	50	0 54	0 58	1 03	1 07	1 11	1 15
160	4	9	13	18	22	27	31	36	40	44	49	53	0 58	1 02	1 07	1 11	1 16	1 20
170	5	9	14	19	24	28	33	38	42	47	52	57	1 01	1 06	1 11	1 16	1 20	1 25
180	5	10	15	20	25	30	35	40	45	50	55	60	1 05	1 10	1 15	1 20	1 25	1 30

Table II is for interpolating the LMT of moonrise, moonset and the Moon's meridian passage for longitude. It is entered with longitude and with the difference between the times for the given date and for the preceding date (in east longitudes) or following date (in west longitudes). The correction is normally *added* for west longitudes and *subtracted* for east longitudes, but if, as occasionally happens, the times become earlier each day instead of later, the signs of the corrections must be reversed.

xxxii

Table 26. The NP314 table for interpolating latitude of sunrise, moonrise etc.

221

The next tabled latitude above the required latitude value is 62° North.

   Time of sunrise at 60° North: 0759 (on Greenwich meridian)
   Time " at 62° North: 0813 –

   Difference in tabled times: 0014m

The values with which to enter the table are: 1° 25′ (left hand 2° column) and 14m (top-line). The nearest available entry points are 1° 24′ and 15m, which give an figure of 10m.

This is <u>added</u> to the time of sunrise at 60° North as sunrise occurs later the further North a navigator travels in the winter months (visible by table inspection).

Time of sunrise at 60° N Greenwich meridian: 0759
+ interpolation value for 61° 25′:   10 +
Time of sunrise on Greenwich meridian at 61°25′ N: 0809 GMT/UT

Time adjustment for longitude - 34° 17.5′ West

Use Arc to Time Table (**Section 1.16** p.28) or longitude in time: longitude ($\div$ 15).

Using "longitude in time" 34° 17.5′ = 17.5/60 + 34° = 34.2917° ÷ 15 = 2.286hrs
2.2861 hrs = 2 hrs + 0.2861x 60m = 2hr 17.166m = 2hr 17m + 0.166x60s = 2hr 17m 10s

GMT/UT of sunrise at 61° 25′ North, 0° 0.0′ West: 0809 GMT/UT
GMT/UT sunrise adjustment for long. 34° 17.5′ West: 0217 +
GMT/UT sunrise 61° 25′N 34° 17.5′W: **1026** GMT/UT

The GMT/UT of sunrise on the observer's meridian is **1026** GMT/UT.

8.12 *Working up a Planet sight* – how is a sight handled?

All planets, like the Moon and Sun, need their apparent altitude values corrected for refraction, due to the effect on light, of the Earth's atmosphere (**Section 2.5** p.43). Planets, except for Venus, are point light sources and have no readily visible diameter when viewed through a sextant and as such, no semi-diameter correction is needed. Venus has a tiny but just significant diameter and also has phases like our Moon, so needs a small correction. Both Venus and Mars, which are relatively near, so also need a very small correction for parallax.

The planet altitude correction table is relatively straight-forward. The NP 314 Altitude table for the planets is shared with that of the Sun and stars and is entered with apparent

– *navigating with the planets*

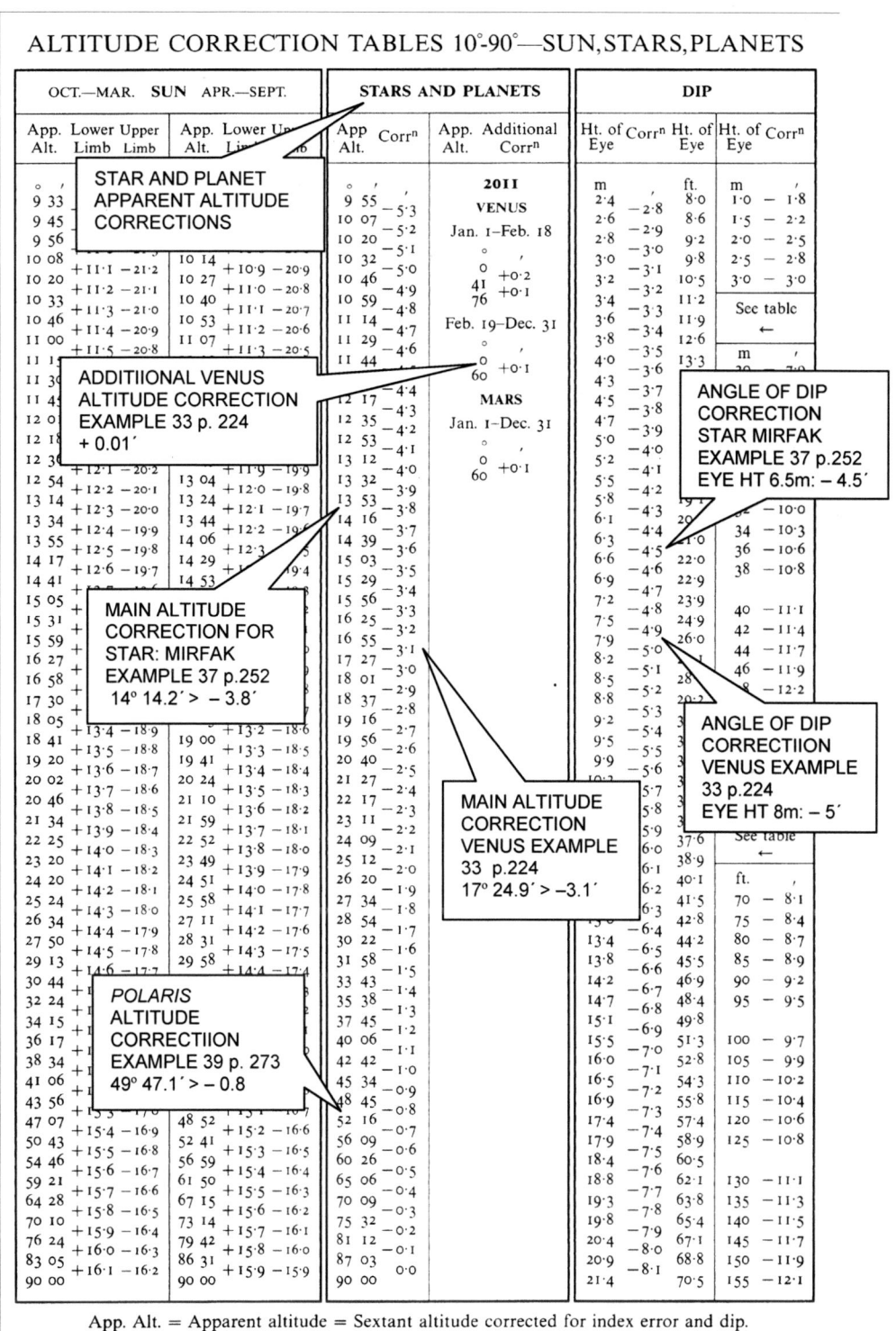

Table 27 NP314-11 Altitude and Dip correction tables for Sun, planets and stars

altitude (H_a); after the correction for dip is applied to the corrected sextant altitude. The altitude table is reproduced in **Table 27** and has a single central column for the refraction corrections for all four planets. Alongside are the corrections for semi-diameter, phase and parallax corrections for Venus, combined in a single figure, together with the parallax correction for Mars. Notice that these values vary through the year, depending upon their orbital positions with respect to Earth and the Sun. Jupiter and Saturn are considerably further away and like the stars, require no parallax correction, simply an adjustment for refraction.

An example follows which will place the observer and the declination of the observed body in the southern hemisphere and uses the Sight Reduction Table (NP 401) method of finding calculated altitude, by first finding a calculated zenith distance, then subtracting this from 90°.

Example 33. An evening twilight sight was taken of Venus on Tuesday November 15th 2011 at 1949 hrs and 45s local time at the chosen latitude of 50° 25′ South and longitude 50° 20′ West. The deck watch proved to be 8 seconds fast. A sextant altitude of 17° 32.4′ was recorded, with an index error "on the arc" of 2.4′, taken at an eye height of 8m. Find the observed altitude, calculated altitude, intercept and azimuth and then plot the position line.

In this example, for clarity, the same 1-8 stages will be used for the Sun and Moon sight reductions in Sections 6 and 7. On this occasion, the Sight Reduction Tables for Marine Navigation (NP 401) will be used (**Section 4.7** p.96) to find both calculated altitude and azimuth. This method uses tabulated data and as such interpolation may be necessary and importantly, accuracy may not be as high as with other methods.

A summary 3-D diagram is drawn below (**Figure 120**) to aid in visualising the problem:

1. **Chosen position**: Latitude 50° 00′ South.

Estimated position: Latitude 50° 25′ South Longitude 50° 20′ West. The chosen longitude is arrived at after the GHA is worked out, such that GHA minus chosen longitude gives a whole number for the LHA.

2. **Observed altitude** (H_o)

$$H_s \pm IE - Dip = H_a + \text{main correction} = H_o$$

Sextant altitude (Hs) =	17° 32.4′
Sextant Index error 2.4 on arc	2.4′ – (on the arc, so taken off)
	17° 30.0′
Dip correction 8m	5.0′ – (**Table 27**) subtracted
Apparent alt. (H_a) =	17° 25.0′
Main corrn.	3.1′ – (**Table 27** - subtracted)
	17° 21.9′
Addit. Venus corrn:	0.1′ + (Feb 19th-Dec 31st)
Observed alt.(H_o)	**17° 22.0′**

– navigating with the planets

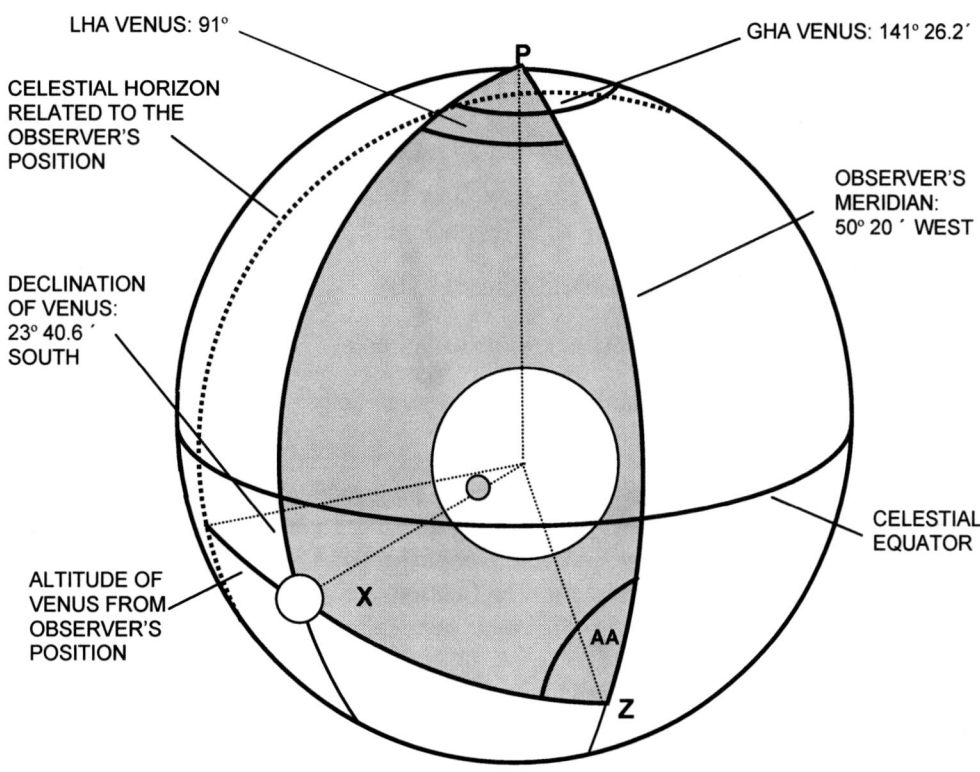

Figure 120. Example 33 drawn on the celestial sphere

3. **Greenwich Hour Angle** of Venus: daily page of almanac: Tuesday Nov. 15th 2011

Local time of sight	19 49 hrs 37s
Time zone of long. 50° 45′ West = + 3:	3 00
Therefore the GMT/UT of observation =	22 49 hrs 37s

Finding the GHA at the time of the sight:

GHA Venus 2249 37s GMT/UT:22 hrs	129° 02.6′ (**Table 24** p.209)
49m 37s	12° 24.3′ (**Table 22** p.195)
	141° 26.9′
v corrn. (− 0.9)	0.7′ −
GHA Venus	141° 26.2′
Chosen longitude of sight	50° 26.2′ − (West)
Local Hour Angle Venus	91° 00.0′

225

4. **Declination** of Venus at time of site: daily page of almanac:

Declination at 22 hrs	23° 40.3′ South (**Table 24** p.209)
d value rate change (0.4)	0.3′+ (inc) (**Table 22** p.195)
Declination	**23° 40.6′ South**

5. **Calculated altitude**: Using the Sight reduction Tables for Marine Navigation there are three specific entry criteria (sometimes called arguments):

    1). a whole number value of LHA

    2). a whole number value of latitude

    3). the declination of the observed body

The chosen position is selected to adjust the estimated position accordingly. In this case, the chosen latitude will be 50° South. The chosen longitude should be within 30′ of the estimated position, aiming to keep the resulting intercept small. In this case the chosen longitude is 50° 26.2′ which, when subtracted from the GHA, results in a whole number for the LHA. Note that both Latitude and Declination are South i.e. they have the SAME NAMES. If Lat and Dec are in different hemispheres then they are said to have CONTRARY NAMES.

The table is then entered at the page for **LHA 91°, SAME NAMES**, the column chosen is **Lat 50°** and the line chosen is that for the whole number of degrees of declination: **Dec 23°**. See **Table 28**.

The following figures are extracted: H_c 16 47.9 d +44.4 z 74.0

The d value is the difference value, used to enter a second table to correct the calculated altitude figure for the minutes (40.6) of declination (**Table 29**).

Hc	16° 47.9′
diff +40	27.1′
4.4	3.0′ +
Calculated altitude =	**17° 18.0′**

6. **Intercept**:   observed altitude (H_o) − calculated altitude (H_c) = intercept

H_o =	17° 22.0′
H_c =	17° 18.0′ −
Intercept =	**+ 4.0′**

The intercept is positive as H_o is greater than H_c and therefore lies <u>towards</u> the geographical position of Venus. The figures are illustrated in **Figure 121** p.228.

– *navigating with the planets*

LATITUDE SAME NAME AS DECLINATION

N. Lat. { L.H.A. greater than 180°......Zn=Z
{ L.H.A. less than 180°.............Zn=360°-

EXAMPLE 33 p.224
CALCULATED ALTITUDE / AZIMUTH:
Hc 16° 47.9´
d 44.4
Z 74.0°

DECLINATION WHOLE NUMBER 23° SOUTH

CHOSEN LATITUDE 50° SOUTH

CHOSEN LOCAL HOUR ANGLE 91°

L.H.A. 91°,

Table 28. NP 401 Sight Reduction Table: LHA 91° - Lat SAME NAME as Dec - part of

7. **Azimuth**: The sight reduction table gives the value of Z to be 74°. This is adjusted to give Z_n, the Azimuth according rules dependent on the LHA of the observed body and the Latitude from which it was observed. The rules are:

In North latitudes: LHA > 180° ... $Z_n = Z$
                     LHA < 180° ....$Z_n = 360° - Z$

In South latitudes: LHA > 180° ... $Z_n = 180° - Z$
                     LHA < 180° ... $Z_n = 180° + Z$

In this case LHA < 180° and the latitude was South:

$$LHA = 180° + 74° = 254°$$

8. **Plot the azimuth and intercept**

Azimuth, intercept and position line plotted in **Figure 121**.

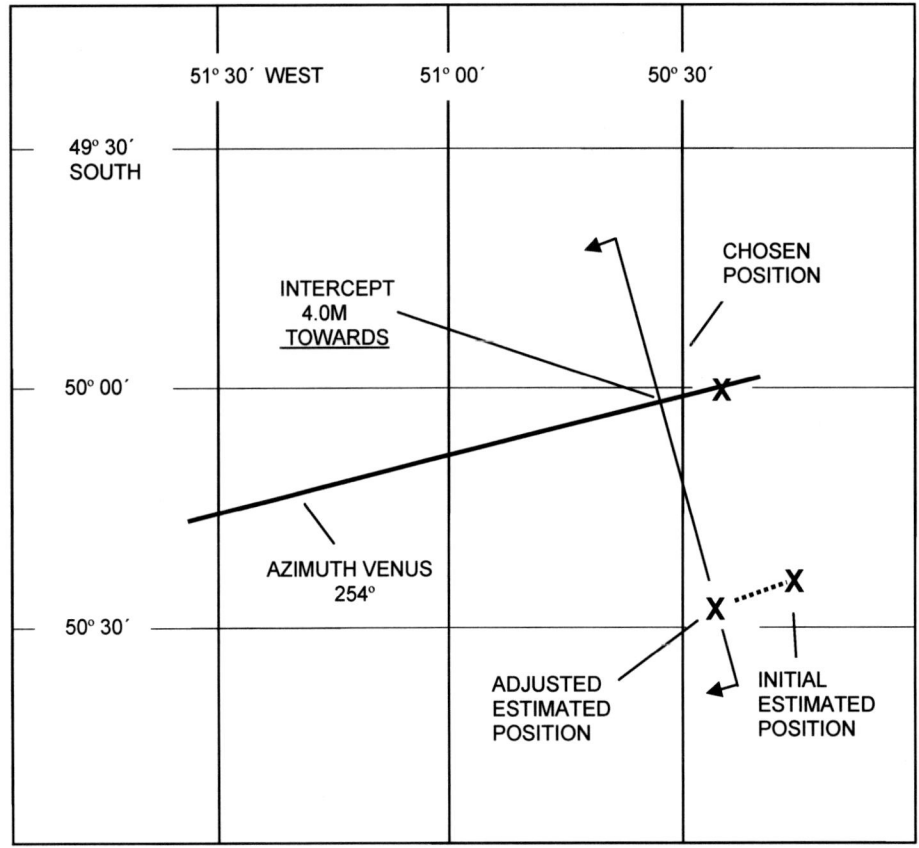

Figure 121. Plotting of the intercept on the azimuth of Venus for Example 33

NP 401 INTERPOLATION TABLE

Dec. Inc.	Altitude Difference (d)															Double Second Diff. and Corr.
	Tens					Decimals	Units									
	10'	20'	30'	40'	50'		0'	1'	2'	3'	4'	5'	6'	7'	8' 9'	
39.0	6.5	13.0	19.5	26.0	32.5	.0	0.0	0.7	1.3	2.0	2.6	3.3	3.9	4.6	5.3 5.9	18.5 1.1
39.1	6.5	13.0	19.5	26.0	32.6	.1	0.1	0.7	1.4	2.0	2.7	3.4	4.0	4.7	5.3 6.0	20.2 1.2
39.2	6.5	13.0	19.6	26.1	32.6	.2	0.1	0.8	1.4	2.1	2.8	3.4	4.1	4.7	5.4 6.1	22.0 1.3
39.3	6.5	13.1	19.6	26.2	32.7	.3	0.2	0.9	1.5	2.2	2.8	3.5	4.1	4.8	5.5 6.1	23.7 1.4
39.4	6.6	13.1	19.7	26.3	32.8	.4	0.3	0.9	1.6	2.2	2.9	3.6	4.2	4.9	5.5 6.2	25.5 1.5
																27.3 1.6
39.5	6.6	13.2	19.8	26.3	32.9	.5	0.3	1.0	1.6	2.3	3.0	3.6	4.3			
39.6	6.6	13.2	19.8	26.4	33.0	.6	0.4	1.1	1.7	2.4	3.0	3.7	4.3	FIGURE FOR UNITS (4) AND DECIMAL (0.4) OF VALUE 44.4 > 3.0		
39.7	6.6	13.3	19.9	26.5	33.1	.7	0.5	1.1	1.8	2.4	3.1	3.8	4.4			
39.8	6.7	13.3	19.9	26.6	33.2	.8	0.5	1.2	1.8	2.5	3.2	3.8	4.5			
39.9	6.7	13.3	20.0	26.6	33.3	.9	0.6	1.3	1.9	2.6	3.2	3.9	4.5			
40.0	6.6	13.3	20.0	26.6	33.3	.0	0.0	0.7	1.3	2.0	2.7	3.4	4.0			
40.1	6.7	13.3	20.0	26.7	33.4	.1	0.1	0.7	1.4	2.1	2.8	3.4	4.1		6.1	
40.2	6.7	13.4	20.1	26.8	33.5	.2	0.1	0.8	1.5	2.2	2.8	3.5			5.5 6.2	0.9 0.1
40.3	6.7	13.4	20.1	26.9	33.6	.3	0.2	0.9	1.6	2.2	2.9	3.6	4.3	4.9	5.6 6.3	2.8 0.2
40.4	6.7	13.5	20.2	26.9	33.7	.4	0.3	0.9	1.6	2.3	3.0	3.6	4.3	5.0	5.7 6.3	4.6 0.3
																6.5 0.4
40.5	6.8	13.5	20.3	27.0	33.8	.5	0.3	1.0	1.7	2.4	3.0	3.7	4.4	5.1	5.7 6.4	8.3 0.4
40.6	6.8	13.5	20.3	27.1	33.8	.6	0.4	1.1	1.8	2.4	3.1	3.8	4.5	5.1	5.8 6.5	10.2 0.5
40.7	6.8	13.6	20.4	27.2		.7	0.5	1.1	1.8	2.5	3.2	3.8	4.5	5.2	5.9 6.5	12.0 0.6
40.8	6.8	13.6	20.4	27.2	34.0		0.5	1.2	1.9	2.6	3.2	3.9	4.6	5.3	5.9 6.6	13.9 0.7
40.9	6.9		20.5	27.3	34.1		1.3	2.0	2.6	3.3	4.0	4.7	5.3	6.0	6.7	15.7 0.8
																17.6 0.9
41.0	6.	MINUTES OF DECLINATION 40.6'			34.1		FIGURES FOR THE TENS VALUE (40) OF THE DIFFERENCE OF MINUTES OF DECLINATION > 27.1'				4.1	4.8	5.5	6.2		19.4 1.0
41.1	6.				34.2						4.2	4.9	5.6	6.3		21.3 1.1
41.2	6.				34.3						4.3	5.0	5.7	6.4		23.1 1.2
41.3	6.				34.4						4.4	5.0	5.7	6.4		25.0 1.3
41.4	6.9	13.8	20.7	27.6	34.5						4.4	5.1	5.8	6.5		26.8 1.4
																28.7 1.5
41.5	6.9	13.8	20.8	27.7	34.6						4.5	5.2	5.9	6.6		30.5 1.6
41.6	6.9	13.9	20.8	27.7	34.7	.6	0.4	1.1	1.8	2.5	3.2	3.9	4.6	5.3	5.9 6.6	32.3 1.7
41.7	7.0	13.9	20.9	27.8	34.8	.7	0.5	1.2	1.9	2.6	3.3	3.9	4.6	5.3	6.0 6.7	34.2 1.8
41.8	7.0	14.0	20.9	27.9	34.9	.8	0.6	1.2	1.9	2.6	3.3	4.0	4.7	5.4	6.1 6.8	

Table 29. NP 401 Interpolation table for minutes of Declination - part of.

As with the Sun and Moon it is possible to use a Planet sight proforma, which may improve both accuracy and consistency, bit I emphasize again, this should NOT be a substitute for understanding (**Section 6.11** p.150). An example is shown below in **Figure 122**.

Astronavigation from Square One

Planet Sight proforma

Date

Estimated position	Latitude	°	. ′	N / S	Longitude	− 30′	°	. ′	W / E
					+ 30′				

	Day	Hour	Min
Zone time			
Zone			
GMT			
GMT / UT			

GHA Hours		°	. ′
Min Sec		°	. ′
v value ___ : corrn. +			. ′
GHA		°	. ′
Chosen longitude −W+E		°	. ′
LHA		°	. ′

	Day	Hour	Min	Sec
Watch time				
Watch error				
GMT / UT				

Declination Hrs ___	°	. ′
d + / − Min page ___	°	. ′
Declination: N / S	°	. ′

Horizontal Parallax	

Sextant Altitude (Hs)	°	. ′
Index Error on − / off +		. ′
	°	. ′
Height of Eye m dip −		
Apparent Altitude (Ha)	°	. ′
Main Correction +		. ′
Mars/Venus correction		. ′
Observed Altitude Ho	°	. ′

↓

Ho	°	. ′
Hc	°	. ′
Intercept		. ′

Chosen Lat. N / S		
Declination N / S		
SAME / CONTRARY		
Chosen Lat	°	N / S

Azimuth Z		°
Zn		°

Hc	°	. ′
d +/−		. ′ +
Hc	°	. ′

←

Chosen position N / S   ° W / E   °   . ′
Intercept   . ′   towards / away
Azimuth   ° T

Figure 122. An example of a Planet sight proforma

Section 9: The stars in general and *Polaris* in particular

In this section the general features of stars are considered in relation to their usefulness in navigation and then one particularly important star, ***Polaris*** is looked at in detail for its role in the fixing of latitude. We live in an age of light pollution and we seldom see the sky as perhaps God intended, but the astro-navigating sailor still has a good chance. There are of course thousands of stars in the sky at night but this number is soon whittled down when the key features of a navigational star are understood. In fact a little light pollution is not a bad thing for the novice star gazer, although to suggest this is probably a heresy; it can help in pattern recognition by obscuring some less bright stars and mimicking twilight.

By the end of Section 9, the following KEY FACTS should be understood:-

- **Stars useful for navigation need to be bright enough to be visible at dawn and dusk.**

- **Brightness varies on a scale of apparent magnitude.**

- **Stars are vast distances from our solar system and their positions remain fixed relative to one another on the celestial sphere.**

- **The pattern of stars visible in the night sky change predictably over the year.**

- **The First point of Aries is used as the reference point to which the East-West position of all stars is related, each by their Sidereal Hour Angle (SHA).**

- **The GHA of Aries is listed for every hour in the astro-nautical almanac.**

- **The GHA of a star = GHA Aries + SHA of the star.**

- **The declination of each star is fixed and does not change through the year.**

- **The fixed positions of stars render some permanently invisible to an observer from certain latitudes.**

- **At increasing latitudes stars will tend to become circumpolar; visible throughout the night, rotating around the pole.**

- **Sight planning and the use of Sight Reduction Tables can identify the best stars available for a navigational fix.**

- **The apparent altitude of all stars needs correction only for the effects of refraction and not for either semi-diameter or parallax.**

- ***Polaris*** **is useful in the northern hemisphere for the estimation of latitude.**

9.1 Stars in General

A star is an unbelievably vast sphere of gas atoms, so highly energised that they form an ionised plasma. The gas is mostly composed of hydrogen which, by atomic fusion in the dense core of the star, produces helium. This process releases huge quantities of energy as light and heat. The first recorded star map is reported to be of ancient Egyptian origin[9.1] and the birth of the astrology, based on the perceived patterns in the stars had a similarly early birth, probably in Mesopotamia, modern day Iraq[9.2].

Giordano Bruno (1548-1600), the Italian priest, scholar and subsequent martyr, suggested that stars were really other suns and that the universe was infinite[9.3]. The first measurements of the great distance between stars were made by in 1838 by Friedrich Bessel (1784-1846)[9.4]. William Herschel (1738-1822), by counting stars along lines of sight, found more in certain directions and suggested the existence of massive groups of stars that we call galaxies, our own being the Milky Way. Herschel also described the first binary star where two stars have attracted one another by gravitational forces to form a double star system.[9.5].

With the development of spectroscopy, the study of emitted light wavelengths, a comparison of star properties and importantly their content, became possible. Stars form by gravitational effects within the interstellar medium, perhaps instigated by shockwaves from distant star activity. As material coalesces, the central dense matter heats and reaches the condition to begin the process of nuclear fusion which becomes self perpetuating[9.6]. This process can last several billion years, depending upon the original mass of hydrogen and the rate at which it fuses.

When the hydrogen is exhausted the star expands and cools forming a "red giant". Helium fusion then begins and the star volume decreases. Eventually the helium is consumed, converted into Carbon and Oxygen which themselves can further fuse producing heavier elements until the metal iron is produced. At this stage the fusion process ceases as iron nuclei are too tightly bound to fuse with each other to produce a heavier element. The star then begins to shrink to form a dense "white dwarf". In very large stars at the stage of iron production, a sudden collapse may occur which initiates an enormous explosion of the star, called a supernova[9.7].

9.2 The celestial sphere – is it a useful model for stars?

The universe we are told is still expanding; the stars are moving away from each other at unimaginable speeds, but fortunately for the navigator, their angular positions relative to one another from our viewpoint on Earth, stay constant. The patterns of **constellations** we see are formed by stars that appear to lie closely together from our particular viewpoint on Earth. In reality they are not usually in close proximity to one another at all. Constellations we recognise, if viewed from another point in the universe, would appear completely different, if not disappear altogether. New constellations would become apparent, depending upon the relative positions of stars with respect to the observer.

From a navigational viewpoint however, the position of each star is fixed on the celestial sphere. This is a wholly artificial but supremely useful way of considering the positions of the stars as viewed from Earth, each useful star having a known declination and hour angle; their celestial latitude and longitude, all at an infinite but irrelevant distance from Earth. For a star the hour angle is called the *sidereal* **hour angle** (Latin: *sidus* – a star), which is the fixed angular relationship between each star and the First Point of Aries, the stellar reference meridian, introduced in **Section 1.7** p.11.

9.3 *The characteristics of navigationally useful stars* – which star to find?

For a star to be useful for position fixing it must be readily identifiable. This usually means that it is part of or near or directionally related to a recognizable constellation; a group of stars that together produce a distinctive pattern. A useful star must also be visible at the appropriate time i.e. when the horizon is visible, either at dawn or dusk, so the star must be quite bright. Lastly the navigator must have access to the tabulated data about a particular star to be able to exploit its navigational secrets.

The star once chosen is then used in the standard way, just as the Sun, Moon and planets; its sextant altitude being converted to an observed altitude and then compared with a calculated altitude from a chosen position, to produce an intercept on the star's azimuth.

Ideally, all the major stars visible in the navigator's particular hemisphere should be known, some of which will be resident in the opposite celestial hemisphere. This does not necessitate knowing and recognising hundreds of stars. Only a relatively small number are bright enough to be visible at dawn and dusk when the horizon is also defined. The approximate spatial relationship between the star-groups should be known, as well as recognising individual stars, as only portions of the sky may be usefully visible due to cloud cover or the brightness of the Moon.

2011 NOVEMBER 15, 16, 17 (TUES., WED., THURS.)

UT	ARIES GHA	VENUS −3.8 GHA	VENUS Dec	MARS +0.9 GHA	MARS Dec	JUPITER −2.9 GHA	JUPITER Dec	SATURN +0.8 GHA	SATURN Dec	STARS Name	SHA	Dec
15 00	53 44.3	159 ...	S23 31.1	259 09.1	N12 32.2	22 22.5	N11 10.7	210 40.0	S 7 11.4	Acamar	315 18.7	S40 15.4
01	68 46.8	174 ...	31.5	274 10.3	31.8	37 25.3	10.6	225 42.2	11.5	Achernar	335 27.0	S57 10.6
02	83 49.3	189 ...	31.0	289 11.6	31.4	52 28.1	10.5	240 44.4	11.6	Acrux	173 11.1	S63 09.7
03	98 51.7	(ARIES - THE STAR'S REPRESENTATIVE)			31.0	67 30.8	10.4	255 46.6	11.7	Adhara	255 13.2	S28 59.2
04	113 54.2				30.6	82 33.6	10.3	270 48.8	11.8	Aldebaran	290 50.3	N16 32.0
05	128 56.6				30.2	97 36.3	10.2	285 51.0	11.9			
06	143 59.1				29.8	112 39.1	N11 10.2	300 53.2	S 7 12.0	Alioth	166 22.1	N55 53.5
07	159 01.6	264 16.0	34.1	4 18.0	29.4	127 41.8	10.1	315 55.4	12.1	Alkaid	153 00.2	N49 15.1
08	174 04.0	279 15.2	34.5	19 19.2	29.0	142 44.6	10.0	330 57.6	12.2	Al Na'ir	27 45.1	S46 54.3
T 09	189 06.5	294 14.3	34.9	34 20.5	28.6	157 47.3	09.9	345 59.8	12.3	Alnilam	275 47.2	S 1 11.7
U 10	204 09.0	309 13.4	35.3	49 21.8	28.2	172 50.1	09.8	1 02.0	12.4	Alphard	217 57.2	S 8 42.6
E 11	219 11.4	324 12.5	35.8	64 23.1	27.8	187 52.8	09.7	16 04.2	12.5			
S 12	234 13.9	339 11.6	S23 36.2	79 24.3	N12 27.4	202 55.6	N11 09.6	31 06.4	S 7 12.6	Alphecca	126 12.3	N26 40.6
D 13	249 16.4	354 10.7	36.6	94 25.6	27.0	217 58.4	09.5	46 08.6	12.7	Alpheratz	357 44.5	N29 09.7
A 14	264 18.8	9 09.8	37.0	109 26.9	26.6	233 01.1	09.4	61 10.8	12.8	Altair	62 09.5	N 8 54.2
Y 15	279 21.3	24 08.9	47.1	124 28.2	26.2	248 03.9	09.3	76 13.0	12.9	Ankaa	353 16.5	S42 14.5
16	294 23.7	39 08.0	37.8	139 29.4	25.8	263 06.6	09.2	91 15.2	12.9	Antares	112 28.0	S26 27.4
17	309 26.2	54 07.1	38.3	154 30.7	25.4	278 09.4	09.1	106 17.4	13.0			
18	324 28.7			169 32.0	N12 25.0	293 12.1	N11 09.0	121 19.6	S 7 13.1	Arcturus	145 57.1	N19 07.2
19	339 31.1	(HOURLY VALUES OF GHA ARIES)		184 33.3	24.6	308 14.9	08.9	136 21.8	13.2	Atria	107 31.4	S69 02.9
20	354 33.6			199 34.5	24.2	323 17.6	08.8	151 24.0	13.3	Avior	234 18.3	S59 32.7
21	9 36.1			214 35.8	23.8	338 20.4	08.7	166 26.2	13.4	Bellatrix	278 32.9	N 6 21.6
22	24 38.5			229 37.1	23.4	353 23.1	08.6	181 28.4	13.5	Betelgeuse	271 02.2	N 7 24.5
23	39 41.0			244 38.4	23.0	8 25.9	08.6	196 30.6	13.6			
16 00	54 43.5	159 00.8	S23 41.1	259 39.7	N12 22.6	23 28.6	N11 08.5	211 32.8	S 7 13.7	Canopus	263 56.2	S52 42.0
01	69 45.9	173 59.9	41.5	274 40.9	22.2	38 31.4	08.4	226 35.0	13.8	Capella	280 35.7	N46 00.5
02	84 48.4	188 59.0	41.9	289 42.2	21.8	53 34.1	08.3	241 37.2	13.9	Deneb	49 32.4	N45 19.8
03	99 50.9	203 58.1	42.3	304 43.5	21.4	68 36.9	08.2	256 39.4	14.0	Denebola	182 35.0	N14 30.2
04	114 53.3	218 57.2	42.7	319 44.8	21.0	83 39.6	08.1	271 41.6	14.1	Diphda	348 56.8	S17 55.2
05	129 55.8	233 56.3	43.1	334 46.1	20.6	98 42.4	08.0	286 43.8	14.2			
06	144 58.2	248 55.4		349 47.4	20.2	113 45.1	N11 07.9	301 46.0	S 7 14.3	Dubhe	193 53.3	N61 40.9
W 07	160 00.7	(EXAMPLE 38 p.259 GHA ARIES 0700 160° 00.7')				128 47.9	07.8	316 48.2	14.4	Elnath	278 13.7	N28 37.0
E 08	175 03.2					143 50.6	07.7	331 50.4	14.5	Eltanin	90 47.1	N51 29.5
D 09	190 05.6					158 53.4	07.6	346 52.6	14.6	Enif	33 48.2	N 9 56.0
N 10	205 08.1					173 56.1	07.5	1 54.8	14.7	Fomalhaut	15 25.1	S29 33.5
E 11	220 10.6	323 50.9	45.3	64 55.7	18.5	188 58.9	07.4	16 57.0	14.8			
S 12	235 13.0	338 50.0	S23 45.7	79 55.0	N12 17.9	204 01.6	N11 07.3	31 59.2	S 7 14.9	Gacrux	172 02.6	S57 10.6
D 13	250 15.5	353 49.1	46.3	94 56.3	17.5	219 04.4	07.2	47 01.4	15.0	Gienah	175 53.7	S17 36.4
A 14	265 18.0	8 48.2	46.7	109 57.6	17.1	234 07.1	07.2	62 03.6	15.1	Hadar	148 50.2	S60 25.7
Y 15	280 20.4	23 47.3	47.1	124 58.9	16.7	249 09.9	07.1	77 05.8	15.2	Hamal	328 01.7	N23 31.3
16	295 22.9	38 46.4	47.4	140 00.2	16.3	264 12.6	07.0	92 08.0	15.3	Kaus Aust.	83 45.6	S34 22.7
17	310 25.4	53 45.5	47.8	155 01.4	15.9	279 15.4	06.9	107 10.2	15.4			
18	325 27.8	68 44.6	S23 48.2	170 02.7	N12 15.5	294 ...	(EXAMPLE 37 p.252 SHA OF THE STAR MIRFAK 308° 41.5' DECLINATION 49° 54.3')		15.5	Kochab	137 21.1	N74 06.3
19	340 30.3	83 43.7	48.6	185 04.0	15.1	309 ...			15.6	Markab	13 39.3	N15 16.4
20	355 32.7	98 42.8	49.0	200 05.3	14.7	323 ...			15.7	Menkar	314 15.9	N 4 08.3
21	10 35.2	113 41.9	49.3	215 06.6	14.3	338 ...			15.8	Menkent	148 09.3	S36 25.6
22	25 37.7	128 41.0	49.7	230 07.9	13.9	353 ...			15.9	Miaplacidus	221 39.8	S69 45.7
23	40 40.1	143 40.1	50.1	245 09.2	13.5	8 ...			16.0			
17 00	55 42.6	158 39.2	S23 50.5	260 10.5	N12 13.1	23 ...			16.1	Mirfak	308 41.5	N49 54.3
01	70 45.1	173 38.3	50.9	275 11.7	12.7	39 ...			16.2	Nunki	76 00.0	S26 16.8
02	85 47.5	188 37.4	51.2	290 13.0	12.3	54 40.1	06.1	242 30.1	16.3	Peacock	53 21.2	S56 41.9
03	100 50.0	203 36.5	51.6	305 14.3	11.9	69 42.9	06.0	257 32.3	16.4	Pollux	243 28.9	N27 59.7
04	115 52.5	218 35.5	52.0	320 15.6	11.5	84 45.6	05.9	272 34.5	16.5	Procyon	245 00.7	N 5 11.6
05	130 54.9	233 34.6	52.3	335 16.9	11.1	99 48.3	05.8	287 36.7	16.6			
06	145 57.4	248 33.7	S23 52.7	350 18.2	N12 10.7	114 51.1	N11 05.7	302 38.9	S 7 16.7	Rasalhague	96 07.8	N12 33.3
07	160 59.9	263 32.8	53.1	5 19.5	10.3	129 53.8	05.6	317 41.1	16.8	Regulus	207 44.7	N11 54.4
T 08	176 02.3	278 31.9	53.4	20 20.8	09.9	144 56.6	05.5	332 43.3	16.9	Rigel	281 12.8	S 8 11.2
H 09	191 04.8	293 31.0	53.8	35 22.1	09.6	159 59.3	05.4	347 45.5	16.9	Rigil Kent.	139 54.0	S60 52.9
U 10	206 07.2	308 30.1	54.2	50 23.3	09.2	175 02.1	05.3	2 47.7	17.0	Sabik	102 14.2	S15 44.3
R 11	221 09.7	323 29.2	54.5	65 24.6	08.8	190 04.8	05.2	17 49.9	17.1			
S 12	236 12.2	(EXAMPLE 37 p.252 GHA ARIES 1900HRS: 341° 29.4')			08.4	205 07.6	N11 05.1	32 52.1	S 7 17.2	Schedar	349 41.4	N56 36.5
D 13	251 14.6				8.0	220 10.3	05.0	47 54.3	17.3	Shaula	96 23.8	S37 06.7
A 14	266 17.1				7.6	235 13.0	05.0	62 56.5	17.4	Sirius	258 34.5	S16 43.9
Y 15	281 19.6				7.2	250 15.8	04.9	77 58.7	17.5	Spica	158 32.7	S11 13.3
16	296 22.0				6.8	265 18.5	04.8	93 00.9	17.6	Suhail	222 53.2	S43 28.7
17	311 24.5	53 ...		56.7	155 32.4	06.4	280 21.3	04.7	108 03.1	17.7			
18	326 27.0	68 22.8	S23 57.0	170 33.7	N12 06.0	295 24.0	N11 04.6	123 05.3	S 7 17.8	Vega	80 40.0	N38 48.0
19	341 29.4	83 21.9	57.4	185 35.0	05.6	310 26.8	04.5	138 07.5	17.9	Zuben'ubi	137 07.0	S16 05.4
20	356 31.9	98 21.0	57.7	200 36.3	05.2	325 29.5	04.4	153 09.7	18.0		SHA	Mer.Pass.
21	11 34.3	113 20.1	58.1	215 37.6	04.8	340 32.3	04.3	168 11.9	18.1			h m
22	26 36.8	128 19.2	58.4	230 38.9	04.4	355 35.0	04.2	183 14.1	18.2	Venus	104 17.4	13 25
23	41 39.3	143 18.3	58.8	245 40.2	04.0	10 37.7	04.1	198 16.3	18.3	Mars	204 56.2	6 41
	h m									Jupiter	328 45.2	22 22
Mer.Pass. 20 17.8		v −0.9	d 0.4	v 1.3	d 0.4	v 2.7	d 0.1	v 2.2	d 0.1	Saturn	156 49.4	9 52

Table 30. NP 314-11 Aries, Star and Planet and star data Nov. 15th-17th

There are around sixty stars that are useful for navigation. NP 314 gives data on 57 navigational stars on its daily pages, 27 with northern declinations and 30 with southern declinations. Depending upon the star's declination, only a proportion may be visible to an observer at any given latitude (**Section 9.10**). They are listed in alphabetical order, with SHA and Dec. (**Table 30**). At a separate point, in the end-pages of NP 314, the 57 are listed again both alphabetically and in order of given number, based on their sidereal hour angle (as one increases the other decreases) and their magnitude, giving SHA and Dec in whole numbers (**Section 9.13** – choosing stars).

NP 314 also lists in 6 pages of detail, a total of 173 stars, the 57 main stars alongside details of 120 or so other lesser stars that complete constellations, including the zodiacal signs ranged around the ecliptic (the Sun's path through the heavens over the course of a year). In this table all the stars are named with by their constellation and a letter of the Greek alphabet (e.g. Ursae Majori α, β, γ, ε) and their given number if they are a main star. Each is given a monthly mean SHA and declination, the small but not quite insignificant differences adjust for the effect of Earth's orbital variations detailed in **Section 5.12** p.130.

NP 314 -1 1 has detailed star charts of northern and southern hemispheres showing the named constellations a well as the brighter navigational stars. The northern hemisphere star chart is reproduced in **Figure 123** p.236. Emphasis is needed though, that learning lists of stars is not necessary and there are means available to make star recognition, especially at twilight, a relatively straight-forward, if not easy, matter (See **Section 9.13** p.248).

9.4 Brightness and Magnitude – what is their relationship to the stars?

For a star to be navigationally useful it needs to be bright enough to be seen and identified during twilight, which immediately rules out several thousand less bright ones! The bright stars are relatively easy to recognise. The brightness of a heavenly body is called its **apparent magnitude**: how bright it appears from Earth. The main star data table in NP 314 includes the brightness factor.

Apparent magnitude is based on a comparison scale which somewhat confusingly, has the Sun's brightness as –26.7 and all the useful stars having a brightness value of less than 3! A magnitude scale was first devised by the Greeks; they looked at the stars and saw 6 levels of brightness. The brightest stars they termed to be of the 1st magnitude and the least of the 6th magnitude, each magnitude being half as bright as the one preceding it.

This scale was usefully modified in the 19th century by the astronomer Norman Pogson (1829–1891)[9.8], who suggested that a magnitude 1 star should be defined as having a level of brightness of 100 times a magnitude 6. He developed the brightness factor, called Pogson's ratio, representing the six star magnitudes, from 1 to 6, by levels of brightness on a logarithmic scale: 2.5^5, 2.5^4, 2.5^3, 2.5^2, 2.5^1 and 2.5^0. These values, when calculated and rounded up, reveal the figures of 99.6, 39.7, 15.8, 6.3, 2.5 and 1, fitting well with Pogson's 100:1 ratio; a magnitude 1 star having a brightness of 100 times that of a magnitude 6 star.

Astronavigation from Square One

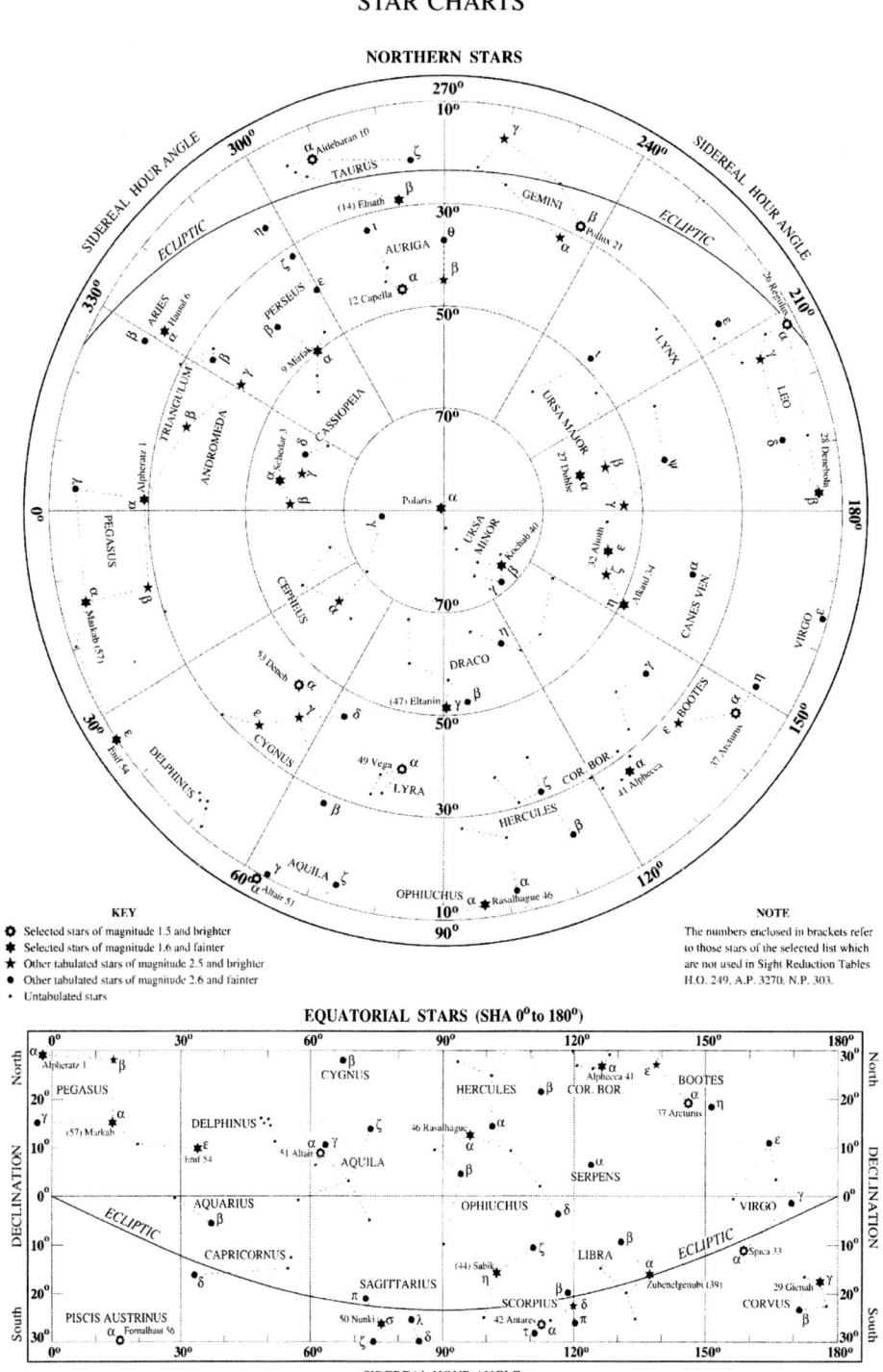

Figure 123. The standard northern hemisphere star chart from NP-314

Polaris, the North Star, was initially used as the reference point in the scale, being of magnitude 2. Then the brighter star *Vega* was used as the reference with its near zero magnitude, but was in turn superseded when astrophysicists, in their search for accuracy, arrived at a calculated zero. *Sirius*, the brightest star visible from Earth (except for our Sun) has a magnitude of –1.4, giving it a brightness of over 800 times ($2.5^{7.4}$) that of a star of magnitude 6. A table of comparative magnitudes was drawn up and eventually enlarged to include the full Moon (–12.6) and Sun (–26.7). These figures should begin to make some sense if "relative brightness" is worked out in an example.

Example 34. The difference between the Sun and Moon's brightness in terms of apparent magnitudes is –26.7 – (–12.6) = –14.1. Therefore the difference in brightness can be mathematically calculated and expressed using Pogson's ratio as: $2.5^{14.1} \approx 400{,}000$. This indicates that the Sun is four hundred thousand times brighter than the full Moon; reason enough not to stare!

9.5 The celestial sphere's movement – how does it move relative to Earth?

A careful series of night observations from any latitude will reveal that the celestial sphere, as shown by star motion, appears to rotate a fraction more than 360° each 24 hours. This is most obvious when looking at the same area of sky over a few weeks. For example, if a group of stars are observed from the same position at the same time of night over the period of a month, they will be seen to advance at the given time, from East to West by 30° (**Figure 124**).

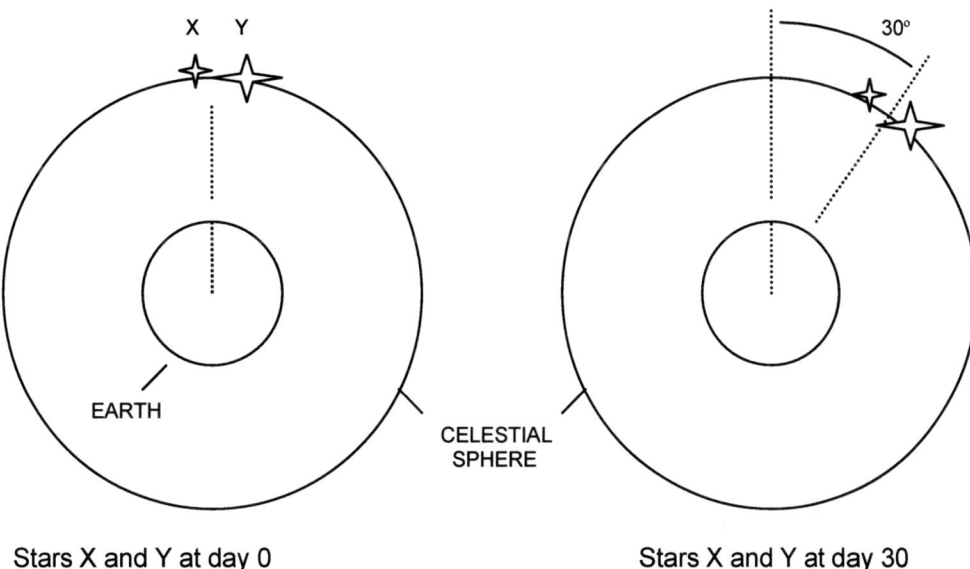

Stars X and Y at day 0   Stars X and Y at day 30

Figure 124. The cumulative movement of the celestial sphere over 30 days

Astronavigation from Square One

This apparent movement is of course, a consequence of the Earth's orbit around the Sun at a rate of just less than one degree per 24 hours as detailed in **Section 5.9** (p.120). Remember, 1° of angular change is equivalent to 4 minutes of time, so watching the same segment of sky a given star will pass over an observer's viewpoint 4 minutes earlier each night. In effect then, the celestial sphere appears to us on Earth, to rotate 361° in a westerly direction every 24 hours.

9.6 The changing night sky – how and why does the star pattern change?

The Earth moves round the Sun over a year and results in a gradual change in the stars seen in the night sky, as explained in the previous section. Just 1° of the 361° of rotation per 24 hours results in the accumulation of 360° over a year. The infinite scale of this view cannot be readily compressed into a diagram, but **Figure 125** may help in your mind's eye view.

The diagram shows the Earth, viewed from afar, in two positions in its orbit round the Sun, six months apart. It can be seen that at night, when half the Earth's sphere faces away from the Sun, a different pattern of the stars would be seen, depending on the position of the Earth in its annual orbit. The apparent view (that which you see) as an observer from the Earth's geocentric and apparently static position can be reconciled to this wider view. Imagine the Sun rotating through 360° and the stars rotating through 361° per 24 hours (**Figure 126**).

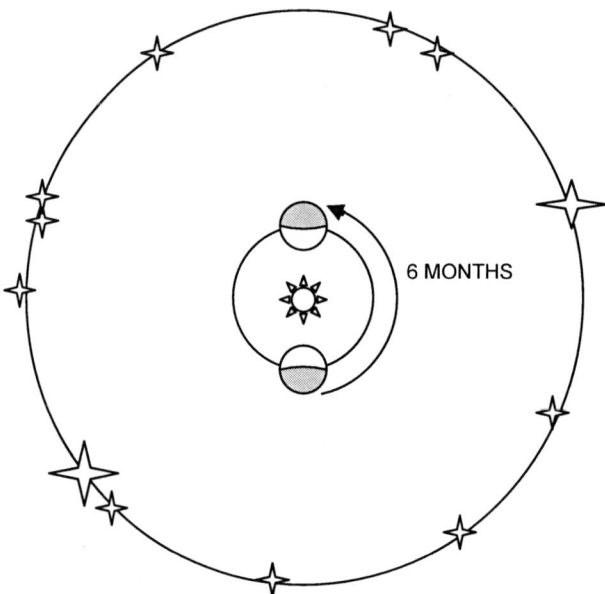

Figure 125. Earth's differing night skies depend upon its solar orbit

– navigating with the Stars

In this way over a year, the Sun will rotate around its background of stars and produce the apparent seasonal change of the stars at night. We have dealt with the details of the Sun's apparent movement across the celestial sphere when viewed from the Earth in **Section 6.3** (p.144), but we need now to look in detail at how the stars appear to move?

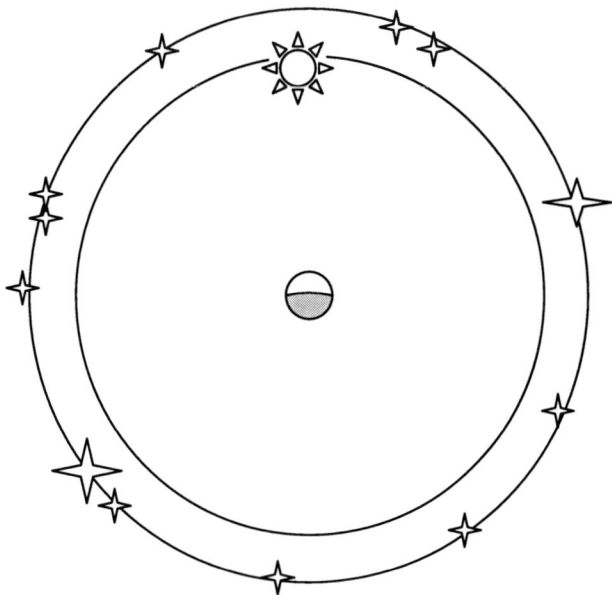

Figure 126. The geocentric view from Earth of the Sun and celestial sphere

9.7 Movement of the stars on the celestial sphere – the star turn?

The stars appear to us to rotate as one, around a North-South axis, in an East-West direction on a daily basis, each star remaining fixed in position on the celestial sphere relative to all others. This is quite unlike the relatively unfettered and seemingly unrelated movements of the Sun, Moon and planets. Each star's individual hour angle (its celestial longitude) changes as the sphere rotates. Its declination (celestial latitude) remains <u>unchanged</u>. Each star therefore lies directly over a parallel of latitude on the Earth's surface and continually follows this latitude line as it rotates with the celestial sphere.

The end result of this movement for an observer in mid-latitudes is that certain constellations, those with declinations near the celestial equator, tend to appear at particular seasons. Other constellations with higher declinations, towards the celestial poles, are present throughout the year. Exactly what stars are seen is dependent on the observer's latitude and the time at which the stars are observed.

Astronavigation from Square One

In the northern hemisphere for instance, the "Orion" constellation appears at night in winter (**Figure 127**). It is readily identified by its central triple line of stars, "Orion's belt", the middle star being *Alnilam*. From Orion's central band there are 4 stars spread like limbs, *Betelgeuse* and *Bellatrix* to the North and *Rigel* and a less bright star *Orionis κ* to the South.

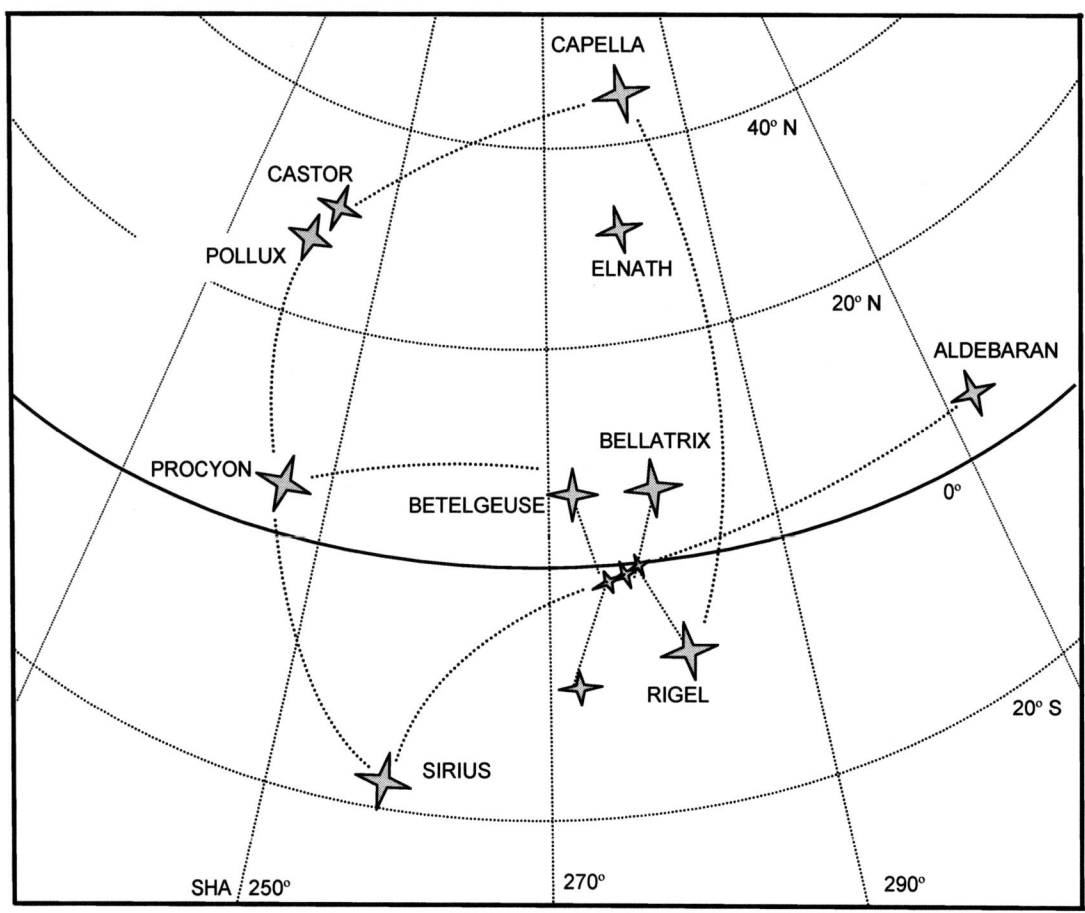

Figure 127. The navigational stars in and around the constellation of Orion

240

– navigating with the Stars

The night sky in the summer months is dominated by the "Summer Triangle" (useful stars *Altair*, *Vega* and *Deneb*). Each is from a different constellation: *Altair* from "Aquila" the Eagle, *Vega* from "Lyra", the Lyre and *Deneb* from "Cygnus", the Swan (**Figure 128**).

Both Orion and the Summer triangle are of course fixed in position on the celestial sphere and rotate around the Earth once each 24 hours, but when one is prominent at night, the other passes overhead in the daytime; it is up there above us, but our view is obscured by the Sun's light. Orion can be seen in the very early morning as summer draws to a close and autumn creeps on. Half a year later the reverse applies, the Earth having rotated that extra 1° each day resulting in 180° over 6 months. Orion will be in the sky, invisible in the middle of the day and replaced instead at night by the stars of the Summer Triangle.

9.8 Leading and trailing stars – recognisable patterns of movement

The orientation of those stars we do recognize is important as certain stars will always lead a constellation and others trail. This gives the navigator a way of orientating the stars; the key to recognition[9.9]. Again, the constellation of Orion is a classic example seen in the winter skies in the northern hemisphere, rising in the East, passing across the sky to the South of northern hemisphere observers and setting in the West. The leading stars are *Bellatrix* to the North and *Rigel* to the South. The northern star of the trailing pair is the bright *Betelgeuse*. In the summer triangle, *Vega* leads and *Deneb* and *Altair* trail. *Duhbe* and *Merak* lead *Ursa Major*.

9.9 Recognizing patterns in the sky – the key to the stars?

Certain stars within a constellation will usefully point to other stars standing alone or in other constellations, producing a useful and reliable route-map of the sky. Continuing with Orion for the present (**Figure 127**); a curved line running North and West, leads to *Aldebaran*, the reddish bull's eye of the constellation called "Taurus".

A similar curved line running southeast from the trailing end of the belt leads to *Sirius*, the brightest star in our sky and the first to be seen as night approaches in the winter and spring months in the northern hemisphere. A curved line running North and East from *Sirius* leads to *Procyon*, another bright star, with a declination slightly greater than *Betelgeuse* and so following a similar path across the sky. The directions from star to star are not exact, but depend upon the latitude of the observer and the time of observation during their nightly procession, but the overall patterns persist.

Continuing to the North and West from *Procyon*, a pair of slightly less prominent stars will be seen; the "Gemini Twins", *Castor* and *Pollux*, themselves lying to the North and East of "Orion". Together they point in the direction of *Capella*, the "she-goat" another bright star lying midway between "Orion" and *Polaris* and with a sidereal hour angle similar to *Rigel*, helping its identification. *Capella* has a declination of 46° North, passing overhead in mid-northern latitudes.

241

Astronavigation from Square One

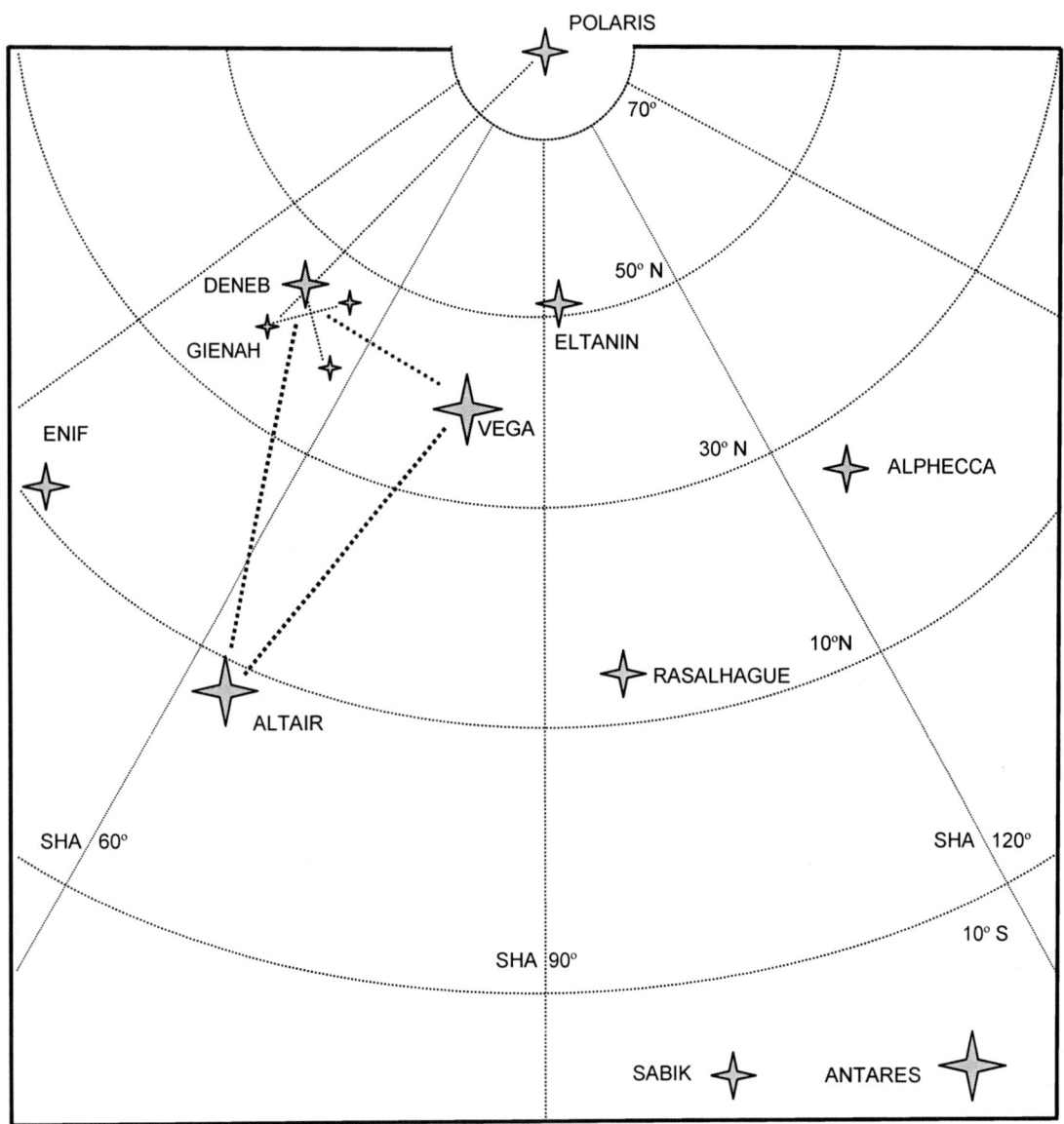

Figure 128. The navigational stars in and around the Summer Triangle

– *navigating with the Stars*

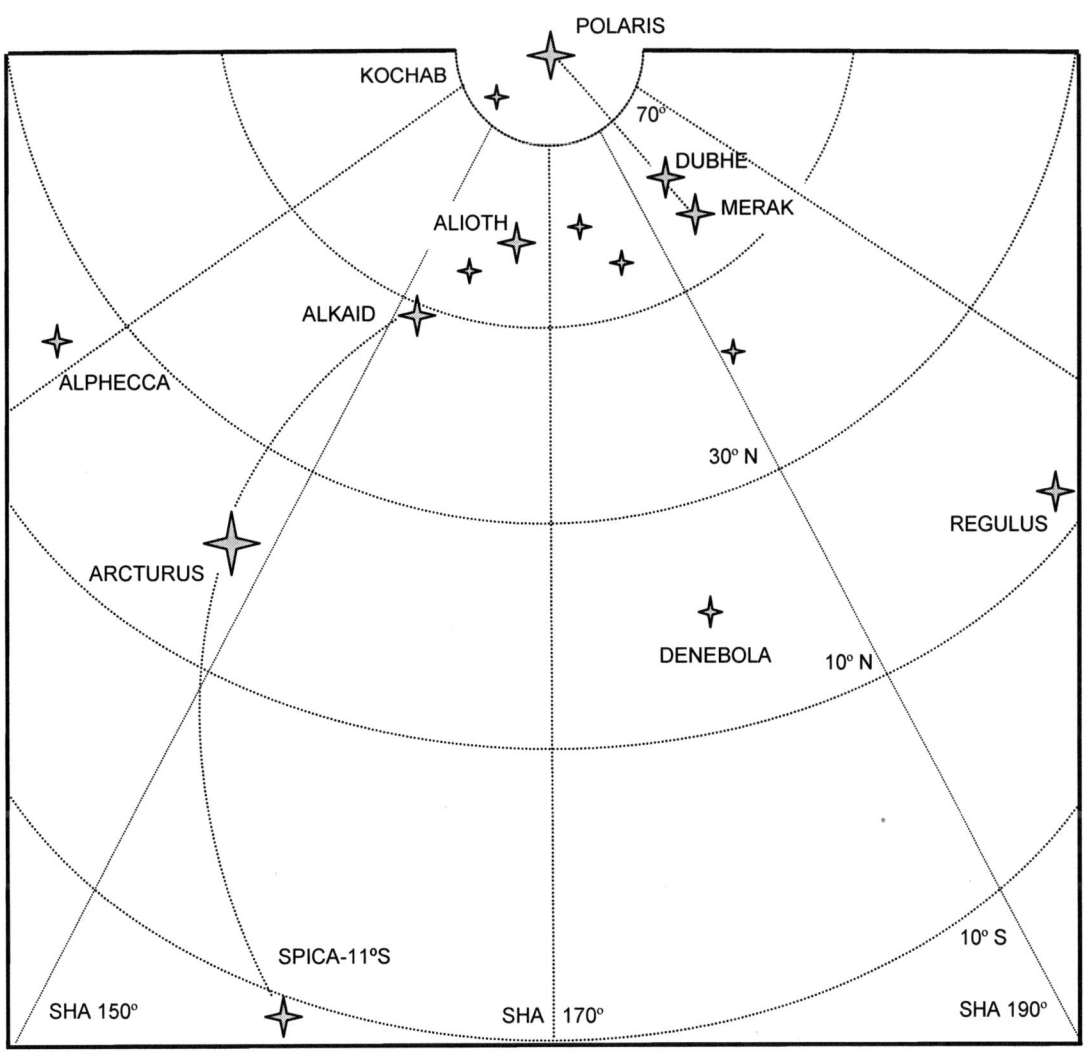

Figure 129. The link between *Polaris*, *Ursa Major*, *Arcturus* and on to *Spica*

Second only to Orion, the next most easily recognised constellation in the northern hemisphere is the "Great Bear", (*Ursa Major*), also known variously as the plough or big dipper (meaning spoon or ladle) (**Figure 129**). Its two leading stars *Dubhe* and *Merak* give the direction to *Polaris* and they are consequently known as the "pointers". A curved line running from the prominent handle of the plough or dipper, running through the stars *Alioth*, *Mizar* and *Alkaid* leads West of South to *Arcturus* and on South to *Spica* at a declination on 11° South.

9.10 Invisible stars – why can we not see all the stars in the celestial sphere?

Depending upon the latitude of the observer, there will be parts of the celestial sphere not visible <u>at any time</u> of the year. A general rule applies: that a heavenly body having a declination in the opposite hemisphere to the observer and a declination greater than 90° minus the observer's latitude, will not be visible at all. They cannot be seen from the observer's latitude at any point during the star's 24 hour period of rotation around the Earth. **Figure 130** helps to clarify the situation.

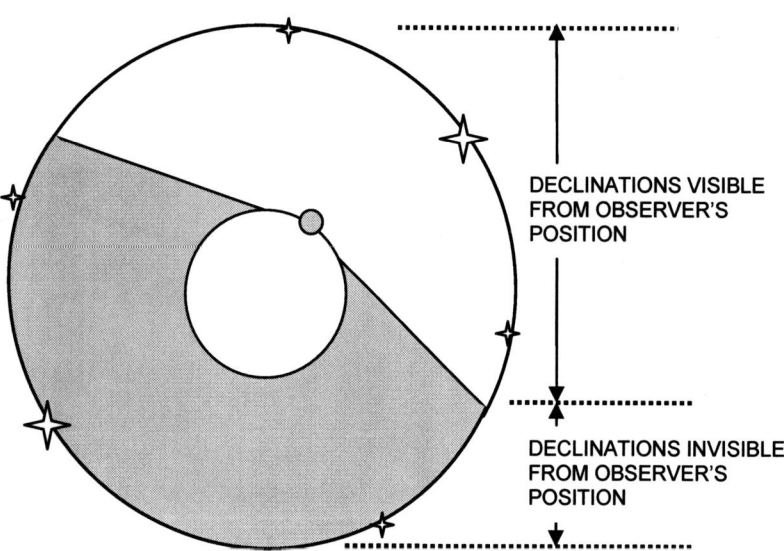

Figure 130. Visible and invisible declinations from a given latitude

Example 35: What is the southerly limit of declination on the celestial sphere visible from an observer's latitude at 50° North?

The visible limit, in the opposite hemisphere is given by: 90° − latitude

$$= 90° − 50°N = \mathbf{40° \ South}$$

Stars with declinations greater than 40° South are never visible from latitude 50° North or above.

9.11 Circumpolar stars – what are they and how are they defined?

To our observer looking towards the northern horizon from a position in the northern hemisphere and if able to watch for a period, the stars would be seen to rotate from East to West, in an anti-clockwise fashion with *Polaris* at the centre of rotation. This would be seen most easily in time-lapse photography. The greater the latitude of the observer, the higher *Polaris* will appear to be and also the greater the number of continually visible stars.

These stars are called circumpolar stars as they will be present throughout the whole night of every night of the year (clouds and pollution permitting!). **Figure 131** shows the orientation of the movement of circumpolar stars with the dominant constellation of *Ursa Major* seen as the observer looks North.

On any given meridian, the declinations of the stars visible to an observer follow a rule. Stars will appear circumpolar if their declination is of the same name as the latitude from which they are observed and is greater than 90° − latitude of observer. The higher the observer's latitude, then the greater will be the number of visible circumpolar stars. For example, the "Great Bear" constellation (*Ursa Major*) is visible all year round above the latitude of 41°N: it is circumpolar at this latitude and above.

If observed at a lesser latitude than 41° N, *Ursa Major* will be seen to rise and set in the ordinary way. Note that circumpolar stars will cross the observer's meridian twice per 24 hour rotation, once above and once below *Polaris*, once in the day and once at night, but to witness both, the observer would need to be at or above the Arctic Circle in the winter months when the Sun does not rise and where the whole of the heavens is circumpolar! A worked example will help understanding.

Astronavigation from Square One

Example 36. At what latitude would an observer be when Arcturus becomes circumpolar?

A star is circumpolar to an observer if its declination is greater than 90° minus the observer's latitude, expressed as: declination = 90 – latitude of observation.

The declination of Arcturus is 15° North:

$\quad\quad$ 15 = 90 – latitude (at which point the star becomes circumpolar)

$\quad\quad$ **latitude** = 90 – 15 = **75° North** (well above the arctic circle).

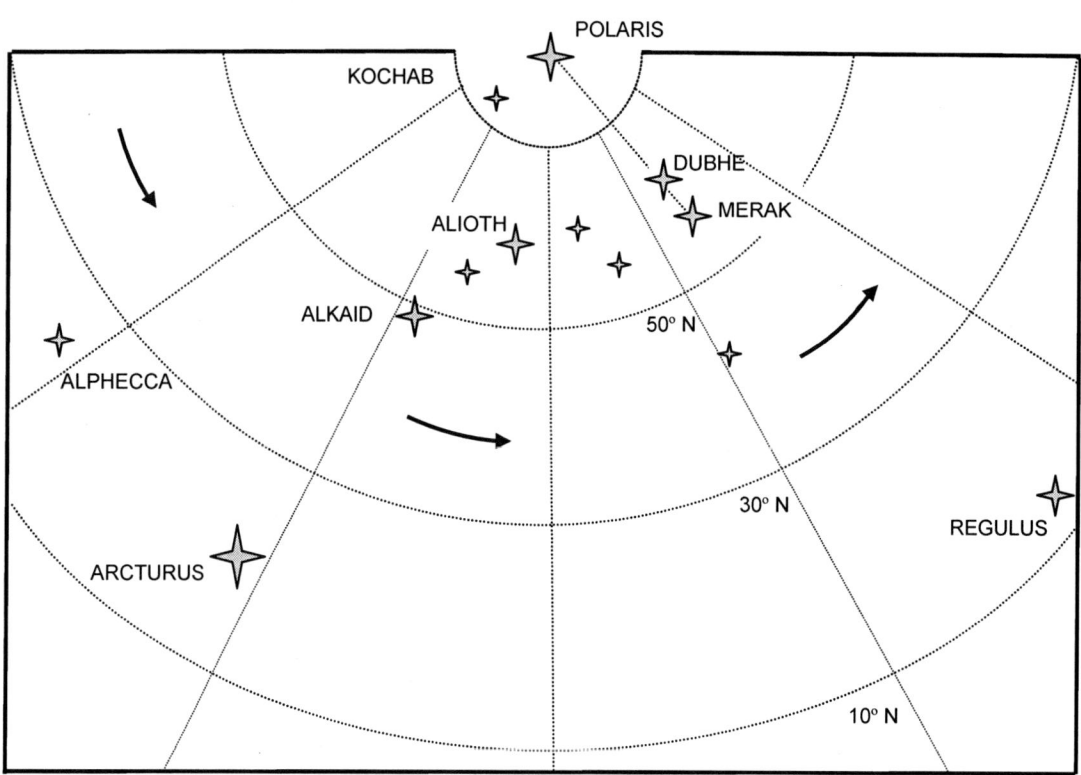

Figure 131. Looking due North at potential circumpolar stars

9.12 Non-circumpolar stars – how do they behave?

Stars which at particular latitude are not constantly in view, will rise, pass across the sky and then set. Stars with northern declinations, rise North of East, pass across the sky and then set North of West. Stars with southern declinations rise South of East, pass across the sky and then set South of West. A star having the same declination as the observer's latitude will pass over the observer's head (**Figure 132**).

Each non-circumpolar star will reach its peak height exactly half way between its time of rising and setting. This peak height corresponds to the meridian passage of the star as it crosses the meridian of the observer. It may, with a little forward planning, be possible to view a star at meridian passage at dawn or dusk twilight, when an horizon is visible. Meridian passages of bodies other the Sun are not useful for latitude estimations, as the Sun is the sole arbiter of time, but they may be used for longitude estimation, as can the Moon (**Section 7.11** p.193).

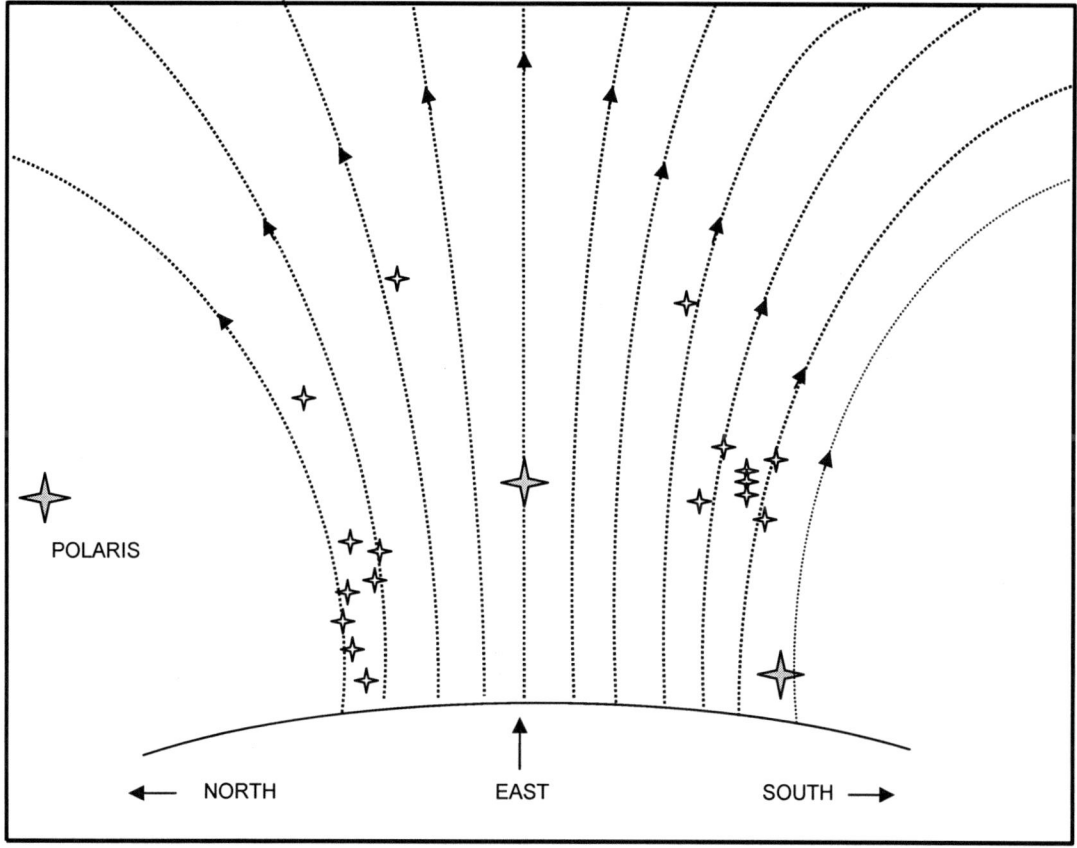

Figure 132. Observer facing East - apparent paths of stars at 50° N

9.13 *Sight planning and choosing stars* – how to pick the most useful stars

What then are the ideal features of a star for sight taking? The first must be visibility as determined by **apparent magnitude**, detailed in sections **9.3** and **9.4**. The top 57 are listed in order alphabetical order on the Aries/planet/star daily table in NP 314 (**Table 30** p.234). The remaining and occasionally useful less bright brethren are added in NP 314's second star list, helpfully arranged in order of Sidereal hour angle: as they appear ranged around the horizon. Their orientation at a moment in time being fixed by the GHA of Aries their representative, added to their Sidereal hour angle (**Section 1.7** p.11).

A further factor to consider early in the planning stage and a major one, is the angle of cut between the eventual plotted position lines. Three lines at angle approximating 60° to one another are ideal, giving the best positional fix.

Of the myriad of stars and the occasional planet above, just how does the navigator start to plan for the task of identifying, sighting and recording those bodies available in the relatively short time at dawn and evening twilight. Random sight taking and then attempted identification after the event is a sure road to disaster.

Having worked through the preceding paragraphs it is plain that being positioned in one or other hemisphere at particular latitude, immediately rules out any chance of seeing stars in the opposite hemisphere that are outside the critical barrier dictated by latitude itself, detailed in paragraph **9.10** p.244.

The next likely arbiter is the season of the year which as detailed in paragraphs **9.5-7**, directly affects the expected star pattern above the navigator during the hours of darkness and twilight. This also brings in to play the local time of twilight, itself varying widely through the year.

The planet diagram in **Section 8.9** p.216, shows the meridian passages of planets and importantly the Sidereal hour angles of their star background. Knowing the time of twilight enables the navigator to extract from the diagram the SHA range of stars that will be passing across the navigator's meridian at a given time. From this, it is then reasonable to extrapolate to an approximate assumption about the stars with greater SHA values that will have already passed meridian passage and to those with lesser SHAs that have yet to pass. The former should be visible to the West of the observer's meridian and the latter to the East.

Begin with deriving the range of Greenwich hour angle of Aries during the likely observation period. GHA is then converted to Local hour angle Aries by adjusting for the observer's longitude in the usual way (+ E / –W). Once the LHA Aries range is broadly known for the specific twilight period, there are several methods of finding the available stars, from the traditional to the modern. Ease of use is important. The range includes star globes, star charts, star-finders, navigational calculators and computers and not least, specifically designed star sight reduction tables.

1. Star Globes

The traditional method uses a star globe: a boxed 3D model of the hemispheres with marked stars, on which can be set the observer's latitude and the LHA Aries. The approximate altitude and azimuth of specific stars available can then be read off. Planetary positions can also be temporarily marked in. Star globes are relatively expensive and now not widely found.

2. Star charts

Star charts of varying complexity are found in most astro-navigational publications. Those found in NP 314 (**Figure 123** p.236) are detailed but clear, giving good indications of a star's sidereal hour angle and declination and therefore which stars should pass through the observer's sky at night. Most importantly the charts correlate exactly with the large star table in NP 314, referred to in para. **9.3** p.233.

3. Star Finders

The standard Star-Finder, model 2102-D is made by Weems and Plath of Annapolis U.S.A (**Figure 133**). Its use is taught in nautical colleges. It is easy to use, cheap (£30), durable and stores flat in a protective case. This device has far-reaching advantages in that it selects the navigational stars visible at a given range of latitude and gives their approximate altitude and azimuth. This diminishes the need to identify stars in the sky and the need to use the sextant to follow them down to the horizon. Rapid sights can be taken by simply setting the approximate altitude on the sextant, to look along the given azimuth, using a hand-bearing compass and sight the star that appears in the sextant's mirror. This method works well in conjunction with the Rapid sight Reduction Tables (**NP 303** vol.1) (**Example 38** p. 259).

Model 2102-D consists of an opaque white plastic disc base plate with central pivot, on which are marked the main 57 navigational stars. On the edge of the plate is circumferentially printed a scale of LHA Aries 0-360°. One of a range of 9 concentric discs, each with a small central hole, can be applied to the base plate and press fitted over the central pivot peg. The discs are arranged for observers in both the northern or southern hemispheres.

The upper disc is transparent and on which is eccentrically printed a blue graticule (a ruled grating) of an "azmuthal equidistant projection" of azimuth and altitude. The 9 discs cover latitudes from 5°- 85° North and South at 10° intervals. The graticule curves are labelled for altitude, increasing from 0-80° in 10° intervals towards its centre and for azimuth, 0-360° in 10° intervals, marked around its edge.

The correct graticule is selected for the observer's approximate latitude. An arrow runs from the centre of the graticule to the edge of the upper disc, which is aligned with the required point of LHA Aries 0-360° on the base plate scale.

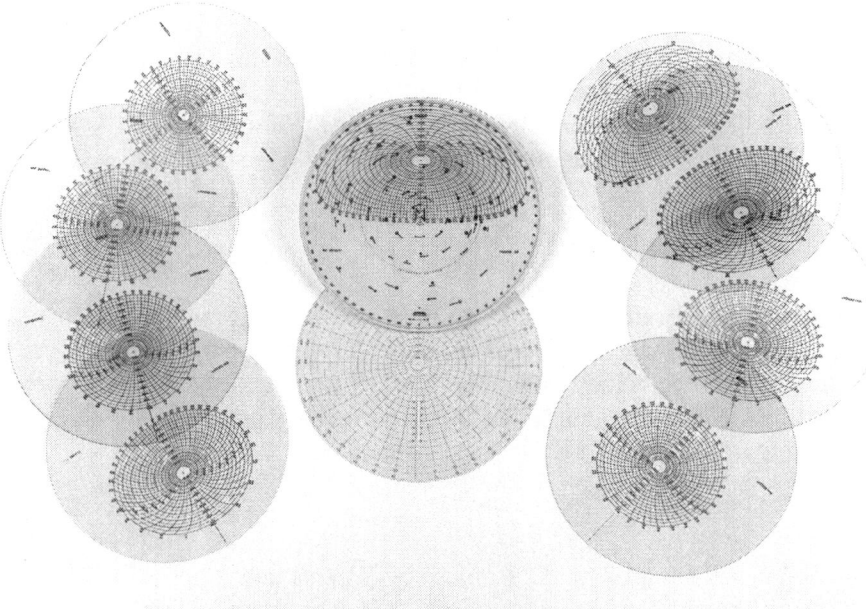

Figure 133. Weems and Plath 2102-D Star-Finder – showing the base and latitude discs

The graticule is then seen to overlay a proportion of the stars printed on the base. The overlain stars are those which should be visible above the observer. For each it is possible to read off an azimuth and altitude from the curves of the graticule. The sextant can then be preset with the approximate altitude of chosen star and the skies scanned much more efficiently in the direction of its approximate azimuth. The positions of the Sun, Moon and planets can also be plotted using a secondary red printed disc.

It is also possible to use the star-finder to identify a star, using the red disc applied over the blue graticule. The red disc has a slot which can be rotated over the altitude position of an unidentified star and from the edge of the slot can be read the approximate declination of the star. The LHA of the star can be converted to SHA and the star can then be sought in star tables with SHA and Dec.

$$\text{SHA Star} = \text{LHA Star} - \text{LHA Aries}.$$

4. Navigational computers/calculators – programs are available into which the parameters of observer's latitude, star altitude and azimuth are entered and from which is given the SHA and declination and hence the identity of a star. For the derivation of calculated altitude and azimuth, several generations of navigation calculators have been designed and built, the current popular models being the Texas instruments StarPilot T89 and TI voyage 200 models (both approx. $400). StarPilot software is also available for a PC.

"Navpac" is the navigating software produced by the Nautical Almanac Office of the UK Hydrographic Office primarily for the Royal Navy but made available to other users[9,11]. It provides "simple and efficient methods for calculating the positions of the Sun, Moon, navigational planets and stars over several years to a consistent precision with the aid of a pocket calculator, personal computer or laptop". "AstroCalc" is a similar program available from www.pangolin.co.nz.

There are several "astro" websites that run algorithms for live almanac data which show the sky overhead for a given latitude and longitude with the Sun, Moon, planets and navigational stars in real time. A particular favourite of mine is a Brazilian site: http//www.tecepe.com.br.

5. Sight Reduction Tables

Rapid Sight Reduction Tables for Navigation NP 303 – vols. 1, 2 & 3, were detailed in **Section 4.7 p.99**. Volume 1 can be used for star sight planning as it selects the 7 most useful stars at all declinations, available at a given LHA Aries and for the observer's latitude. The seven best stars are chosen for their particular azimuth, magnitude, altitude, continuity in latitude and hour angle. They give the required angle of cut in order to produce a good position fix. A fully worked example is given in paragraph **9.15 p.259**.

9.14 *Reducing a star sight* – what stages are necessary?

A star is chosen and a timed sighting made and reduced to an observed altitude. The tabulated data for Aries and the Sidereal Hour Angle of the star, with the observer's longitude is then used to determine the local hour angle of the star. The LHA and declination of the star, with the chosen latitude, are then used to work out the calculated altitude (H_c) and azimuth (Z) for the star at the time of the sight from the chosen position.

As stars have no appreciable diameter the sight does not have to be adjusted for semi-diameter and as they are so very far away they do not need to be corrected for parallax, unlike sights of the Sun and Moon and planets. The star's sextant altitude (H_s) is simply amended for index error, then corrected for height of eye (dip) and refraction using the standard tables, to give the observed altitude H_o.

1. Altitude

The NP 314- 11 altitude correction table for the Sun, planets and stars (**Table 27 Section 8.12 p.223**) is entered with apparent altitude (sextant altitude corrected for index error and dip). Observed and calculated altitudes are then compared in the usual way, to produce an intercept on the azimuth of the star. Calculated altitude and azimuth are worked out using sight reduction tables, the ABC formula, ABC tables or a calculator equation, introduced in **Section 4.1-4.14**, using the latitude of the observation, the local hour angle and declination of the chosen star.

2. GHA Aries and SHA Star

NP 314-11 lists the hourly GHA for Aries on a daily basis together with that of the planets. At the foot of the page is given the time of Aries meridian passage at Greenwich for the middle of the three days on each page (**Table 30** p.234). The increments for minutes and seconds of GHA Aries are included with those for the Sun, planets and Moon (**Table 31** p.254 shows the 14 & 15 minute increments and corrections table). The rate of change of GHA Aries is constant at 15° 02.46′ per hour, which totals to 361° in 24 hours (see **Section 9.5-6**). Its v value is zero, meaning that GHA Aries does not change its rate over time, so no correction is needed. The method for converting GHA Aries to LHA Aries follows the same general rule:

$$\text{LHA Aries} = \text{GHA Aries} + \text{longitude East } \underline{or} - \text{longitude West}$$

Once the LHA Aries is calculated, the sidereal hour angle of the star used is added to it to produce the Local Hour Angle of the star.

$$\text{LHA Star} = \text{LHA Aries} + \text{SHA star}$$

3. Declination

The declination of the relevant star is read from the tri-daily page and needs no correction, it is a fixed value; stars on the celestial sphere moving as one. The axis of movement is vertical, through the North and South celestial poles and wholly accounted for by changing GHA. Remember that unlike stars, the Sun, Moon and planets move <u>across</u> the celestial sphere so both their GHA <u>and</u> Dec change with time.

Example 37. While crossing the Atlantic on the "ARC", a yacht takes an evening twilight sight of the star *Mirfak* at 1715hrs 45s local time Thursday November 17th 2011. The estimated position was latitude 20° 05.0′ North 25° 52.0′ West. The deck watch was 3 seconds fast. The recorded sextant altitude was 14° 17.6′, with an "off the arc" index error of 1.2′, taken at an eye height of 6.5m. Find the observed altitude, calculated altitude, intercept, and azimuth. Plot the position line.

The situation is summarised in the diagram below (**Figure 134**).

Remember, finding the calculated altitude of the star solves the **PZX triangle** where ZX is the zenith distance on the celestial sphere, from the observer's zenith to the sighted body.

$$ZX = 90° - \text{Altitude} \quad \text{therefore: Altitude} = 90° - ZX.$$

All the maths does is to solve the PZX triangle: side ZX is found from the two other known sides PZ and PX and the known angle between them (LHA).

$$PZ = 90° - \text{latitude}$$
$$\text{and } PX = 90° - \text{declination of the star.}$$

– *navigating with the Stars*

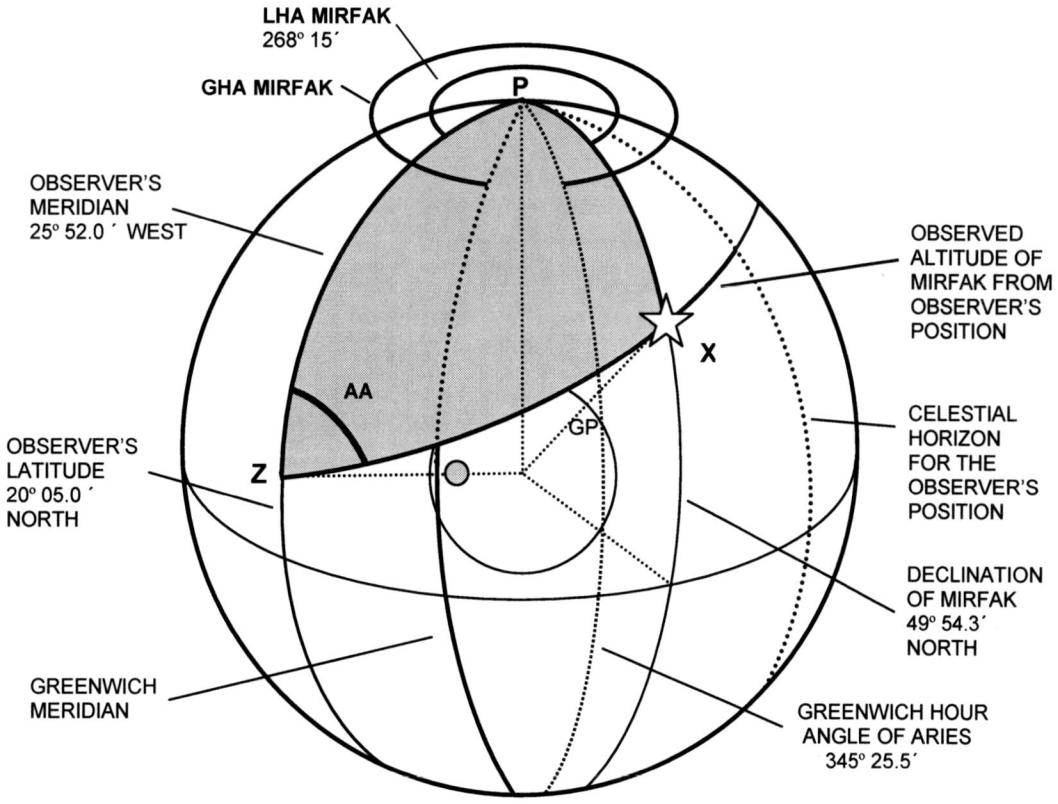

Figure 134. Visualising the solution to Example 37

In this example, for clarity, the same 1-8 stages will be used as those for the Sun, Moon and planet sight reductions in Sections 6, 7 and 8. On this occasion, the calculated altitude will be worked out using the modified cosine formula and azimuth formula, both of which can be easily handled on a standard calculator which has the trigonometric functions of sine, cosine and tangent.

1. **Chosen position**: Latitude 20° 05.0′ North   25° 52.0′ West

2. **Observed altitude** (H_o)

   $H_s \pm IE - Dip = H_a + $ main correction (just refraction for stars) $ = H_o$

   Sextant altitude (Hs) = 14° 17.6′

   Sextant Index error   1.2 ′ + (off the arc, so added on)

Astronavigation from Square One

14ᵐ						INCREMENTS AND CORRECTIONS							15ᵐ

m 14	SUN PLANETS	ARIES	MOON	v or d Corrⁿ	v or d Corrⁿ	v or d Corrⁿ	m 15	SUN PLANETS	ARIES	MOON	v or d Corrⁿ	v or d Corrⁿ	v or d Corrⁿ
s	° ′	° ′	° ′	′ ′	′ ′	′ ′	s	° ′	° ′	° ′	′ ′	′ ′	′ ′
00	3 30·0	3 30·6	3 20·4	0·0 0·0	6·0 1·5	12·0 2·9	00	3 45·0	3 45·6	3 34·8	0·0 0·0	6·0 1·6	12·0 3·1
01	3 30·3	3 30·8	3 20·7	0·1 0·0	6·1 1·5	12·1 2·9	01	3 45·3	3 45·9	3 35·0			3·1
02	3 30·5	3 31·1	3 20·9	0·2 0·0	6·2 1·5	12·2 2·9	02		3 45·5	3 46·1	3 35·2		3·2
03	3 30·8	3 31·3	3 21·1	0·3 0·1	6·3		03		3 46·4	3 35·5			3·2
04	3 31·0	3 31·6	3 21·4	0·4 0·1	6·4		04		3 46·6	3 35·7			3·2
05	3 31·3	3 31·8	3 21·6	0·5 0·1	6·5		05		3 46·9	3 35·9	0·5 0·1	6·5 1·7	12·5 3·3
06	3 31·5	3 32·1	3 21·9	0·6 0·1	6·6		06		3 47·1	3 36·2	0·6 0·2	6·6 1·7	12·6 3·3
07	3 31·8	3 32·3	3 22·1	0·7 0·2	6·7 1·6	12·7 3·1	07	3 46·8	3 47·4	3 36·4	0·7 0·2	6·7 1·7	12·7 3·3
08	3 32·0	3 32·6	3 22·3	0·8 0·2	6·8 1·6	12·8 3·1	08	3 47·0	3 47·6	3 36·7	0·8 0·2	6·8 1·8	12·8 3·3
09	3 32·3	3 32·8	3 22·6	0·9 0·2	6·9 1·7	12·9 3·1	09	3 47·3	3 47·9	3 36·9	0·9 0·2	6·9 1·8	12·9 3·4
10	3 32·5	3 33·1	3 22·8	1·0 0·2	7·0 1·7	13·0 3·1	10	3 47·5	3 48·1	3 37·1	1·0 0·3	7·0 1·8	13·0 3·4
11	3 32·8	3 33·3	3 23·1	1·1 0·3	7·1 1·7	13·1 3·2	11	3 47·8	3 48·4	3 37·4	1·1 0·3	7·1 1·8	13·1 3·4
12	3 33·0	3 33·6	3 23·3	1·2 0·3	7·2 1·7	13·2 3·2	12	3 48·0	3 48·6	3 37·6	1·2 0·3	7·2 1·9	13·2 3·4
13	3 33·3	3 33·8	3 23·5	1·3 0·3	7·3 1·8	13·3 3·2	13	3 48·3	3 48·9	3 37·9	1·3 0·3	7·3 1·9	13·3 3·4
14	3 33·5	3 34·1	3 23·8	1·4 0·3	7·4 1·8	13·4 3·2	14	3 48·5	3 49·1	3 38·1	1·4 0·4	7·4 1·9	13·4 3·5
15	3 33·8	3 34·3	3 24·0	1·5 0·4	7·5 1·8	13·5 3·3	15	3 48·8	3 49·4	3 38·3	1·5 0·4	7·5 1·9	13·5 3·5
16	3 34·0	3 34·6	3 24·			13·6 3·3	16	3 49·0	3 49·6	3 38·6	1·6 0·4	7·6 2·0	13·6 3·5
17	3 34·3	3 34·8	3 24·			13·7 3·3	17	3 49·3	3 49·9	3 38·8	1·7 0·4	7·7 2·0	13·7 3·5
18	3 34·5	3 35·1	3 24·			13·8 3·3	18	3 49·5	3 50·1	3 39·0	1·8 0·5	7·8 2·0	13·8 3·6
19	3 34·8	3 35·3	3 25·			13·9 3·4	19	3 49·8	3 50·4	3 39·3	1·9 0·5	7·9 2·0	13·9 3·6
20	3 35·0	3 35·6	3 25·			3·4	20	3 50·0	3 50·6	3 39·5	2·0 0·5	8·0 2·1	14·0 3·6
21	3 35·3	3 35·8	3 25·4	2·1 0·5	8·1 2·0	14·1 3·4	21	3 50·3	3 50·9	3 39·8	2·1 0·5	8·1 2·1	14·1 3·6
22	3 35·5	3 36·1	3 25·7	2·2 0·5	8·2 2·0	14·2 3·4	22	3 50·5	3 51·1	3 40·0	2·2 0·6	8·2 2·1	14·2 3·7
23	3 35·8	3 36·3	3 25·9	2·3 0·6	8·3 2·0	14·3 3·5	23	3 50·8	3 51·4	3 40·2	2·3 0·6	8·3 2·1	14·3 3·7
24	3 36·0	3 36·6	3 26·2	2·4 0·6	8·4 2·0	14·4 3·5	24	3 51·0	3 51·6	3 40·5	2·4 0·6	8·4 2·2	14·4 3·7
25	3 36·3	3 36·8	3 26·4	2·5 0·6	8·5 2·1	14·5 3·5	25	3 51·3	3 51·9	3 40·7	2·5 0·6	8·5 2·2	14·5 3·7
26	3 36·5	3 37·1	3 26·6	2·6 0·6	8·6 2·1	14·6 3·5	26	3 51·5	3 52·1	3 41·0	2·6 0·7	8·6 2·2	14·6 3·8
27	3 36·8	3 37·3	3 26·9	2·7 0·7	8·7 2·1	14·7 3·6	27	3 51·8	3 52·4	3 41·2	2·7 0·7	8·7 2·2	14·7 3·8
28	3 37·0	3 37·6	3 27·1	2·8 0·7	8·8 2·1	14·8 3·6	28	3 52·0	3 52·6	3 41·4	2·8 0·7	8·8 2·3	14·8 3·8
29	3 37·3	3 37·8	3 27·4	2·9 0·7	8·9 2·2	14·9 3·6	29	3 52·3	3 52·9	3 41·7	2·9 0·7	8·9 2·3	14·9 3·8
30	3 37·5	3 38·1	3 27·6	3·0 0·7	9·0 2·2	15·0 3·6	30	3 52·5	3 53·1	3 41·9	3·0 0·8	9·0 2·3	15·0 3·9
31	3 37·8	3 38·3	3 27·8	3·1 0·7	9·1 2·2	15·1 3·6	31	3 52·8	3 53·4	3 42·1	3·1 0·8	9·1 2·4	15·1 3·9
32	3 38·0	3 38·6	3 28·1	3·2 0·8	9·2 2·2	15·2 3·7	32	3 53·0	3 53·6	3 42·4	3·2 0·8	9·2 2·4	15·2 3·9
33	3 38·3	3 38·8	3 28·3	3·3 0·8	9·3 2·2	15·3 3·7	33	3 53·3	3 53·9	3 42·6	3·3 0·9	9·3 2·4	15·3 4·0
34	3 38·5	3 39·1	3 28·5	3·4 0·8	9·4 2·3	15·4 3·7	34	3 53·5	3 54·1	3 42·9	3·4 0·9	9·4 2·4	15·4 4·0
35	3 38·8	3 39·3	3 28·8	3·5			35	3 53·8	3 54·4	3 43·1	3·5 0·9	9·5 2·5	15·5 4·0
36	3 39·0	3 39·6	3 29·0	3·6			36	3 54·0	3 54·6	3 43·3	3·6 0·9	9·6 2·5	15·6 4·0
37	3 39·3	3 39·9	3 29·3	3·7			37	3 54·3	3 54·9	3 43·6	3·7 1·0	9·7 2·5	15·7 4·1
38	3 39·5	3 40·1	3 29·5	3·8			38	3 54·5	3 55·1	3 43·8	3·8 1·0	9·8 2·5	15·8 4·1
39	3 39·8	3 40·4	3 29·7	3·9			39	3 54·8	3 55·4	3 44·1	3·9 1·0	9·9 2·6	15·9 4·1
40	3 40·0	3 40·6	3 30·0	4·0			40	3 55·0	3 55·6	3 44·3	4·0 1·0	10·0 2·6	16·0 4·1
41	3 40·3	3 40·9	3 30·2	4·1			41		3 55·9	3 44·5	4·1 1·1	10·1 2·6	16·1 4·2
42	3 40·5	3 41·1	3 30·5	4·2 1·0	10·2 2·5	16·2 3·9	42	3 55·5	3 56·1	3 44·8	4·2 1·1	10·2 2·6	16·2 4·2
43	3 40·8	3 41·4	3 30·7	4·3 1·0	10·3 2·5	16·3 3·9	43	3 55·8	3 56·4	3 45·0	4·3 1·1	10·3 2·7	16·3 4·2
44	3 41·0	3 41·6	3 30·9	4·4 1·1	10·4 2·5	16·4 4·0	44	3 56·0	3 56·6	3 45·2	4·4 1·1	10·4 2·7	16·4 4·2
45	3 41·3	3 41·9	3 31·2	4·5 1·1	10·5 2·5	16·5 4·0	45	3 56·3	3 56·9	3 45·5	4·5 1·2	10·5 2·7	16·5 4·3
46	3 41·5	3 42·1	3 31·4	4·6 1·1	10·6 2·6	16·6 4·0	46	3 56·5	3 57·1	3 45·7	4·6 1·2	10·6 2·7	16·6 4·3
47	3 41·8	3 42·4	3 31·6	4·7 1·1	10·7 2·6	16·7 4·0	47	3 56·8	3 57·4	3 46·0	4·7 1·2	10·7 2·8	16·7 4·3
48	3 42·0	3 42·6	3 31·9	4·8 1·2	10·8 2·6	16·8 4·1	48	3 57·0	3 57·6	3 46·2	4·8 1·2	10·8 2·8	16·8 4·3
49	3 42·3	3 42·9	3 32·1	4·9 1·2	10·9 2·6	16·9 4·1	49	3 57·3	3 57·9	3 46·4	4·9 1·3	10·9 2·8	16·9 4·4
50	3 42·5	3 43·1	3 32·4	5·0 1·2	11·0 2·7	17·0 4·1	50	3 57·5	3 58·2	3 46·7	5·0 1·3	11·0 2·8	17·0 4·4
51	3 42·8	3 43·4	3 32·6	5·1 1·2	11·1 2·7	17·1 4·1	51	3 57·8	3 58·4	3 46·9	5·1 1·3	11·1 2·9	17·1 4·4
52	3 43·0	3 43·6	3 32·8	5·2 1·3	11·2 2·7	17·2 4·2	52	3 58·0	3 58·7	3 47·2	5·2 1·3	11·2 2·9	17·2 4·4
53	3 43·3	3 43·9	3 33·1	5·3 1·3	11·3 2·7	17·3 4·2	53	3 58·3	3 58·9	3 47·4	5·3 1·4	11·3 2·9	17·3 4·5
54	3 43·5	3 44·1	3 33·3	5·4 1·3	11·4 2·8	17·4 4·2	54	3 58·5	3 59·2	3 47·6	5·4 1·4	11·4 2·9	17·4 4·5
55	3 43·8	3 44·4	3 33·6	5·5 1·3	11·5 2·8	17·5 4·2	55	3 58·8	3 59·4	3 47·9	5·5 1·4	11·5 3·0	17·5 4·5
56	3 44·0	3 44·6	3 33·8	5·6 1·4	11·6 2·8	17·6 4·3	56	3 59·0	3 59·7	3 48·1	5·6 1·4	11·6 3·0	17·6 4·5
57	3 44·3	3 44·9	3 34·0	5·7 1·4	11·7 2·8	17·7 4·3	57	3 59·3	3 59·9	3 48·4	5·7 1·5	11·7 3·0	17·7 4·6
58	3 44·5	3 45·1	3 34·3	5·8 1·4	11·8 2·9	17·8 4·3	58	3 59·5	4 00·2	3 48·6	5·8 1·5	11·8 3·0	17·8 4·6
59	3 44·8	3 45·4	3 34·5	5·9 1·4	11·9 2·9	17·9 4·3	59	3 59·8	4 00·4	3 48·8	5·9 1·5	11·9 3·1	17·9 4·6
60	3 45·0	3 45·6	3 34·8	6·0 1·5	12·0 2·9	18·0 4·4	60	4 00·0	4 00·7	3 49·1	6·0 1·6	12·0 3·1	18·0 4·7

Table 31. NP 314-11 14 & 15 minute increment and correction tables

– navigating with the Stars

	14° 18.8′	
Dip correction 6.5m	4.5′	– (**Table 27** p.223 - always subtracted)
Apparent alt. (H_a)	14° 14.3′	
Refraction corrn.	3.8′	– (**Table 27** p.223 - subtracted)
Observed alt.(H_o) =	**14° 10.5′**	

3. **Greenwich Hour Angle** of Aries: daily page of almanac: Nov. 17th 2011

Local time of sight 1715 hrs 45s corrected to:	1715 hrs 42s
Time zone of longitude 25° 53.5′ West is GMT/UT	+2 00 (22.5° - 37.5° W)
Therefore the GMT of observation =	19 07 hrs 42s

GHA Aries at 1915hrs 42s:	19 hrs	341° 29.4′	(**Table 30**)
	15m 42s	3° 56.1′ +	(**Table 31**)
GHA Aries 1915hrs 42s		345° 25.5′	
Longitude of sight: subtract West		25° 52.0′ –	
Local Hour Angle of Aries		**319° 33.5′**	
SHA Mirfak Nov. 17th: daily page		308° 41.5′ +	(**Table 30**)
		628° 15.0′	
– 360 °		360° 00.0′ –	
Local Hour Angle Mirfak		**268° 15.0′**	

4. **Declination** of Mirfak: daily page of almanac:

Declination 49° 54.3′ (**Table 30**)

5. **Calculated altitude**: Given the observer's latitude and the Local Hour Angle and Declination of Mirfak, with the modified cosine formula:

sin Calc. Altitude = (*cos* Lat x *cos* Dec x *cos* LHA) ± (*sin* Lat x *sin* Dec)

Lat and Dec are added in the 2nd brackets if they are of the <u>same</u> name and subtracted if <u>different</u> names. In this case both latitude and declination are North.

Substituting: Lat	= 20° 05.0′ North =	20.0833°
Dec	= 49° 54.3′ North=	49.9050°
LHA	= 268° 15.0′ =	268.2500°

sin Calc. Alt. = (*cos* 20.0833 x *cos* 49.9050 x *cos* 268.25) + (*sin* 20.0833 x *sin* 49.9050)

Astronavigation from Square One

$$= (0.9392 \times 0.6441 \times -0.0305) + (0.3434 \times 0.7650)$$
$$= -0.0185 + 0.2627$$

sin Calc. Alt.	=	0.2442
Calc. Altitude	=	*arcsin* 0.0.2442
	=	14.1346° (0.1346° x 60′ = 8.1′)
Calculated altitude	=	**14° 08.1′**

6. Intercept: observed altitude (H_o) − calculated altitude (H_c) = intercept

$$H_o = 14°\ 10.5'$$
$$H_c = \underline{14°\ 08.1'} -$$
$$+2.4'$$

Intercept = + 2.4 ′

The intercept is positive as H_o is more than H_c and therefore lies <u>towards</u> the geographical position of Mirfak.

7. Azimuth: Using the equation for calculator:

sin Azimuth	=	(*sin* LHA x *cos* Dec) ÷ *cos* Calc. Alt.
sin Az	=	(*sin* 268.2500 x *cos* 49.9050) ÷ *cos* 14.1346
	=	(−0.9995 x 0.6441) ÷ 0.9697
	=	− 0.6438 ÷ 0.9697
sin Az	=	− 0.6639
Azimuth	=	*arcsin* −0.6639
Azimuth	=	41.5980° = **41° 35.9 ′** (0.5980 x 60 = 35.9′)

To convert the quadrantal azimuth to a true bearing requires a referral to the rules applying to the ABC Tables (**Section 4.8** p.101) for deriving the azimuth: where C = A ± B:
A is named opposite to latitude except when LHA value is between 90-270°. LHA is 319° so in this case the A is named South. B is named same as declination, North in this case.
C = A±B: same names +, opposite names −.   C = A−B in this case.
Using *Nories Nautical Tables*, for LHA 319° and Latitude of 20°, A = 0.42, LHA 319° Dec 50° B = 1.82. C takes the name of the larger which is the name of B: North. C then gives the same prefix to Azimuth. The suffix to the azimuth depends upon the LHA size: in this case it is East, as the LHA is between 180-360°.

– navigating with the Stars

The Azimuth is therefore: **North 41° 35.9′ East**

The true bearing of the **Azimuth** is therefore: N Az E = 000° + 41° 35.9′ = **41° 35.9′**

8. Plot the azimuth and intercept

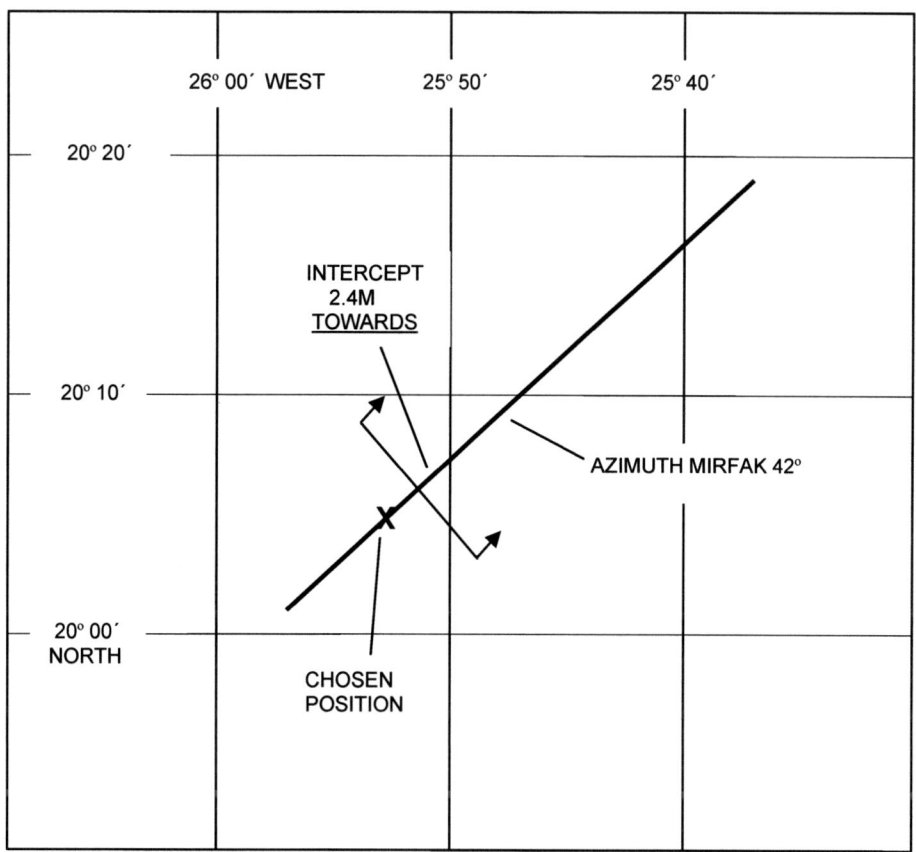

Figure 135. Plotting the intercept on the azimuth of Mirfak for Example 37

As with the Sun and Moon and planets, a Planet sight proforma could be used, which may improve both accuracy and consistency, but I emphasise again, it should NOT be a substitute for an understanding of the process of handling the sight. See **Figure 136** below.

Astronavigation from Square One

Star Sight proforma

Date						
					− 30′	
Estimated position	Latitude	° . ′	N / S	Longitude	° . ′	W / E
					+ 30′	

	Day	Hour	Min
Zone time			
Zone			
GMT			
GMT / UT			

GHA Aries Hours		° . ′
Min Sec		° . ′
GHA Aries:		° . ′
SHA Star +		° . ′
GHA Star		° . ′
Chosen longitude −W+E		° . ′
LHA Star		° . ′

	Day	Hour	Min	Sec
Watch time				
Watch error				
GMT / UT				

Declination star	° . ′ N / S

Chosen Lat	N / S
Declination	N / S
SAME / CONTRARY	
CHOSEN LAT	° N / S

Sextant Altitude (Hs)	° . ′
Index Error on − / off +	. ′
	° . ′
Height of Eye m dip −	. ′
Apparent Altitude (Ha)	° . ′
Main Correction +	. ′
Observed Altitude Ho	° . ′

↓

Ho	° . ′
Hc	° . ′
Intercept	. ′

Azimuth Z	°
Zn	°

Hc	° . ′
d +/−	. ′ +
Hc	° . ′

Chosen position N / S	° W / E ° . ′
Intercept	towards / away
Azimuth	° T

Figure 136. Example of a Star Sight Proforma

9.15 Rapid Sight Reduction Tables

Rapid sight Reduction Tables, volumes 1, 2 and 3 were originally developed for use by air navigators using bubble sextants, where positions change much more quickly than those of Earth-bound navigators. They became popular with sea-going navigators although rightly, professional mariners might take issue with their accuracy. They are stated to enable the derivation of a position with an accuracy of 0.5 nautical miles, where two or more position lines intersect.

Volume 1 is designed for use without reference to a nautical almanac, containing tables for the computation of GHA Aries, dip, refraction and time conversion for an epoch of 5 years. The current publication is designed for use from 2008-2013. The bulk of the volume contains tables for whole degrees of latitude from 89° North to 89° South, giving the calculated altitude to (1′) and azimuth (in whole degrees) of seven of a range of 41 first and second magnitude stars. Their positions are fixed for a mean equinox called J2010.0.

Corrections for precession and nutation though small (**Section 5.12** p.130) would need to be made to increase accuracy and a table for this is included, along with tables for computing latitude based on the altitude of *Polaris*.

The groups of seven selected stars remains the same for 15 consecutive entries of LHA Aries, at 1° intervals for latitudes below 69° and 2° of LHA Aries for those above 69° North and South. Three stars in each group are marked as being the most suitable of the group for a "three star fix" at that particular range of LHA Aries and latitude, in terms of their brightness, wide-spacing of their azimuths and altitude.

As data is tabulated in NP 303 vol.1 for 7 selected stars at a given latitude, then sight planning is an important task prior to the relatively short time available at twilight as outlined in para. **9.13**. Before the event, it is necessary to arrive at an estimated or chosen position such that the Greenwich and Local Hour Angle of Aries can be computed for the likely time of observation. From the seven stars listed, the best can be selected for a sextant sight, depending upon the observer's particular situation and the conditions present at the time. It is then possible to preset the sextant at the likely altitude just prior to the observation and to use a star finder such as the Weems and Plath 2102-D to narrow the arc of the horizon to be examined.

Example 38. A "3 star fix" observation is planned for dawn twilight of Wednesday November 16th 2011 at an assumed latitude of 36°N, on an approach to the island of Madeira. Find the time of dawn twilight and the likely range of LHA Aries. After consulting **NP 303 - vol.1**, sights were taken of the stars Kochab, Spica and Pollux, the observation times and sextant altitudes are tabulated below. The eye height was 4.6m and the sextant index error was 1.9′ "on the arc". Use the results to plot a fix. Use **Tables 30** p.234, **Table 18** p.139 and **Table 32** p.261. (Note that not all minute increment and correction tables are shown).

Astronavigation from Square One

1. **Dawn twilight and Aries range:**
   Proposed time of observation: twilight Wednesday Nov.16th 2011 (**Table 30** p.234):
   Using the daily page of NP 314 -11:
   Civil twilight at latitude North 35° = 0608-0635hrs on the Greenwich Meridian
   This time can be used as local time of twilight on all other meridians

Local Time of twilight at longitude 14° 15´ West	= 0608	- 0635
Zone Time 7.5° – 22.5° = +1	0100	0100 +
Time of twilight at longitude 14° 15´ West	= 0708	- 0735 GMT/UT

2. **GHA/LHA Aries for twilight:**

	07hrs	160° 00.7´	07hrs	160° 00.7´ (**Table 30**)
	08m	2° 00.3´	35m	8° 46.4´
GHA Aries 0708		162° 01.0´	0735	168° 47.1´
LHA Aries = GHA – long W		14° 15.0´ –		14° 15.0´ –
LHA Aries 0708		147° 46.0´	0735	154° 32.1´

   LHA Aries range during the proposed sight period is approximately 147-155° (**Table 32** p.261)

3. **Choice of stars**

   Using *Rapid Sight Reduction Tables NP 303/AP 3270* –epoch 2010.0-Vol.1: the relevant part-page from **NP 303** for latitude 36° North is shown in **Table 32** p.261. Latitude 36° is chosen as it is nearest whole number of latitude to the estimated latitude of 35° 45´.

   At Latitude 36° North the nearest whole number of latitude, the 7 best stars available for sights at LHA Aries 147-155° are in two groups: LHA 147-149:-
   Kochab   *ARCTURUS   SPICA   REGULUS   *SIRIUS   BETELGEUSE   *CAPELLA

   The second group of stars: LHA 150-155:-
   *Kochab   ARCTURUS   *SPICA   REGULUS   PROCYON   *POLLUX   CAPELLA

   Note that the timing of the LHA spans two 15° paragraphs of best stars and in this case, the stars indicated in each are not identical. Capitals indicate first magnitude (the brightest) stars; lower case indicates a second magnitude star. The asterisks indicate the best 3 of the 7 stars available at the chosen LHA Aries range to provide a good 3 star fix; where the relative azimuths are >100° apart..

   To save time and aid in identifying the stars, a Weems and Plath 2102-D star-finder (p.249) was used to help preset the approximate altitudes on the sextant. The 35° North graticule disc was fixed in place and the LHA Aries set with the blue arrow on the base disc scale. A hand bearing compass was used to align the LHA Aries. The sights were taken, corrected for deck watch error and are listed below. It was assumed that there was no significant vessel movement between the sights, although it is possible to allow for this in the calculations.

– *navigating with the Stars*

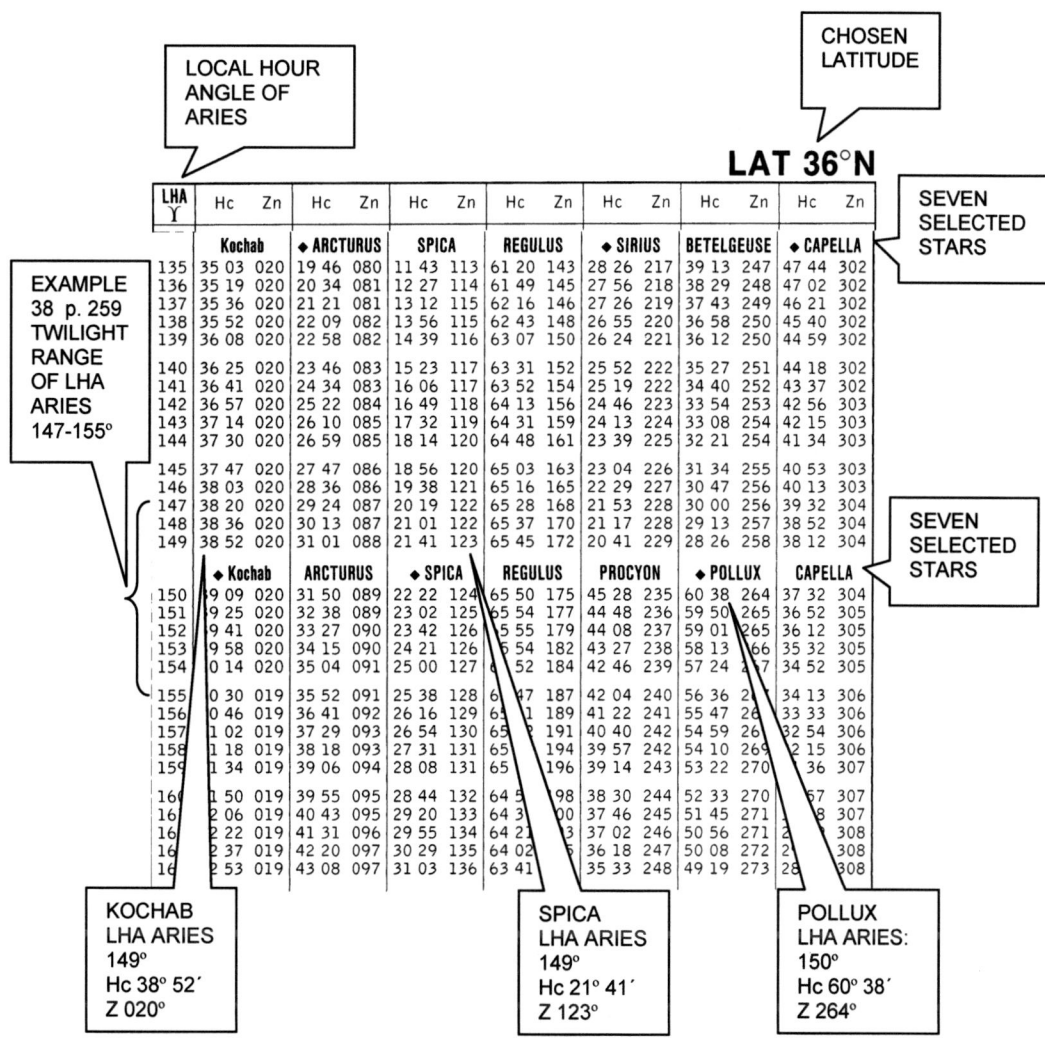

Table 32. Extract of NP 303 Rapid sight Reduction Tables : latitude 36° North

4. Star sights

The following sights were obtained:-

	Kochab	**Spica**	**Pollux**
Sextant altitude (H_s) =	38° 54.4′	21° 59.2′	60° 35.6′
Local time of sight:	0612 23s	0614 41s	0617 13s
Zone time +1	1 +	1 +	1 +
GMT/UT of Sight	0712 23s	0714 41s	0717 13s
GHA Aries 07hrs	160° 00.7′	160° 00.7′	160° 00.7′
	+12m 23s 3° 06.3′	+14m 41s 3° 40.9′	+17m 13s 4° 19.0′
GHA Aries	0712 23s **163° 07.0′**	0714 41s **163° 41.6′**	0717 13s **164° 19.7′**

It is then necessary to convert the GHA Aries to LHA Aries <u>and</u> to arrive at a whole number for the LHA Aries for each sight. This is done by adding for East longitudes, or subtracting for West longitudes, the longitude figure being chosen to be within 30′ of the estimated position, in this case it is 14° 15′ West.

The chosen longitude is tailored to suit each star sight, a slightly different assumed longitude figure is computed for each of the three sights which in turn will give a slightly different point from which each is eventually plotted. This is done to give a <u>whole number</u> value for each of the LHAs for Aries.

GHA Aries	163° 07.0′	163° 41.6′	164° 19.7′
minus longitude West	14° 07.0′ −	14° 41.6′ −	14° 19.7′ −
LHA Aries	**149° 00.0′**	**149° 00.0′**	**150° 00.0′**

This enables an accurate, whole number entry into the *Rapid Sight Reduction* NP 303 table.

Referring back to **Table 32** enables the calculated altitude and true azimuth to be extracted, at the latitude of 36° North; the nearest whole number of degrees to the estimated latitude of 35° 45.0′ North. The three chosen stars were: Kochab *SPICA and *POLLUX, though alternatives could have been used.

The extracted figures for the three stars are:

	Kochab	***SPICA**	***POLLUX**
Calc. altitude (H_c)	38° 52′	21° 41′	60° 38′
True Azimuth (Z_n)	020°	123°	264°

The calculated altitudes derived from the table are then compared with the corrected sextant observations in the usual way to arrive at 3 intercepts, one for each of the chosen stars. The intercepts are then plotted on the azimuths of the stars, each from their slightly different longitude, consequent upon the differing timing of each sight, which in turn had changed slightly the GHA/LHA Aries.

	Kochab	***SPICA**	***POLLUX**
Sextant altitude (H_s) =	38° 54.4′	21° 59.2′	60° 35.6′
Sextant Index error (1.9 on)	1.9′ −	1.9′ −	1.9′ −
	38° 52.5′	21° 57.3′	60° 33.7′
Dip correction 4.6m (−)	3.8′ −	3.8′ −	3.8′ −
Apparent alt.(H_a)	38° 48.7′	21° 53.5′	60° 29.9′
Main corrn.(subtract.)	1.2′ −	2.4′ −	0.5′ −
Observed alt.(H_o) =	38° 47.5′	21° 51.1′	60° 29.4′
Calculated Alt (H_c)	38° 52.0′ −	21° 41.0′ −	60° 38.0′ −
Intercept (H_o–H_c)	**− 4.5′**	**+10.1′**	**− 8.6′**

It simply remains to plot the 3 sights from their individual assumed positions along the 36° North parallel of latitude, by drawing in the azimuth derived from NP 303, marking the intercepts on each and then drawing in the position lines. As can be seen in **Figure 137**, the three position lines overlap to produce a position fix in the style of the "cocked hat" found when fixing position using multiple bearings taken over a short period of time.

The estimated position is revised from its pre-sight position of latitude 35° 45.0′ North and longitude 14° 15.0′ West to a new position fix within the cocked hat of position lines: latitude 36° 00′ North 14° 16′ West.

Astronavigation from Square One

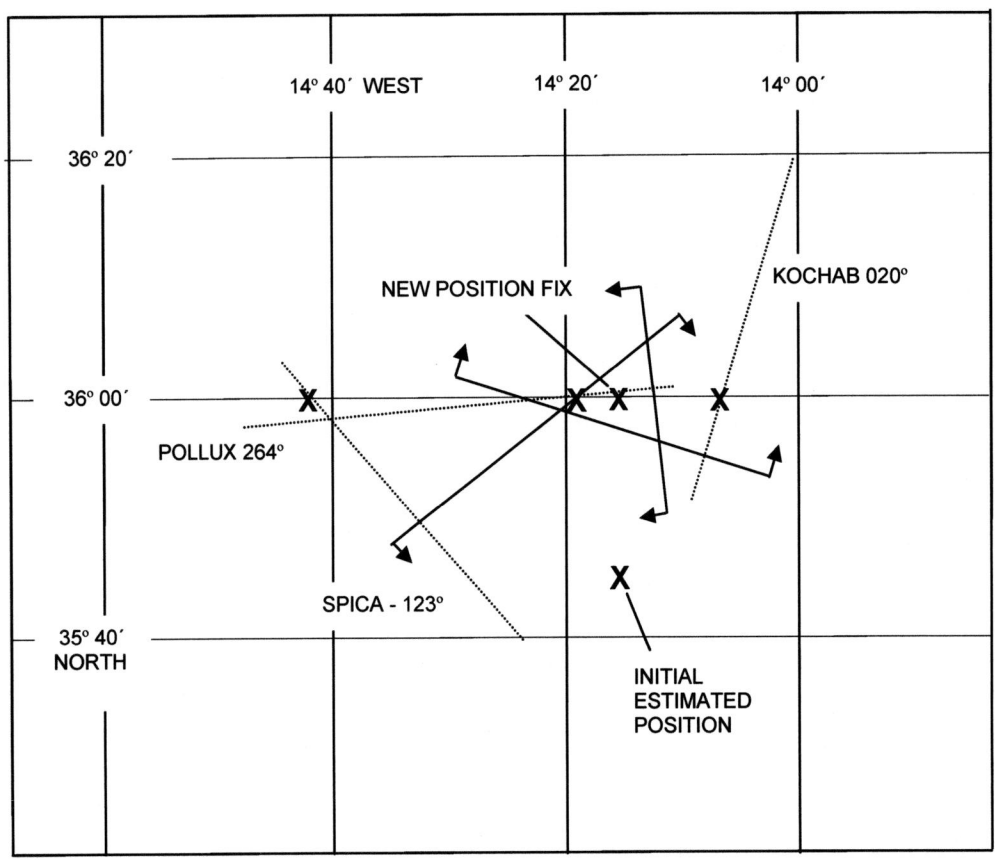

Figure 137. 3-Star fix using *Rapid Sight Reduction Tables* – NP 303-vol 1.

9.16 *Polaris – the North star* – why is it such a useful star?

Polaris is the marker of the North Pole of the celestial sphere and is visible from all latitudes in the northern hemisphere, only disappearing over the horizon as an observer reaches the equator. *Polaris* currently lies almost on the axis of the northern celestial pole, separated from it by a distance of ~0.7°.

In our static Earth and rotating celestial sphere model, *Polaris* has an apparent movement, describing a small and predictable circle some 1.4° in diameter around the pole every 24 hours. Twice in this period its position will be exactly North of an observer; once when it lies above and again when it lies just below the true North Celestial Pole as illustrated in **Figure 138**. At other times it will lie either slightly to the East or to the West of the true pole.

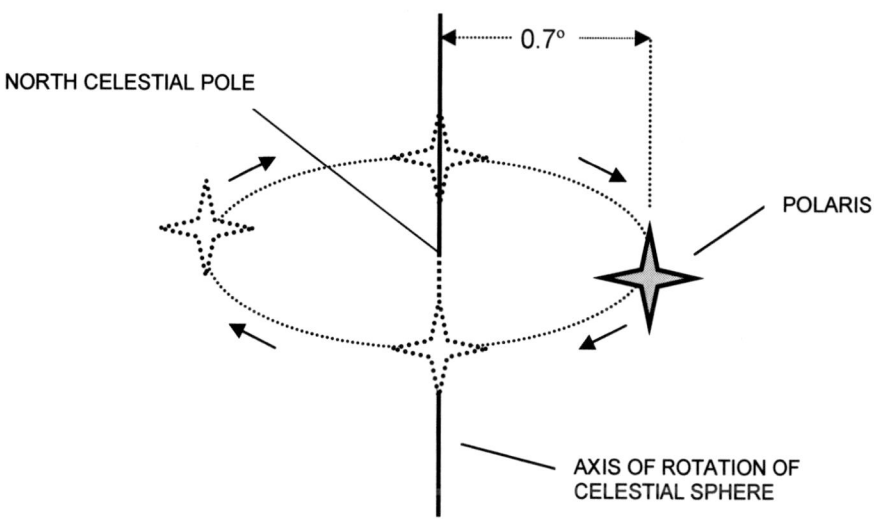

Figure 138. The orbit of Polaris around the celestial sphere's axis

Polaris lies at one end of the *Ursa Minor* (Little Bear) constellation. *Polaris*, with *Kochab* at the opposite end of the constellation are magnitude 2, the rest in the group are of magnitude 3. *Polaris* is by comparison only 48th on the list of star brightness, although an astronomical purist would tell you that *Polaris* is actually a "trinary" system of 3 stars that lie some 430 light years from Earth and it is their combined light that we see.

Polaris can be reliably found in the sky in a number of ways, all relating the specific shape of constellations of stars within which one or more stars point to the direction of *Polaris* and give an indication of its relative distance.

9.17 Finding *Polaris* in the sky - are there easy ways to locate it?

Every navigator experienced or not, MUST be able to instantly recognise the position of *Polaris*. There is nothing more reassuring on a black moonless night passage than being able to glance up and see *Polaris* to check your relative bearing.

From *Ursa Major*: The leading stars *Merak* and *Dubhe* in *Ursa Major* are called the two "pointers". Their useful position has been known for many thousands of years and reference to them appears in Assyrian history[9.10]. A line from *Merak*, through *Dubhe* and then continued for five times their distance apart leads to *Polaris*. All the stars in *Ursa Major* are magnitude 2 in brightness level and are usually readily identifiable against the background of lesser stars (**Figure 129** p.243).

From *Cassiopeia*: The standard alternative if *Ursa Major* is not visible is to use the constellation of *Cassiopeia* (*Andromeda's* mother). Its stars are not particularly bright and only *Schedar* is named (mag. 2.2), but its M-shaped configuration is easily discernible; the leading limb being more upright and its stars slightly brighter than those of its trailing limb. The base of the M lies in the direction of *Polaris* and the length of the base when doubled, gives the distance to *Polaris* when measured at a right angle to the base, from the trailing star.

From the Summer triangle and *Cygnus*: Identifying the summer triangle as outlined in **Section 9.7** (p.239) and illustrated in **Figure 128** (p.242), enables its trailing star *Deneb* to be identified. This is the easiest way of finding the constellation of *Cygnus* (the Swan, also called the "Northern Cross") of which *Deneb* is its most northern and prominent star at the apex of the cross. On the trailing arm of the cross lies *Gienah* (beware: there is a star of the same name in the southern hemisphere); a line from *Gienah*, through *Deneb* and with a length five times the distance between them, leads to *Polaris*.

From *Pegasus*: This constellation (*Pegasus*: the winged horse) shown in **Figure 139**, is prominent in northern skies in the autumn and forms a great square in the sky. Its two leading stars; *Scheat* and *Markab* are the North-pointers and again, five times the distance between them gives the distance to *Polaris*.

From *Auriga*: *Auriga* (the charioteer) is a pentagon formed by 5 prominent stars, the leader being *Capella* (the she-goat) leading her kids; three fainter stars. The trailing pair in Auriga; *Menkalinan* and *Theta-Aurigae* are the North-pointers and again, five times their distance apart gives the approximate distance to *Polaris* (**Figure 140**).

– *navigating with the Stars*

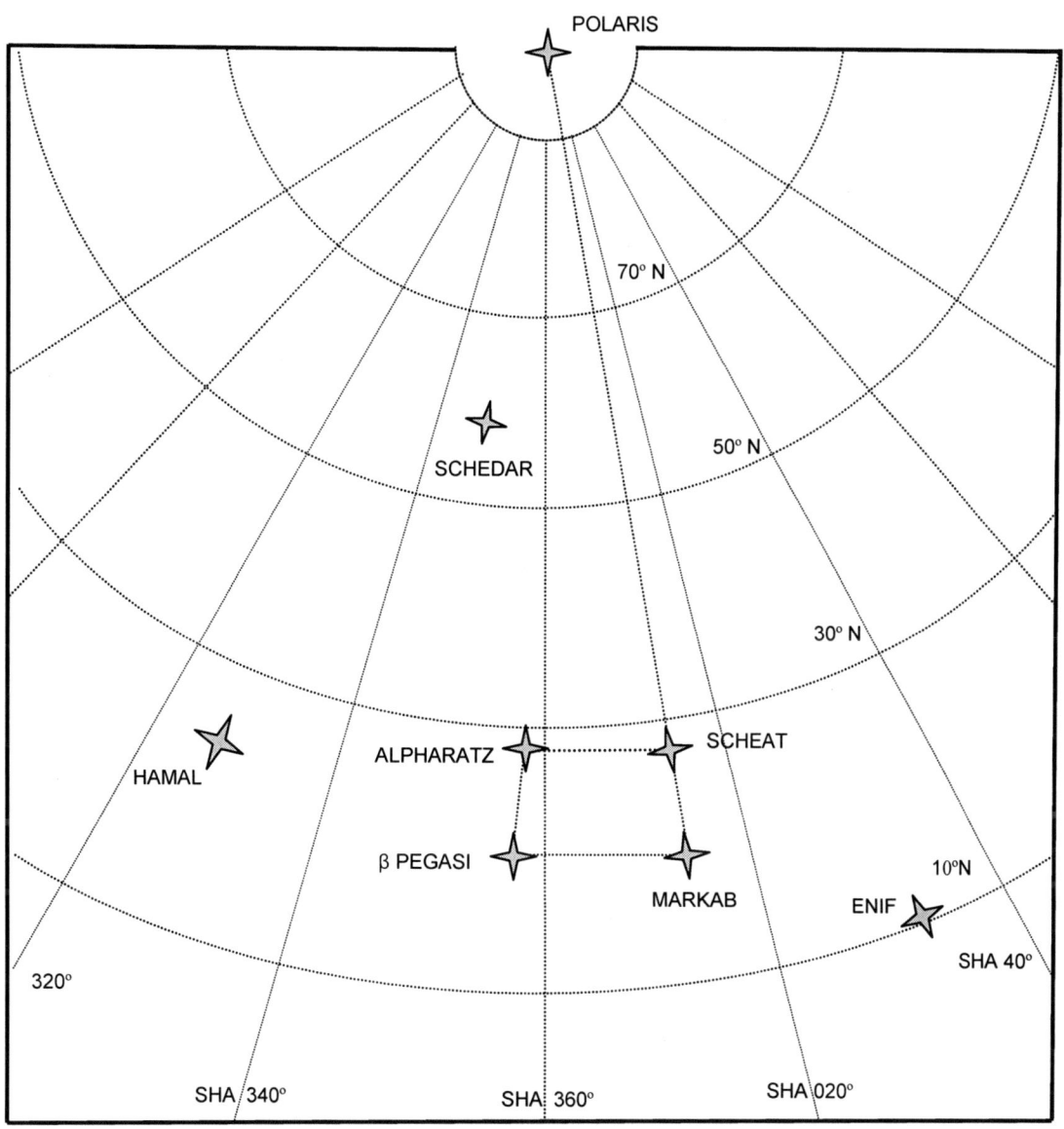

Figure 139. The square of *Pegasus* and its associated stars

Astronavigation from Square One

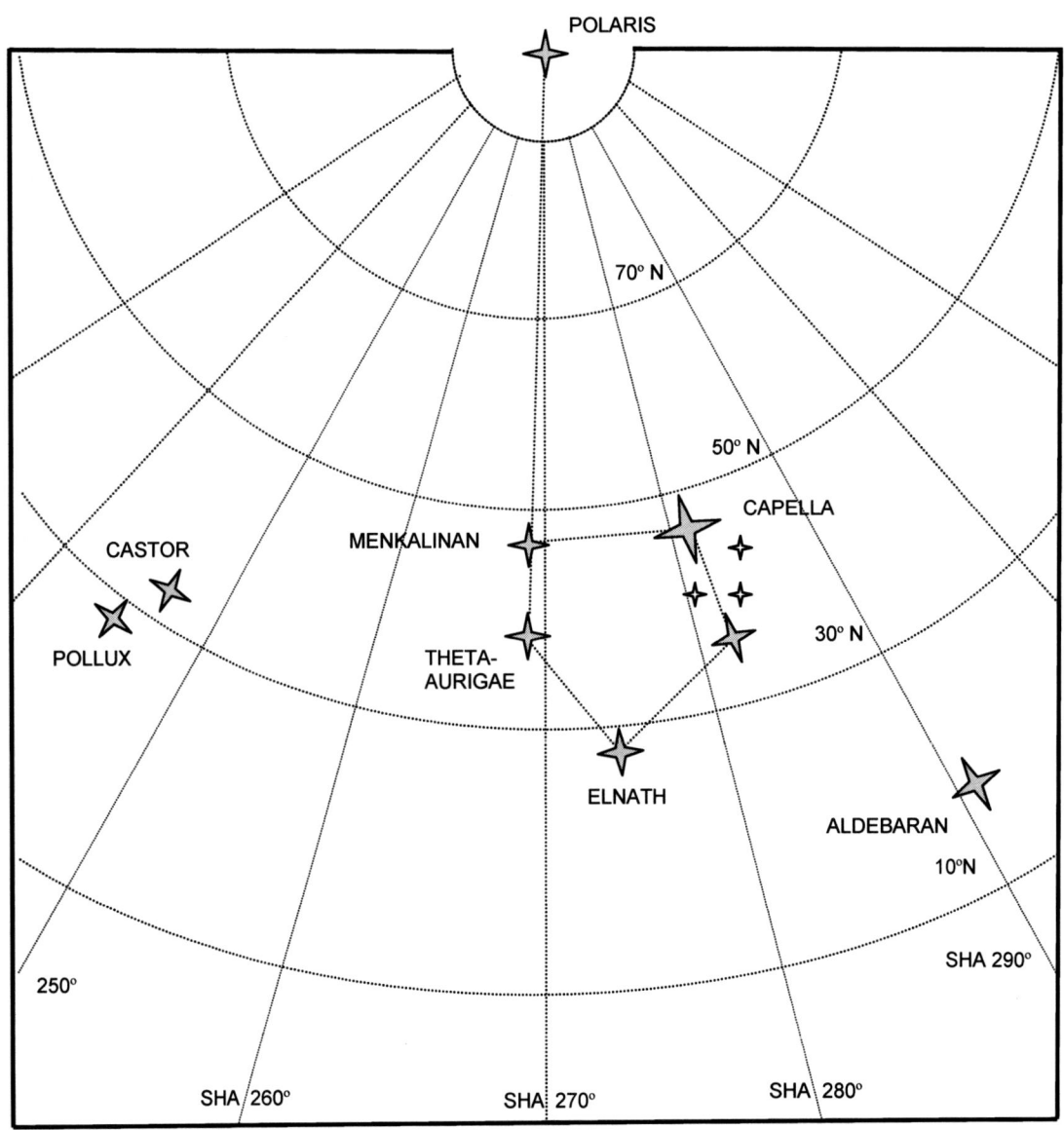

Figure 140. The pentagon of *Auriga* with *Capella* and her kids

9.18 *The special position of Polaris* –how is it used to find latitude?

Observers in the northern hemisphere are fortunate to have the North Celestial Pole marked (very nearly) by *Polaris*. Viewed by an observer at our Earthly North Pole, Polaris would be seen exactly above, at the observer's zenith. Viewed from the equator Polaris would appear at the northern horizon (**Figure 141**), remembering that the distance in the diagram between the Earth's surface and Polaris is hugely foreshortened.

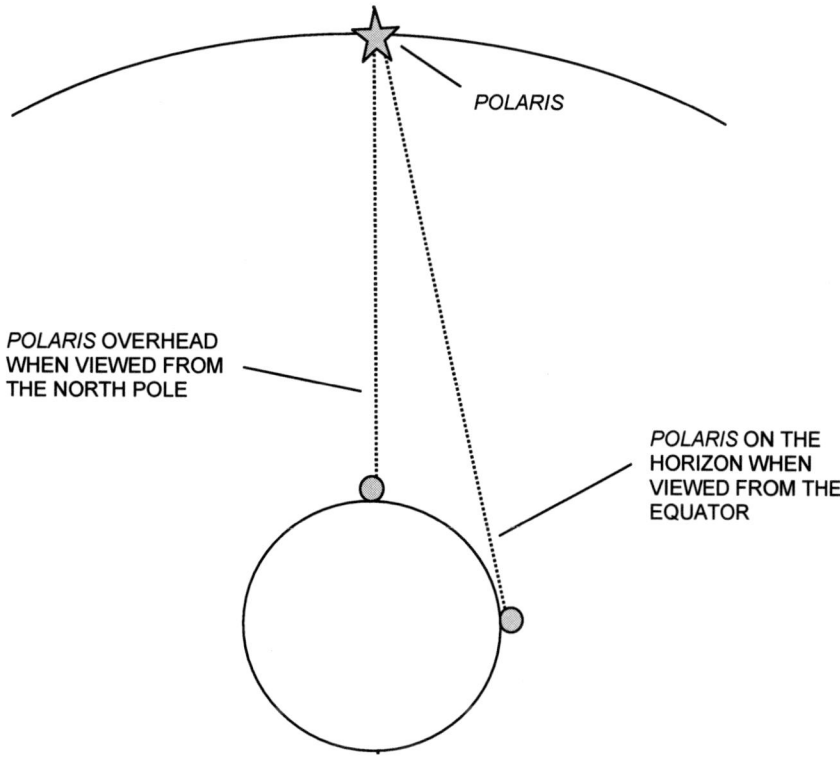

Figure 141. The varying altitude angles of *Polaris* from the Pole and equator

Between these two extremes, the height of Polaris is proportional to the observer's latitude e.g. at 45° N the altitude of *Polaris* will be 45° or so. This is the vital link that makes *Polaris* so useful in determining latitude; a relatively easy alternative to the meridian passage of the Sun (**Section 6.6** p.147), but of course, both Polaris and the horizon need to be visible. Remember at local noon the Sun lies at a particularly specific position for a brief moment; on the same meridian as the observer, exactly on the North-South line. This position is mimicked by Polaris; it also lies on (strictly speaking: almost on) the North-South axis and so can be useful in estimating latitude.

Astronavigation from Square One

9.19 *Polaris altitude equals latitude* – understanding the maths

It is relatively straightforward to prove mathematically that the altitude of *Polaris* equals the observer's latitude. **Figure 142** shows more detail of the angles involved. Consider the situation at the North Pole of the Earth with centre **C** and equator **EQ**: a sight of *Polaris* at position **P** would find the star to be directly above the observer's head, with an altitude A_1 of 90°. At this point on Earth, the observer's latitude must equal 90°. At position **X** a sight would indicate a lesser altitude (A_2). Because of the unimaginable distances involved, the lines of sight from the Earth's centre to Polaris and from the observer on the surface at **X** to Polaris are in effect, parallel.

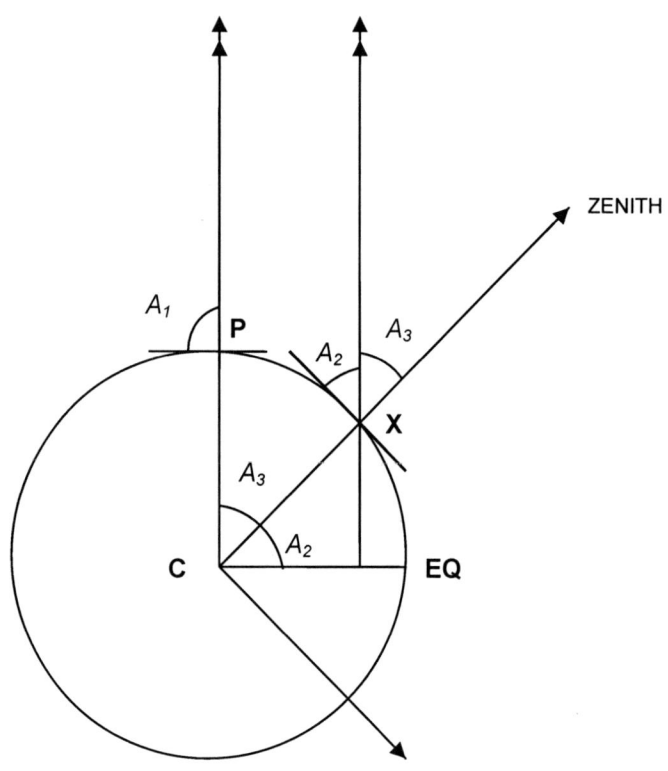

Figure 142. Proving that the altitude of *Polaris* equals the latitude of point of observation

The line from **C** to the observer's zenith lies at a right angle to the plane of the sensible horizon (**Section 2.1** p.37) at **X**. Angle A_3 must therefore $= 90° - A_2$. By following the simple rules governing the angles of right-angled triangles; then angle **PCX** must also equal angle A_3. As angle **PCEQ** is a right angle then angle **XCEQ** $= 90° - A_3 = A_2$. At the centre of the Earth, angle A_2 is the latitude of point **X**, thus proving that the altitude of Polaris measured at point **X** equals the latitude of point **X**.

9.20 *Adjusting for the less than central position of Polaris*

If *Polaris* was sited exactly at the pole then a simple corrected measurement of its altitude from any position in the northern hemisphere to give observed altitude of *Polaris*, would be equal to the observer's latitude. But, *Polaris* currently lays approximately 40′ of arc from the North Celestial Pole and is very slowly getting closer. Unlike the Sun at meridian passage, the local hour angle of Polaris is not always zero, because of its small rotation around the Pole.

The cause of this apparent motion is actually the process of precession, the slow helical motion of the axis of the Earth described in **Section 5.12** (p.130). It is not really *Polaris* that is moving, it merely appears so to an observer on Earth. Accordingly, to be useful to the navigator for an accurate assessment of latitude, an allowance must be made for this motion. Just like all other stars, the rotation by *Polaris* around the axis of the celestial sphere is represented by <u>Aries</u>.

The first step in attaining a latitude value from the altitude of Polaris is the calculation of the local hour angle of Aries at the observer's position, from the GHA Aries for the time of the sight and adjusting it for the longitude of the observer. The position of *Polaris* in relation to the North Celestial Pole varies through the year and from year to year, consistent with the Earth's precession that causes this apparent movement. The *Polaris* Table is therefore published as part of the Ephemeris tables of the astro-nautical almanac and not as a simple fixed correction table.

Using NP 314-11, the LHA Aries is calculated and then used to enter a table from which three values are extracted (**Table 33** p.272). The first value (a_o) is a function of LHA Aries, calculated on mean values of the SHA and Dec of Polaris at Lat. 50° and it determines the column used. The second value (a_1) is a function of LHA and the latitude of the observation and the third (a_2) is based on the month of the observation, correcting for the variation of *Polaris* from its mean position, due in reality to the effect of precession and nutation of the Earth on its axis. The three values are arranged mathematically so that:

Latitude = *Polaris* observed altitude − 1° + a_o + a_1 + a_2

The observed altitude (H_o) is the sextant altitude (H_s) corrected for index error and dip together with the main correction which for a star just corrects for refraction, as stars, being so far away need neither a parallax or semi-diameter correction. The *Polaris* table, at its foot, also gives an azimuth for the star, which increases very slightly with the observer's latitude, from 358°- 001.6°. An example of the process of reducing a Polaris sight follows (**Example 39**).

POLARIS (POLE STAR) TABLES, 2011
FOR DETERMINING LATITUDE FROM SEXTANT ALTITUDE AND FOR AZIMUTH

LHA ARIES	240°–249°	250°–259°	260°–269°	270°–279°	280°–289°	290°–299°	300°–309°	310°–319°	320°–329°	330°–339°
	a_0	a_0	a_0	a_0	a_0	a_0	a_0	a_0	a_0	a_0
°	° ′	° ′	° ′	° ′	° ′	° ′	° ′	° ′	° ′	° ′
0	1 37·8	1 35·0	1 31·1	1 26·2	1 20·5	1 14·2	1 07·3	1 00·2	0 53·1	0 46·1	0 39·5	0 33·4
1	37·6	34·7	30·7	25·7	19·9	13·5	06·6	0 59·5	58·8		38·8	32·9
2	37·3	34·3	30·2	25·2	19·3	12·8	05·9	58·8			38·2	32·3
3	37·1	33·9	29·7	24·6	18·7	12·1	05·2	58·1			37·6	31·8
4	36·8	33·6	29·3	24·0	18·1	11·5	04·5	57·3	50·2	43·4	37·0	31·2
5	1 36·6	1 33·2	1 28·8	1 23·5	1 17·4	1 10·8	1 03·8	0		2·7	0 36·4	0 30·7
6	36·3	32·8	28·3	22·9	16·8	10·1	03·1			2·0	35·8	30·2
7		32·4	27·8	22·3	16·1	09·4	02·4			1·4	35·2	29·7
8			27·3	21·7	15·5	08·7	01·6			0·7	34·6	29·2
9				21·1	14·8	08·0	00·9			0·1	34·0	28·7
10	1			1 20·5	1 14·2	1 07·3	1 00·2	0		9·5	0 33·4	0 28·2

Lat.	a_1	a_1	a_1	a_1	a_1	a_1	a_1	a_1	a_1	a_1	a_1	a_1
°	′	′	′	′	′	′	′	′	′	′	′	′
0	0·6	0·5	0·5	0·4	0·4	0·3	0·3	0·3	0·3	0·4	0·4	0·4
10	·6	·5	·5	·4	·4	·4	·4	·4	·4		·4	·5
20	·6	·5	·5	·5	·4	·4	·4	·4	·4		·5	·5
30	·6	·5	·6	·5	·5	·5	·5	·5	·4		·5	·5
40	0·6	0·6	0·6	0·5	0·5	0·5	0·5	0·5	0·5	0·5	0·5	0·6
45	·6	·6	·6	·6	·6	·6	·6	·6	·6	·6	·6	·6
50	·6	·6	·6	·6	·6	·6	·6	·6	·6	·6	·6	·6
55	·6	·6	·6	·6	·6	·7	·7				·6	·6
60	·6	·6	·7	·7	·7	·7	·7				·7	·7
62				0·7	0·7	0·8	0·8	0·			0·7	0·7
64				·7	·8	·8	·8	·8			·7	·7
66				·8	·8	·8	·9	·9			·8	·7
68				0·8	0·9	0·9	0·9	0·9			0·8	0·8

Month	a_2	a_2	a_2	a_2	a_2	a_2	a_2	a_2	a_2	a_2	a_2	a_2
	′	′	′	′	′	′	′	′	′	′	′	′
Jan.	0·4	0·4	0·4	0·5	0·5	0·5	0·5	0·6	0·6	0·6	0·7	0·7
Feb.				·3	·3	·4	·4	·4		·5	·6	·6
Mar.				·3	·3	·3	·3	·3			·4	·5
Apr.				0·3	0·3	0·2	0·2	0·2			0·3	0·3
May				·4	·3	·3	·3	·2	·2	·2	·2	·2
June	·7	·7	·6	·5	·5	·4	·4	·3	·3	·2	·2	·2
July	0·8	0·8	0·8	0·7	0·6	0·6	0·5	0·5	0·4	0·4	0·3	0·3
Aug.	·9	·9	·9	·8	·8	·7	·7	·6	·6	·5	·5	·4
Sept.	·9	·9	·9	·9	·9	·9	·8	·8				·6
Oct.	0·8	0·8	0·9	0·9	0·9	0·9	0·9					0·8
Nov.	·6	·7	·8	·8	·9	·9	1·0	1·0				1·0
Dec.	0·5	0·5	0·6	0·7	0·8	0·8	0·9	1·0				1·1

Lat.					AZIMUTH							
°	°	°	°	°	°	°	°	°	°	°	°	°
0	0·3	0·4	0·5	0·6	0·6	0·7	0·7	0·7	0·7	0·6	0·6	0·5
20	0·3	0·4	0·5	0·6	0·7	0·7	0·7	0·7				0·5
40	0·4	0·5	0·6	0·7	0·8	0·9	0·9	0·9				0·7
50	0·4	0·6	0·7	0·8	0·9	1·0	1·1					0·8
55	0·5	0·6	0·8	0·9	1·1	1·1	1·2	1·2				0·9
60	0·5	0·7	0·9	1·1	1·2	1·3	1·4	1·4	1·3	1·3	1·2	1·0
65	0·6	0·9	1·1	1·3	1·4	1·5	1·6	1·6	1·6	1·5	1·4	1·2

Latitude = Apparent altitude (corrected for refraction) $-1° + a_0 + a_1 + a_2$

The table is entered with LHA Aries to determine the column to be used: each column refers to a range of 10°. a_0 is taken, with mental interpolation, from the upper table with the units of LHA Aries in degrees as argument; a_1, a_2 are taken, without interpolation, from the second and third tables with arguments latitude and month respectively. a_0, a_1, a_2, are always positive. The final table gives the azimuth of *Polaris*.

Table 33. Part of NP 314-11 *Polaris* ephemeris table

– *navigating with the Stars*

Example 39. A yacht, returning from a transatlantic trip towards Falmouth was approaching the Scilly Isles without electronic navigation aids after an electrical storm. The captain was intent on getting a last latitude fix in the evening, not wanting to repeat the error of the homeward bound British Mediterranean fleet in 1707. Whilst under the direction of the distinguished Admiral Sir Clowdisley Shovell, the fleet ran aground on the Scilly Isles with great loss of life.

A sighting of *Polaris* was made during twilight at 1709hrs 56s local time on Tuesday November 15th 2011, at an estimated latitude of 49° 50′ North and longitude 6° 30′ West, at an eye height of 3m. The deck watch was 3s slow. The sextant altitude was 49° 52.2′ (index error of 2.1′ on the arc). Find the latitude from which the observation of *Polaris* was taken? (Use **Table 30** p.234, **Table 33** p.272, **Table 27** p.223. and **Table 18** p. 139).

1. **Observed Polaris altitude** (H_o)

$H_s \pm IE - Dip = H_a$ + main correction (just refraction for stars) = H_o

Sextant altitude = 49° 52.2′

Sextant Index error ___2.1′__ + (on the arc, so taken off)

49° 50.1′

Dip correction 3.0m ___3.0′__ – (**Table 27** - always subtracted)

Apparent alt.(H_a) 49° 47.1′

Main corrn. (refrac.) ___0.8′__ – (**Table 27** - subtracted)

Observed alt.(H_o) = **49° 46.3′**

2. **Chosen position**: Latitude 49° 50.0 ′ North 06° 30′ West

3. **Greenwich Hour Angle** of Aries: daily page of almanac: Nov. 15th 2011

Local time of sight 1709 hrs 56s corrected to:1709 59s (+3s).

Time zone of longitude is <u>same</u> as GMT/UT +0____ (7.5° East-7.5° West)

Therefore the GMT/UT of observation =   1709 59s GMT/UT

Finding the GHA of Aries, the star's representative, 15th Nov. 2010

GHA Aries:17 hrs	309° 26.2′
9m 59s	__2° 30.2′__ +
GHA Aries 1707hrs 59s	311° 56.4′
Longitude of sight:	__6° 30.2′__ – (subtract West)
Local Hour Angle of Aries=	**305° 26.2′**

273

Astronavigation from Square One

4. *Polaris* correction table

The LHA Aries is then used to enter the *Polaris* Table (**Table 33** p.272).

The figure a_0 for 305°	=	1° 03.8′
and that for 306°	=	1° 03.1′
Difference	=	0° 00.7 x 26.2/60 = 0.3
Therefore a_0 value at LHA Aries 305°	=	1° 03.8′
minus that for 26.2′	=	0.3′ – (decreasing)
a_0 for 305° 26.2′	=	1° 03.5′
a_1 value (lat. corrn. = 50°)	=	0.6′ +
a_2 value (month -Nov)	=	1.0′ +
***Polaris* total corrn($a_0 + a_1 + a_2$)**	=	**1° 05.1′**

Latitude = *Polaris* observed altitude – 1° + $a_0 + a_1 + a_2$

	=	49° 46.3′
		1° 00.0′ –
		48° 46.3′
		1° 05.1′ +
Latitude of observation	=	**49° 51.4′ North**

Azimuth of *Polaris* - read from the lower aspect of the *Polaris* table which has a small adjustment for the latitude from which the star is observed:

Azimuth *Polaris* = 1.1°

It is possible to use a proforma for a *Polaris* sight, subject to the usual caveats. An example follows in **Figure 143**.

– *navigating with the Stars*

Polaris Sight proforma

Date	

Estimated position	Latitude ° . ′	N/S	Longitude ° . ′	W/E

	Day	Hour	Min
Zone time			
Zone			
GMT			
GMT / UT			

GHA Aries Hours	° .
Min Sec	° .
GHA Aries	° .
Chosen longitude –W+E	° .
LHA Aries	° .

	Day	Hour	Min	Sec
Watch time				
Watch error				
GMT / UT				

Sextant Altitude (Hs)	° . ′
Index Error on – / off +	. ′
	° . ′
Height of Eye m dip –	. ′
Apparent Altitude (Ha)	° . ′
Main Correction +	. ′
Observed Altitude Ho	° . ′

Observed Altitude Ho	° . ′	
Minus 1°	1° 00.0′	–
	° . ′	
a(0) LHA Aries	° . ′	+
a(1) LHA / Lat	. ′	+
a(2) LHA / Month	. ′	+
Latitude	° . ′	

LHA Aries	° . ′
Latitude	° . ′
Azimuth	° . ′

Chosen position N / S   °   W / E   ° . ′
Azimuth   ° T
Latitude

Figure 143. an example of a *Polaris* sight proforma

9.21 Summarizing navigation using the stars

This section has shown how, by the recognition of pattern of stars and their movements in the sky, together with the help of a star-finder, the navigator can identify the significantly useful stars. Given that stars are only of navigational help in the half-light of dawn and dusk when the horizon is apparent, then the navigator needs to be able to find the major stars that might be expected to be found above in the heavens. This amounts to several constellations and 60 or so named stars. The key to success is as usual, by continued familiarization and practice.

As an example, the major constellations seen in northern skies were described together with their seasonal nature and relevance to the identification of Polaris. The navigational maths for stars is straightforward and is based on the knowledge of GHA Aries and Sidereal Hour Angle for the particular star. The use of *Rapid sight Reduction Tables – NP 303* was introduced for finding position using a three star fix.

The special peculiarities of the movements of *Polaris* were described and using the patterns of the constellations, ways of identifying its position were explored. Relative to an observer, its position is found by the calculation of the local hour angle of Aries for the moment of altitude observation together with small adjustments for the observer's latitude and the month of observation. A worked example was shown.

Appendix: Understanding spherical trigonometry

An understanding of the key equation used to solve the PZX triangle requires a knowledge of the mathematics used to describe planes, angles, perpendiculars, vectors, sines, cosines, tangents and radians. It is difficult to think in 3 dimensions, especially when angles, lines and curves are all involved at once. I frequently have to begin thinking about the problem from square one and cannot progress from one point to another without fully understanding what has gone before, but then I can make steady progress. I make no apology for my technique, for without it I am stumped.

For this section it is helpful to have access to a scientific calculator which has the trigonometric functions of sine, cosine and tangent and to make frequent use of a pencil and paper.

A.1 Relating angles to the sides of triangles – sines, cosines and tangents?

An understanding of spherical trigonometry (Greek: *sphaira* – a sphere, *trigonon* – a triangle and *metron* - a measure) begins with the mathematical description of a simple right-angled triangle. It has sides **a**, **b** and **c**, with the angle **A** between sides **b** and **c**, lying opposite to side **a**. Angle **C**, lies between sides **a** and **b**, opposite to side **c** and the third angle **B** is a right angle, lying opposite to side **b**, as in **Figure A1**.

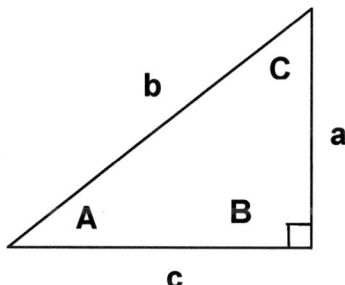

Figure A1. Describing a right-angled triangle

The size of the angle between sides **b** and **c**, angle **A** and that between sides **a** and **b**, angle **C**, are governed by the lengths of the sides of the triangle. As sides **a** and **b** lengthen, angle **A** increases in size and angle **C** decreases. Similarly, if sides **b** and **c** lengthen, then angle **A** decreases and angle **C** increases in size. The relationship between changing side length and angle size can be described mathematically. It is not a simple linear relationship, i.e. if one doubles then the other does not necessarily also double. Angle **A** can be expressed as a ratio of the lengths of the sides of the triangle. The ratio of **a/b**, the opposite side over the hypotenuse, is called the **sine** of angle **A** (**Figure A2**)

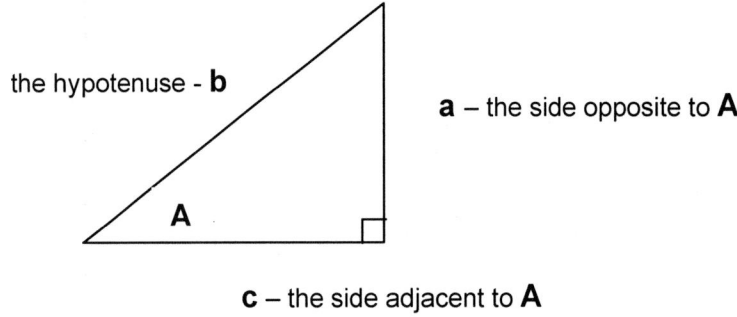

sine of angle **A** = ratio of sides **a** and **b**

sine **A** = **a/b**

Figure A2. Describing the sine of an angle

The relationship between Angle **A** and the ratio of side lengths can be displayed as a graph (**Figure A.3**). It can be seen that as the ratio of sides **a** and **b** (the **sine**) increases then so does the size of angle **A**. The shape of the graph is a curve indicating a nonlinear mathematical relationship. A linear relationship would be seen as a straight line.

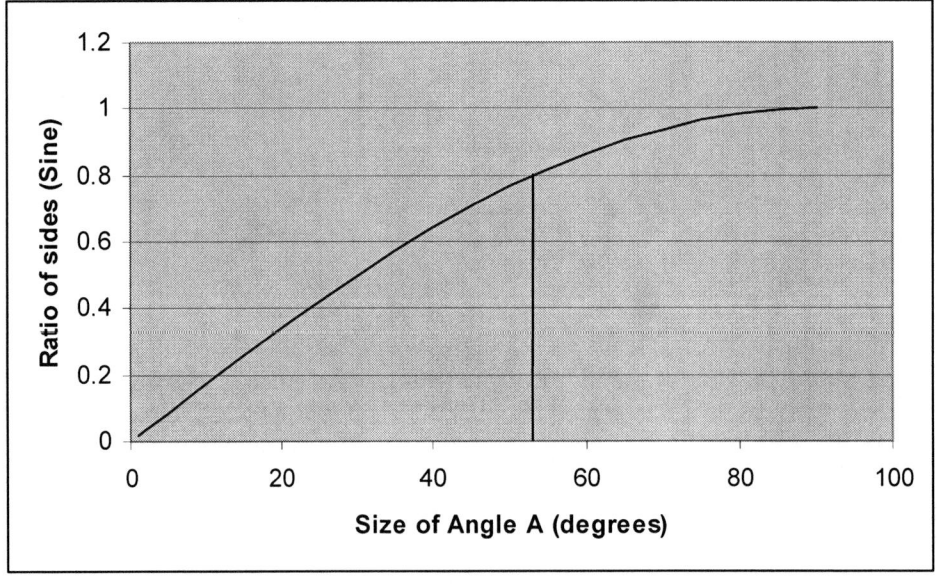

Fig A.3 The ratio of sides a/b compared to angle A: the Sine ratio

– appendix

Using the graph it is possible, given the sizes of sides **a** and **b**, to read off the size of angle **A** and vice versa e.g. an **a/b** sides ratio of 0.8 gives an Angle **A** size of 53 °. A table of sine values and their corresponding angle used to be a common site in schools and colleges, but these have all but disappeared since the advent of scientific calculators which have inbuilt programs for several "trigonometric" functions, one of which is the **sine** of the angle. They still appear in nautical publications such as *Norie's Nautical Tables*.

All that has been said regarding angle **A** and its relationship to sides **a** and **b** can equally be said of angle **C** and sides **c** and **b**: sine of angle **C** = ratio of sides **c** to **b**, so sine **C** = **c/b**.
Each angle of the triangle is related to the side opposite to it in the following way, known as the **sine rule**:

$$\frac{a}{\sin A} = \frac{b}{\sin B} = \frac{c}{\sin C}$$

Sin is an abbreviation of sine. Putting some figures into the equation will help to clarify and check the validity of the **sine rule**. The simplest of right-angled triangles is the 3x4x5 triangle in **Figure A.4**.

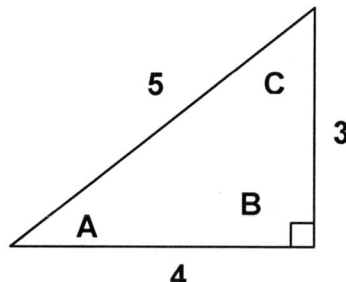

Figure A4 The simplest of right angled triangles

Sin **A** must equal 3/5, the equivalent of sides **a/b** = 0.6. *arcsin* 0.6 = 36.9°. So angle **A** is 36.9°. The abbreviation *"arcsin"*, meaning "arcsine" is used as to indicate the intent to convert from the sine ratio to its corresponding angle. On standard scientific calculators there is a combination of keys that will convert a sine ratio into its angular equivalent. It is usually the *"sin"* key pressed in combination with one other.
Similarly, Sin **C** equals 4/5, the equivalent of sides **c/b** = 0.8. *arcsin* 0.8 = 53.1° = angle **C**. These angles, together with the right angle must equal 180° (36.9° + 53.1° + 90° = 180°). Substituting these figures into the **sine rule**:

$$\frac{3}{\sin 36.9°} = \frac{4}{\sin 53.1°} = \frac{5}{\sin 90°}$$

$$\frac{3.0}{0.6} = \frac{4.0}{0.8} = \frac{5.0}{1.0}$$

5 = 5 = 5   -proving the validity of the sine rule.

A.2. The cosine ratio of the angles of a right-angled triangle – the other side?

Returning to our original triangle, the ratio of sides **c** and **b** can also be related to angle **A** and this ratio is called the **cosine** of angle **A**.

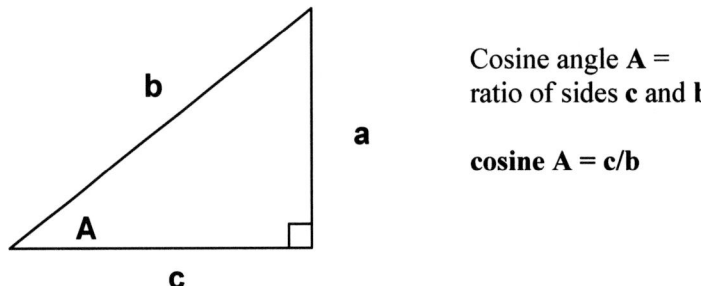

Cosine angle **A** = ratio of sides **c** and **b**

cosine A = c/b

Figure A5 The cosine ratio of angle A

The **cosine** relationship at Angle **A**, like the sine, can also be displayed in the form of a graph (**Figure A.6**). It can be seen that, in contrast to the sine graph in **Figure A3**, as the ratio **c/b** <u>increases</u> then the size of angle **A** <u>decreases</u>.

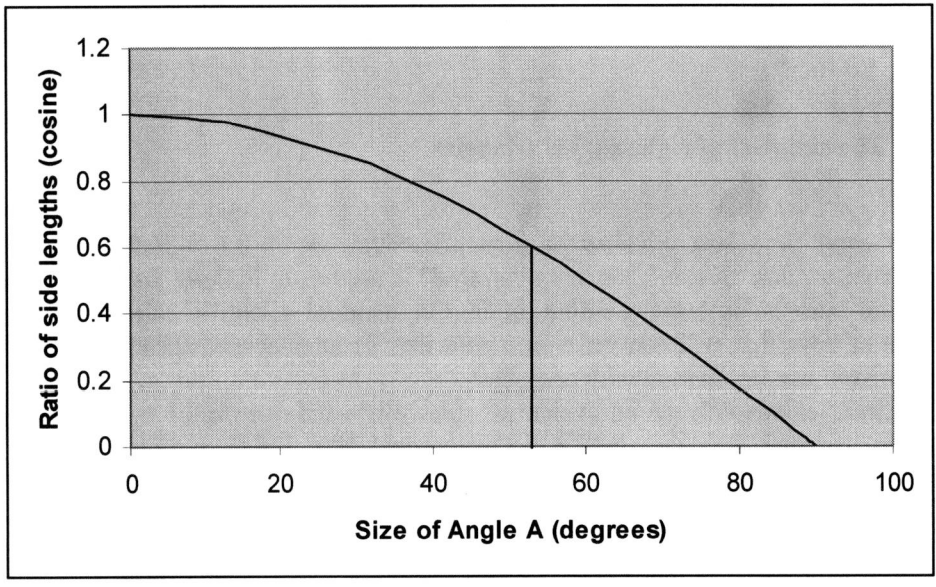

Figure A6. The ratio c/b compared to size of angle A – the Cosine ratio

— *appendix*

As with the sine ratio, by using the graph of cosine values it is possible, given the size of angle **A**, the ratio of sides **c** and **b** can be found and vice versa; if the length of sides **c** and **b** are known, then the angle can be found. Using the graph, a ratio of **c/b** of 0.6 corresponds to an angle of 53°. There is a general **cosine rule** linking sides and angles

$$a^2 = b^2 + c^2 - 2bc \cos A$$

There are two other variants, each based on one of the triangle's other sides and its corresponding angle:

$$b^2 = a^2 + c^2 - 2ac \cos B \text{ and}$$
$$c^2 = a^2 + b^2 - 2ab \cos C$$

These equations have similarities to the equation for the solving of the sides and angles of a spherical triangle, introduced in **Section 2.16** p.54.
If the side lengths and angles of the 3 x 4x 5 triangle are inserted into the first of the above equations the validity can be confirmed:

$$a^2 = b^2 + c^2 - 2bc \cos A$$

$3^2 = 5^2 + 4^2 - (2 \times 5 \times 4 \times 0.8)$ $(c/b = 4/5 = 0.8 = \cos A)$
$9 = 25 + 16 - 32$
$9 = 9$

Similar outcomes will be found by substituting into the corresponding equations.

A.3. The tangent ratio of the sides of a right angled triangle – just touching?

Returning again to our triangle, the ratio of sides **a** and **c** can also be related to angle **A**, this ratio is called the **tangent** of angle **A**.

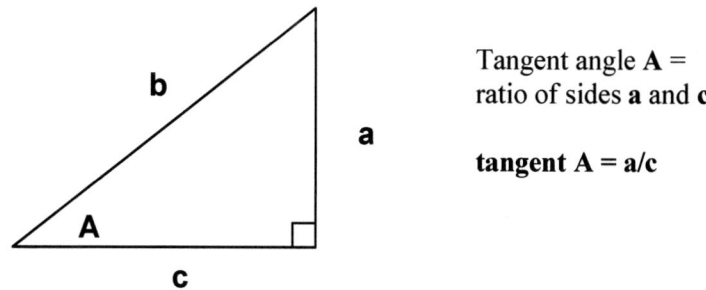

Tangent angle **A** = ratio of sides **a** and **c**

tangent A = a/c

Figure A7. Tangent ratio of angle A

281

The **tangent** relationship at Angle **A**, like the sine and cosine, can also be displayed in the form of a graph (**Figure A8**). It can be seen that the curve is a complex one and that the ratio exceeds 1 when Angle A exceeds 45° proceeding to infinity as the angle reaches 90°.

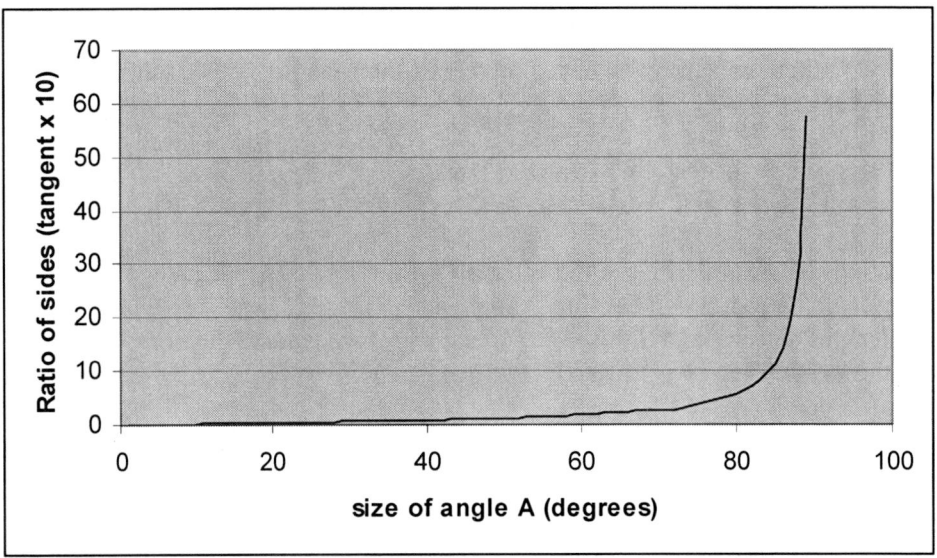

Figure A8. The ratio a/b to the size of angle A – the Tangent ratio

The word tangent derives from circular geometry, where a straight line just touching the circumference of a circle is called a tangent to that circle. If a triangle is drawn in the circle with the radius of the circle as the hypotenuse, the angle between the tangent line and the shortest side of the triangle, side **a**, also equals angle **A**.

In answer to the question at the head of **Section A1**; "What are sines, cosines and tangents?" We can now say that they are ratios of the lengths of the sides of triangles and are used to describe the size of their corresponding angles. The ratio produced by dividing the side opposite to the angle by the long side (the hypotenuse) is the sine of the angle. The ratio produced by dividing the side adjacent to the angle by the hypotenuse is the cosine of the angle and the ratio produced by dividing the opposite side by the adjacent side is the tangent of the angle. There is beautiful mathematical symmetry in this. Taking just two sides of a right-angled triangle, one angle varies as the sine of their ratio and the other angle varies as the cosine.

– *appendix*

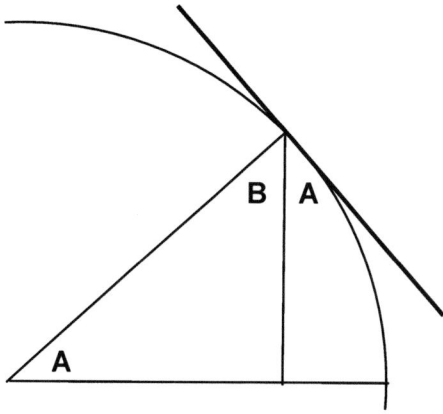

Figure A9. Showing the triangular relationship of the tangent to a circle

It can be shown that the sine of an angle divided by its cosine, gives the tangent of that angle:

$$\frac{\sin A}{\cos A} = \tan A$$

e.g. if angle A = 60° $\frac{\sin 60°}{\cos 60°} = \tan 60°$

$$\frac{0.866}{0.5} = 1.732$$

A.4 The mathematics of describing 3 dimensional space

A **vector**[App.1] is a quantity with both magnitude (meaning size) and a direction in space. A vector has points at which it starts and finishes that can be defined in 3 dimensions, by coordinates such as the latitude and longitude convention that is used on the Earth's surface.

In **Figure A.10** below, two vectors **a** and **b**, arising from the same point and at an angle **C** to one another. They can be represented by a single further element in the same plane. This element called the **dot product** of vectors **a** and **b**, can be written as (**a·b**). Its value is given by the product of **a** and **b** and the cosine of the angle between them.

$$\text{Dot product}(\mathbf{a·b}) = (a)(b) \cos C.$$

In short, the dot product allows one vector to be described in terms of the other. The dot product itself is not considered to be a true vector; it is a scalar, having only magnitude, but to the astro-navigator, it does little harm to look briefly upon it as a vector.

283

Astronavigation from Square One

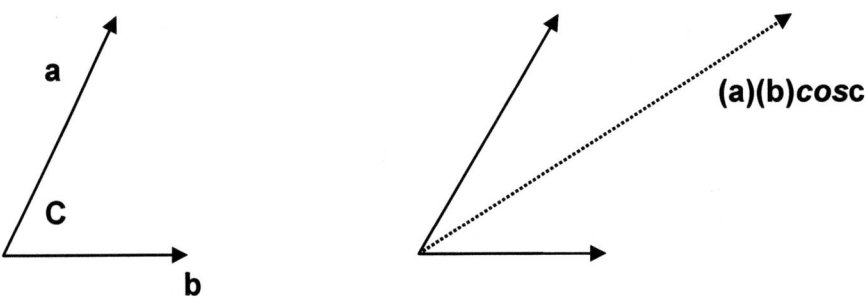

Figure A.10 Defining the dot product of a pair of vectors

For example if vector **a** = 5 and **b** = 4 and the angle between them is 60°, then they can be represented by a third "vector"; 5 x 4 x cos 60° = 20 x 0.5 = 10.

The dot product varies as the cosine of the angle and as cos 0° = 1 and cos 90° = 0, then the smaller the angle between the vectors, the larger is their representative dot product.

A.5 Defining the cross product of two vectors

Vectors **a** and **b** can also be represented by a further vector from their combined origin in a plane <u>perpendicular</u> to the plane in which **a** and **b** lie, as in **Figure A.11**

This vector is a true vector with both magnitude and direction, called the **cross product** and is represented by the mathematical expression (**a**x**b**). Its value is given by the product of **a** and **b** and the sine of the angle between them.

$$\text{cross product } (\mathbf{a}\times\mathbf{b}) = (a)(b)sinC$$

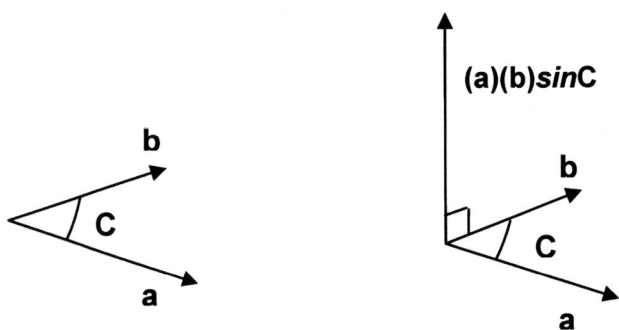

Figure A.11 Defining the cross product of a pair of vectors

– appendix

For example if vector **a** = 5 and **b** = 4 and the angle between them is 60°, then they can be represented by a third vector; 5 x 4 x sin 60° = 20 x 0.86 = 17.2, at right angles to the plane their origins, at their point of origin.

As this third vector varies as the sine of the angle and, as sine 0° = 0 and sine 90° = 1, the larger the angle between the vectors, the larger is their cross product.

Figure A.12 shows two vectors **a** and **b** with their dot product (**a·b**) and cross product (**axb**). The dot product is in the same plane as **a** and **b** and the cross product is perpendicular to the plane of **a** and **b**.

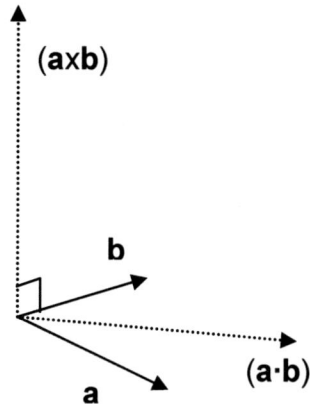

Figure A.12 The dot and cross products of vectors a and b

A.6 Describing 3 vectors in terms of dot and cross products

If three vectors from a point are drawn such as vectors **a**, **b** and **c** in **Figure A.13**, and curved lines; arcs, are drawn from the ends of vectors **a** and **c** to meet at the end of vector **b**, we arrive at a shape that begins to resemble the spherical geometrics at which this section is aimed.

The plane in which vectors **a** and **b** lie and the plane in which vectors **b** and **c** lie form an angle (**B**), along the whole axis of vector **b**, perhaps seen most easily at the vertex (the end) of vector **b**. The cross product of the vectors that demarcate each of the two planes: **a** and **b** (cross product **axb**), and vectors **b** and **c** (cross product **bxc**), perpendicular to their parent vectors are added in **Figure A.14**.

Astronavigation from Square One

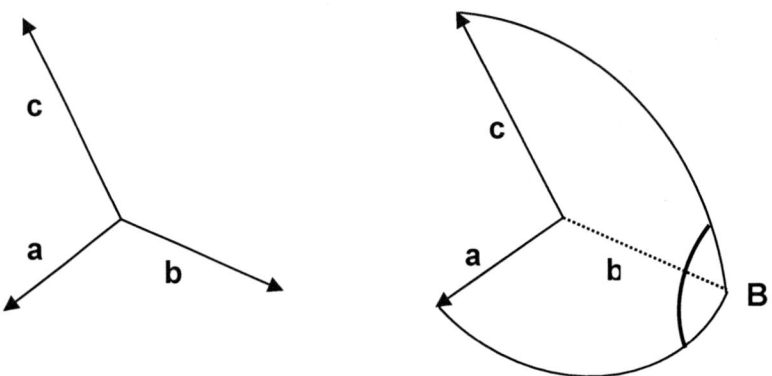

Figure A.13 The 3 dimensional view of vectors a, b and c

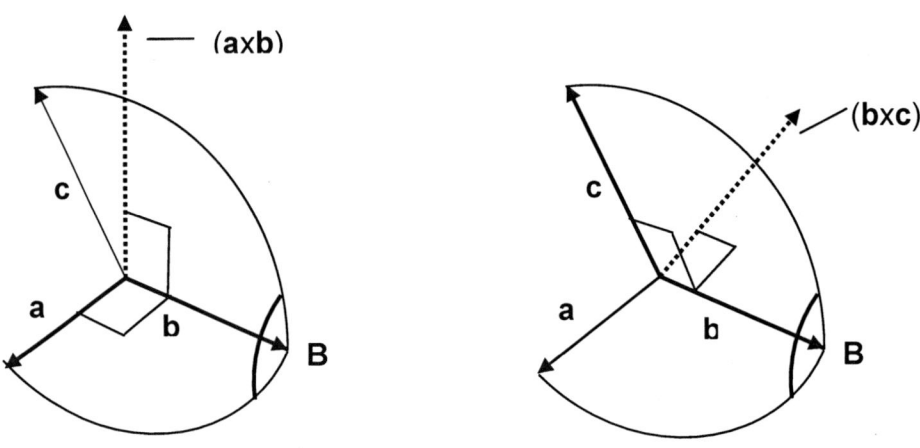

Figure A.14 Cross products of the vectors a and b, and b and c

The two planes generated by vectors **a** and **b**, and by vectors **b** and **c** are represented by their respective cross products, **axb** and **bxc**, but significantly they do not lie on their respective planes; they lie perpendicular to them; a fact difficult to illustrate on the page. It is however, possible to dissect out a 2-dimensional view of the planes and cross products to the planes, based at angle **B**, looking straight along vector **b** (**Figure A.15**).

– *appendix*

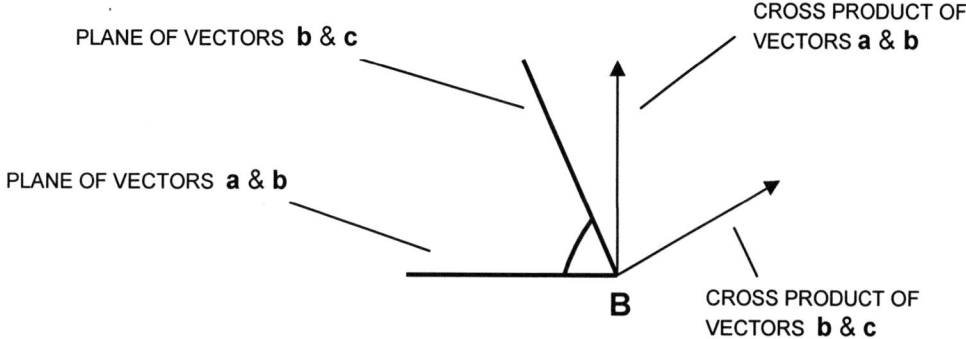

Figure A.15. A 2-dimensional view of vector planes and cross products

Looking closely at the angles dissected out from **Figure A.15**, it is possible to see that the angle between the cross products of the planes (labelled **B*** in **Figure A16**) is one and the same as that between those planes; it equals angle **B**. It must be so as each cross product arises at a right-angle from its respective plane, the planes are fixed at angle **B** and if angle **x** is added to either **B** or **B***, the resulting angle is a right angle.

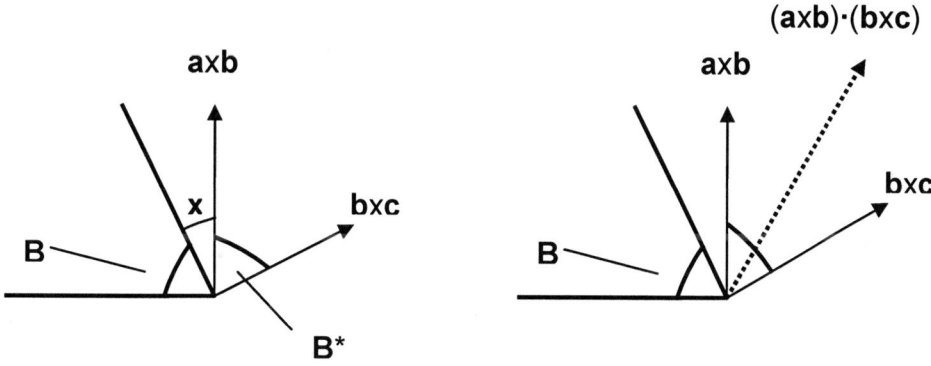

Figure A16. Forming the dot product of the cross products

Given that cross products **axb** and **bxc** are themselves vectors, they can be represented by a dot product formed from the product of vectors **axb** and **bxc** together with the cosine of the angle between them, angle **B**. The dot product is drawn in the right-hand figure of **Figure A.16**:

$$(\mathbf{axb})\cdot(\mathbf{bxc}) = (\mathbf{axb})\,(\mathbf{bxc})\,cos\mathbf{B}$$

287

Astronavigation from Square One

A.7 Describing the full spherical triangle

A spherical triangle such as that formed in **Figure A.17** has three primary vectors running from the centre of the sphere (**O**) to its surface, all of equal length, being that of the radius of the sphere[App 2].

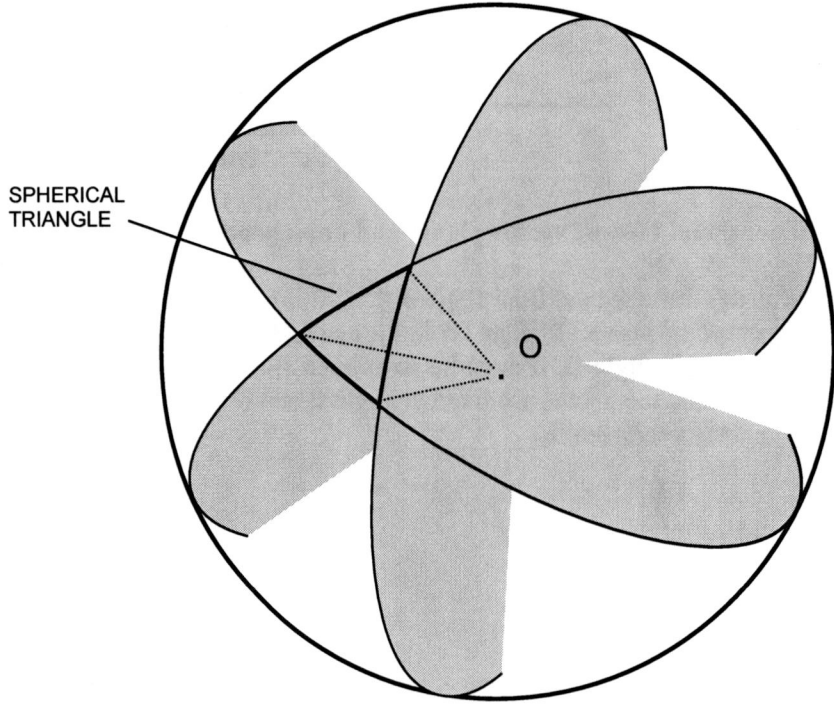

Figure A.17. A spherical triangle formed from 3 great circles

To further consider the triangle formed it is necessary to extract it from the sphere and to look at it face-on as in **Figure A18**. The vectors when taken in pairs describe 3 planes. Vectors **a**, **b** and **c** arise from the common central point. At this point, three angles are formed by the vector origins; **c** by the intersection of vectors **a** and **b**, **a** at the intersection of **b** and **c** and **b** at the intersection of **a** and **c**, as in **Figure A.18**.

– *appendix*

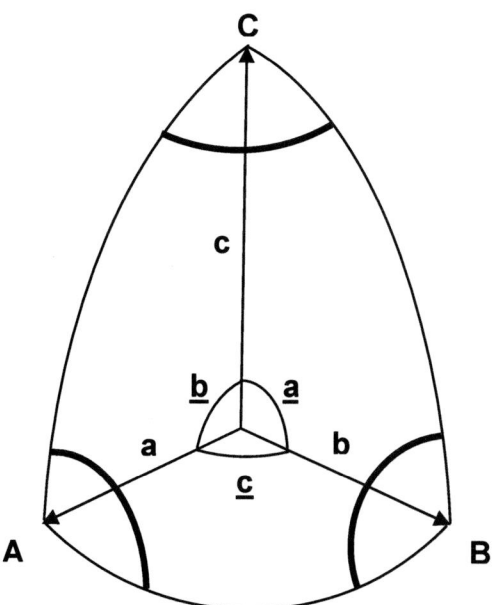

Figure A.18. Describing the spherical triangle

Pairs of planes intersect at each of vectors a, b and c, giving rise to angles A, B and C. The angles formed at O, at the vector origins can be measured in radians rather than degrees and so we must pause briefly to explain this unit of measure.

A.8 Describing the radian as a unit of angular measure – another complication?

The length of the circumference of a circle of diameter d is given by πd, where π (pi) is a constant, having a value of 3.142. Since diameter equals 2 x radius, then the circumference is equally well given by 2 x π x r, more usually written as $2\pi r$. For each and every circle the circumference is therefore 2π times its radius. The length of an arc of circumference equal to the length of the radius of any and every circle is called one radian[App3].

There are 2π radians or 2 x 3.142 = 6.284 radians in the whole circumference. This makes one radian equivalent to $360/2\pi \approx 57.3°$. For a sphere with a radius of 1, then the length of the surface arc equals the angle subtended at the centre, both being measured in radians, as in **Figure A.19**. This is vital to the understanding of the derivation of equations for solving the spherical triangle, as will be seen in the next paragraph.

Astronavigation from Square One

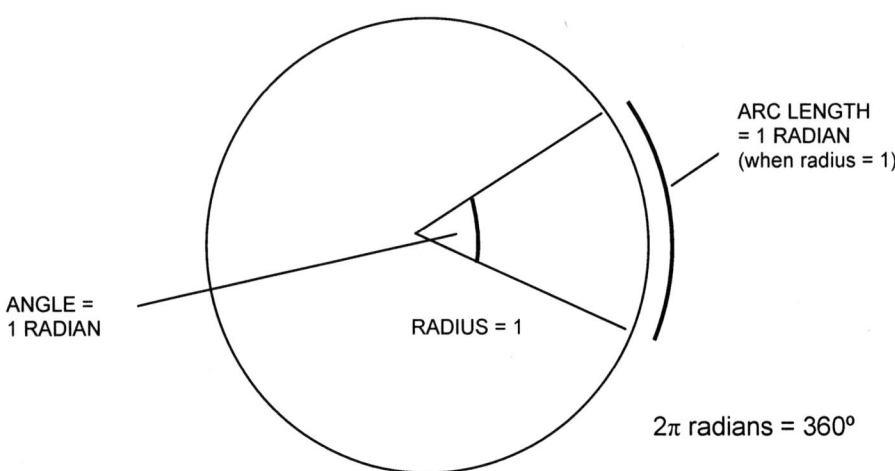

Figure A.19. Defining the radian as a unit of angular measure

For circles with radii greater than 1, then the length of the surface arc subtended by a specific angle in radians is given by that angle multiplied by the radius. For example converting degrees to radians as the angular measure, a central angle of 30° between two radii = 30/360 x 2π = 30/360 x 2 x 3.142 = 0.52 radians. If the sphere's radius is 100m then the surface arc length subtended by 0.52 radians = 0.52 x 100m = 52m.

A.9 Deriving the equations to solve the spherical triangle – the end in sights?

Returning to the description of our spherical triangle, if our central angles $\underline{a}$, $\underline{b}$ and $\underline{c}$ are defined in radians, then their respective surface arcs (**AB**, **BC** and **CA**), when measured in radians, will be $R\underline{a}$, $R\underline{b}$ and $R\underline{c}$, where **R** is the radius of the sphere. As described in **Section A4**, two vectors such as **u** and **v**, from the centre of a sphere to its surface, can be defined by their dot product: $\mathbf{u \cdot v} = [u][v] \cos \theta$, where θ is the angle between the vectors. The pairs of vectors **a** and **b**, **b** and **c** and **c** and **a**, can be described by their respective dot products, $\mathbf{a \cdot b}$, $\mathbf{b \cdot c}$ and $\mathbf{a \cdot c}$:

$$\mathbf{a \cdot b} = [a][b] \cos \underline{c} = [R][R] \cos \underline{c} = [R]^2 \cos \underline{c}$$

$$\mathbf{b \cdot c} = [b][c] \cos \underline{a} = [R][R] \cos \underline{a} = [R]^2 \cos \underline{a}$$

$$\mathbf{a \cdot c} = [a][c] \cos \underline{b} = [R][R] \cos \underline{b} = [R]^2 \cos \underline{b}$$

But if the radius of the sphere = 1, then **R** can be eliminated from the equations:

$$\mathbf{a \cdot b} = \cos \underline{c} \quad [1]$$
$$\mathbf{b \cdot c} = \cos \underline{a} \quad [2]$$

– appendix

$$\mathbf{a} \cdot \mathbf{c} = \cos \underline{b} \qquad [3]$$

The angles **A**, **B** and **C** of the spherical triangle can be calculated using the dot product of the normal (the perpendiculars) to the pair of planes from which they arise as described in **Section A.6**. The normals are given by the <u>cross products</u> of the vectors to the vertices as described in **Section A.5** and illustrated in **Figure A.20**.

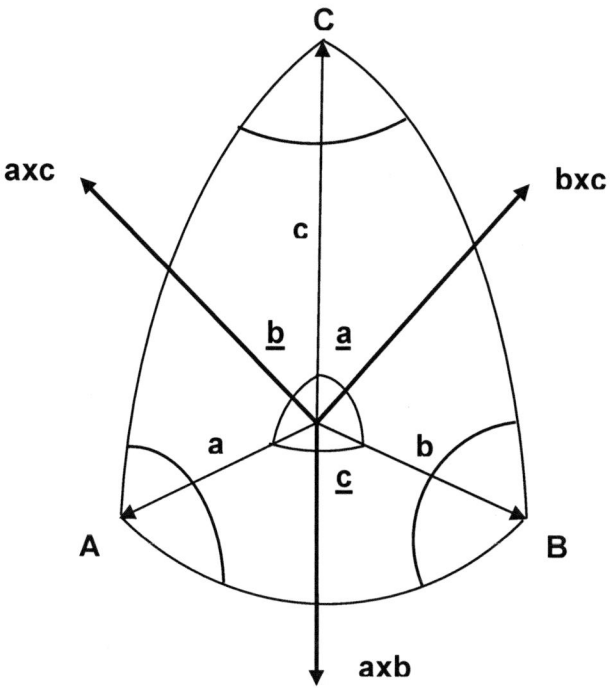

Figure A.20. Showing the cross products of the vectors to A, B and C

Following the general equation that a dot product of vectors **u** and **v** at an angle of $\theta = \mathbf{uv}\cos\theta$ and their cross product $= \mathbf{uv}\sin\theta$, then the cross products of the vertex angles **A**, **B** and **C** are:

For angle **A**: $(\mathbf{a} \times \mathbf{b}) \cdot (\mathbf{a} \times \mathbf{c}) = ([\mathbf{a}][\mathbf{b}]\sin \underline{c})\,([\mathbf{a}][\mathbf{c}]\sin \underline{b})\cos \mathbf{A}$

Likewise the angle **B**: $(\mathbf{a} \times \mathbf{b}) \cdot (\mathbf{b} \times \mathbf{c}) = ([\mathbf{a}][\mathbf{b}]\sin \underline{c})\,([\mathbf{b}][\mathbf{c}]\sin \underline{a})\cos \mathbf{B}$

and Angle **C**: $(\mathbf{a} \times \mathbf{c}) \cdot (\mathbf{b} \times \mathbf{c}) = ([\mathbf{a}][\mathbf{c}]\sin \underline{b})\,([\mathbf{b}][\mathbf{c}]\sin \underline{a})\cos \mathbf{C}$

But as the radius of the sphere is assumed to be 1 then: $\mathbf{a} = \mathbf{b} = \mathbf{c} = 1$, it follows that:

For angle **A**: $(\mathbf{a} \times \mathbf{b}) \cdot (\mathbf{a} \times \mathbf{c}) = \sin \underline{c}\,\sin \underline{b}\,\cos \mathbf{A}$ [4]

For angle **B**: $(a \times b)\cdot(b \times c) = \sin \underline{c} \, \sin \underline{a} \, \cos B$ [5]

For angle **C**: $(a \times c)\cdot(b \times c) = \sin \underline{b} \, \sin \underline{a} \, \cos C$ [6]

It is possible using mathematical means to transform the dot product of two cross products, using a derivative of Lagrange's formula, also called triple product expansion[App 4]. Using this theory, the dot product of two cross products $(a \times b)\cdot(a \times c)$, expands to:

$$a \cdot [b \times (a \times c)]$$

which itself expands to: $a \cdot [a(b\cdot c) - c(a\cdot b)]$

which then expands to: $a\cdot a(b\cdot c) - a\cdot c(a\cdot b)$

Given that in vector theory, a dot product of itself is 1, as the angle between a vector and itself is 0, and cosine $0° = 1$, then $a\cdot a = 1$.

It follows that: $(a \times b)\cdot(a \times c)$ transforms to $(b\cdot c) - (a\cdot c)(a\cdot b)$

As $b\cdot c$, $a\cdot c$ and $a\cdot b$ are dot products, this transformation expresses the dot product of two cross products in terms of dot products alone. Equations [1], [2] and [3] above, derived values for the dot products $a\cdot b$, $b\cdot c$ and $a\cdot c$ which were respectively: $\cos \underline{c}$, $\cos \underline{a}$ and $\cos \underline{b}$.

Therefore as: $(a \times b)\cdot(a \times c) = (b\cdot c) - (a\cdot c)(a\cdot b)$

then: $(a \times b)\cdot(a \times c) = \cos \underline{a} - (\cos \underline{b})(\cos \underline{c})$

But having also found that $(a \times b)\cdot(a \times c) = \sin \underline{c} \, \sin \underline{b} \, \cos A$ [4]

then: $\cos \underline{a} - \cos \underline{b} \cos \underline{c} = \sin \underline{c} \, \sin \underline{b} \, \cos A$

$\cos \underline{a} = \cos \underline{b} \cos \underline{c} + \sin \underline{c} \, \sin \underline{b} \, \cos A$ [7]

Then similarly, following from equations [5] and [6] above:

$\cos \underline{b} = \cos \underline{a} \cos \underline{c} + \sin \underline{a} \sin \underline{c} \cos B$ [8]

and $\cos \underline{c} = \cos \underline{a} \cos \underline{b} + \sin \underline{a} \sin \underline{b} \cos C$ [9]

These three equations [7], [8] and [9] solve the spherical triangle, to give the size of the three angles at the vertices and the lengths of the three sides.

At this point, the reader can return to **Section 2.16** (p.54) where the spherical triangle is solved in an example.

– *appendix*

A10. Transforming the cosine formula to an easily usable form for calculator

From **Section 2.6** p.57-8:

The basic cosine formula for an unknown side (**a**) of the PZX triangle, given the other two sides (**b** and **c**) and their contained angle (**A**) is:

$$\cos \underline{a} = \sin \underline{c} \sin \underline{b} \cos A + \cos \underline{b} \cos \underline{c}$$

a = 90° – altitude, **b** = 90° – latitude, **c** = 90° – declination, **A** = local hour angle

Substituting these values into the equation produces the following:

$\cos 90°$ – altitude = $[\sin (90° - \text{lat}) \sin(90° - \text{dec}) \cos \text{LHA}] + [\cos (90° - \text{lat}) \cos (90° - \text{dec})]$

But, as $\cos 90° - x° = \sin x°$ and $\sin 90° - x° = \cos x°$, substituting into the above gives:

$$\sin \text{altitude} = [\cos \text{lat} \times \cos \text{dec} \times \cos \text{LHA}] + [\sin \text{lat} \times \sin \text{dec}]$$

This equation is easily handled on a calculator with trigonometric functions.

Note that the equation is adapted to allow for the conditions where latitude and declination are in opposite hemispheres.

$$\sin \text{altitude} = [\cos \text{lat} \times \cos \text{dec} \times \cos \text{LHA}] - [\sin \text{lat} \times \sin \text{dec}]$$

A11. *The use of plotting sheets and defining Departure*

When navigating relatively close to land, where large scale charts are available, it is usual to plot position lines to produce a fix directly on to those charts. When crossing oceans, navigating away from land masses, large scale charts are not published and only small scale charts are available.

The small scale of ocean charts does not permit the accurate plotting of astro-navigational azimuths and position lines. It is usual in these circumstances to use plotting sheets whose latitude and longitude scales can be decided on and marked in by the navigator.

Remember that standard charts are Mercator projections, where the longitude scale changes with latitude; distances between meridians of longitude becoming progressively smaller as latitude increases. The distance between adjacent meridians of longitude at a given latitude is called the **departure**. The ratio of longitude to latitude changes as the cosine of latitude e.g. at latitude 0°, the equator, as cosine 0° = 1, then the ratio of longitude to latitude is 1. At the equator then, both 1° of latitude and longitude are equal to 60nm. At the Pole, where latitude is 90° and cosine 90° = 0, then 1° of longitude is 0nm; not very surprisingly. At 51° North, the latitude of London, 1° of longitude = 60 x cosine 51° = 60 x 0.6293 = 37.8nm.

This cosine relationship between latitude and longitude, the **departure**, is the basis of the scale seen in the lower right corner of commercially available plotting sheets. A plotting sheet comprises a sheet of paper with a central vertical longitude meridian scale marked off by horizontal lines across the sheet at latitude intervals of 60 minutes as in **Figure A.21**. A compass rose is usually also present together with a longitude scale, the latter based on the cosine ratio above.

The navigator chooses the latitude nearest his likely position and marks this in on the central horizontal line. Dividers are then used on the longitude scale at the point corresponding to the chosen latitude, to measure the width of 1° of longitude, which is then marked of each side of the central longitude meridian. This essentially produces a mini-Mercator projection. **Figure A.21** shows the chosen latitude of 50°N with the departure for both 1° West and 1° East drawn in. Graph paper can be used instead of a formal plotting sheet, using a chosen ratio of horizontal to vertical squares equivalent to the cosine ratio of the chosen latitude.

– *appendix*

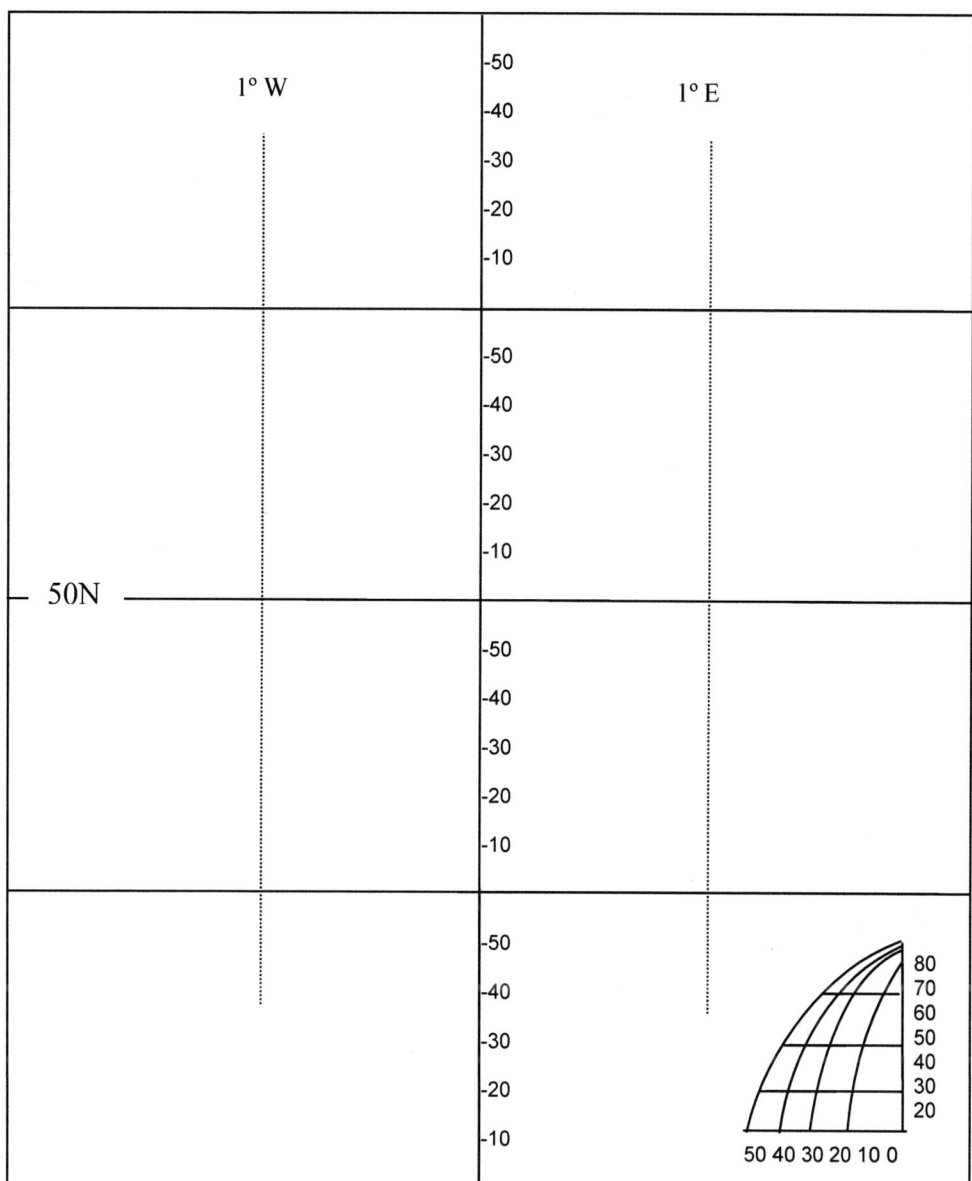

Figure A.21. The features of a plotting sheet with departure at latitude 50°N 0°W

Bibliography: Astro-navigation texts, naval and general reference books

Celestial Navigation; Tom Cunliffe 2000. ISBN 978-1-89866075-0. Published by Fernhurst Books, Arundel UK

Ocean Yachtmaster- Coursebook for Ocean Navigation students-3rd edition Pat Langley-Price and Philip Ouvry 2007. ISBN 978-0-7136-8265-6. Published by Adlard Coles Nautical, London

RYA Astro Navigation Handbook; Tim Bartlett 2009 ISBN 978-1-906435-09-7 Published by The Royal Yachting Association, Southampton UK

Reed's Sextant Simplified 7th Edition by Dag Pike 2003. ISBN 0-7136-6705-2. Published by Adlard Coles Nautical, London

Norie's Nautical Tables A G Blance 2007 ISBN 978 0 85288 945 9. Published by Imray Laurie Norie and Wilson, St Ives, Cambs. UK

The Chambers Dictionary 1998. ISBN 978-0550140005. Published by Chambers Harrap Publishers Ltd, Edinburgh UK

The Oxford Companion to Ships and the Sea, Edited by I.C.B Dear & Peter Kemp, 2006 ISBN 978-0-19-920568-4. Published by Oxford University Press, Oxford

Astro-navigation Data Sources

The Nautical Almanac NP 314-2010, 2009 (annual publication) ISBN 978-0-7077-40843 Published by HM Nautical Almanac Office, the United Kingdom Hydrographic Office

Rapid Sight Reduction Tables for Navigation NP303, 2008 ISBN 978-0-7077-40683 Published by United Kingdom Hydrographic Office

Sight Reduction Tables for Marine Navigation NP401 Vols 1-6, Published by the Hydrographer of the Navy, Taunton, Devon

Reeds Astro Navigation Tables by Lt Cdr Harry J Baker, 2011 (annual publication) ISBN 978-14081-23362 Published by Adlard Coles Nautical, London

Specific references in the text: Section 1: The basics of astro-navigation

1.1. *Dating the oldest Egyptian star map* by Ove von Spaeth, 1999. Centaurus International Magazine of the History of Mathematics, Science and Technology **42** (3): 159-179

1.2. *History of the Prime Meridian -Past and Present* by Jeremy Paul 1999.

– bibliography

http://gpsinformation.net/main/greenwich..html (accessed 12.02.08)

1.3. The nautical mile: Section 4.1 Table 8 in the *International System of Units* 8th ed. (2006) by the Bureau International des Poids et Mesures. http://www.bipm.org/en/si/si_brochure/chapter4/table8.htm (accessed 2007-12-04)

1.4. *Physics* by Ken Dobson, David Grace and David Lovett 1999. Published by Collins Educational, London

1.5. *"Copernicus Nicolaus"- Encyclopedia Britannica Online*, Encyclopedia Britannica. (accessed 2008-07-04

1.6. *John Harrison and Longitude:* the National Maritime Museum Greenwich, England: http://www.nmm.ac.uk/server/show/conWebDoc.355/viewPage/2 (accessed 2008-03-05)

1.7 *Neville Maskelyne* http://www.makingthemodernworld.org.uk/people/BG.0143 (accessed 2008-06-03)

Section 2: Down to the details

2.1. *Admiral Marcq St Hilaire: Line of Position Navigation: Sumner and Saint-Hilaire, the Two Pillars*, by Michel Vanvaerenbergh and Peter Ifland. Published by Unlimited Publishing Limited, 2003. ISBN:1588320685

Section 3: The Sextant and its use

3.1. *John Hadley1 682–1744*: http://www.questia.com/library/encyclopedia/hadley-john.jsp (accessed 2008-04-15)

3.2. *Angus MacDonald*: www.freepatentsonline.com/4421407.html (accessed 2008-03-27).

3.3. *Scientific Instruments of the Seventeenth and Eighteenth Centuries and Their Makers* by Daumas, Maurice, Portman Books, London 1989 ISBN 978-0713407273

Section 4: Table, tables, tables

4.1. *The Nautical Almanac NP 314-2010*, 2008. ISBN 978-0-70-774-0843 Published by HM Nautical Almanac Office, the United Kingdom Hydrographic Office

4.2. *Reeds Astro Navigation Tables* by Lt Cdr Harry J Baker, 2011 (annual publication) Published by Adlard Coles Nautical, London

4.3. *Astro-navigation made easy: using a pocket calculator* by Francois Meyrier, 2003. ISBN 0-7136-6222-0. English translation by David Fairhall. Published by Adlard Coles Nautical, London.

Section 5: The Sun and Time – their important link

5.1. *The beginning of time: a public lecture* by Stephen Hawking, http://www.hawking.org.uk/lectures/bot.html (accessed 2008-05-03)

5.2. *Blue sky and Rayleigh scattering* http://hyperphysics.phy-astr.gsu.edu/Hbase/atmos/blusky. html (accessed 2008-07-03)

5.3. *Saving the Phenomena: The Background to Ptolemy's Planetary Theory*, by Bernard R. Goldstein, *Journal for the History of Astronomy*, 28 (1997): 1-12

5.4. *History of legal time in Britain* by Joseph Myers, 2007 http://www.srcf.ucam.org/~jsm28/british-time/(retrieved 2008 -01-14)

5.5. *Greenwich Mean Time* http://wwp.greenwichmeantime.com (accessed 2008-01-16)

5.6. *NIST-F1 Cesium Fountain Atomic Clock* http://tf.nist.gov/cesium/fountain.htm (accessed 2008-07-03)

5.7. *Visions of the Future: Almanacs, Time, and Cultural Change, 1775-1870* by Maureen Perkins 1996, published by Clarendon Press

5.8. *Kepler's laws of planetary motion* http://en.wikipedia.org/wiki/Kepler_laws_of_planetary_motion (retrieved 2008-01-16).

5.9. *Eccentricity and semi-major axis* Young, Charles Augustus (1902). Manual of Astronomy: A Text Book. Ginn & company, 324–7.

5.10. *Isaac Newton's* Principia *1687*, Translated by Andrew Motte 1729. http://members.tripod.com/~gravitee/axioms.htm (retrieved 2008-06-10)

5.11. *Round the World in eighty days*, Jules Verne 1873. http://www.online-literature.com/

Section 6: Navigating with the Sun

61. *Gypsy Moth circles the world*; Francis Chichester 1967, Published by Hodder and Stoughton, London

6.2. *Shackleton's Boat Journey*; F A Worsley. 1999. Published by Pimlico, London

Section 7: Navigating with the Moon

7.1. *Moon*. Spudis, P.D. 2004. World Book Online Reference Center, NASA. (accessed on 2007-02-18)

7.2. *Neville Maskelyne* http://www.makingthemodernworld.org.uk/people/BG.0143 (accessed 2008-06.3)

– bibliography

7.3. *The Weak Friction Approximation and Tidal Evolution in Close Binary Systems*, Alexander, M. E. (1973). Astrophysics and Space Science **23**: 459–508.

7.4. *Moon*.by Spudis, P.D. 2004. World Book Online Reference Center, NASA. (accessed on 2008-03-14)

7.5. *Variations in Albedo Affect the Moon's Brightness* by Mike Luciuk, 2004 http://www.asterism.org/tutorials/tut26-1.htm (accessed 2008-02-21)

7.6 *RYA Astro Navigation Handbook*; Tim Bartlett 2009. The Royal Yachting Association, Southampton UK

7.7. *Eclipses*: Thieman, J.; Keating, S. Eclipse 99, Frequently Asked Questions. NASA. (accessed on 2008-04-24)

Section 8: Navigating with the Planets

8.1. *IAU 2006 General Assembly: Result of the IAU Resolution votes*. International Astronomical Union 2006. (retrieved on 2007-11-329).

8.2 *Greek Mythography in the Roman World* by Alan Cameron, 2005. Oxford University Press. ISBN 0195171217.

8.3. *2004 and 2012 Transits of Venus* http://eclipse.gsfc.nasa.gov/transit/venus0412.html (accessed 2008-07-2)

Section 9: The stars in general and *Polaris* in particular

9.1. *Dating the oldest Egyptian star map* by Ove von Spaeth, 1999. Centaurus International Magazine of the History of Mathematics, Science and Technology **42** (3): 159-179

9.2. *The history of astrology-another view* by Robert Hand http://www.zodiacal.com/articles/hand/history.htm (accessed 2008-06-01)

9.3. *The Acentric Labyrinth. Giordano Bruno's Prelude to Contemporary Cosmology* by Ramon G. Mendoza PhD, 1995, ISBN 1-85230-640-8

9.4. "Bessel, Friedrich Wilhelm". Dictionary of Scientific Biography 2-97-102. 1980 New York: Charles Scribner and Sons,. ISBN 0684101149.

9.5. "*William Herschel*" by Michael Hoskin 2008 New dictionary of Scientific Biography, 3-289-291 published by Scribners

9.6. *What makes the Sun shine* Raymond Davis Jnr. 2002 Nobel Foundation http://nobelprize.org/nobel_prizes/physics/laureates/2002/illpres/sun.html (accessed 2006-08-30)

9.7. "*The Historical Supernovae*". *Supernovae: A survey of current research* by Clark, D. H.; Stephenson, F. R. (June 29, 1981).; Proceedings of the Advanced Study Institute: 355–370, Cambridge, England, Reidel Publishing Co.

9.8. Norman Pogson Oxford Dictionary of National Biography. http://www.oxforddnb.com/index/101022438 (accessed 2008-05-02)

9.9. *Emergency Navigation* by David Burch 1990. Published by International Marine Publishing Company, Camden, Maine, USA ISBN 0-87742-204-4

9.10. *Lebanon's History Phoenician Beginnings* http://www.ghazi.de/phonecia.html (accessed 2008-04-03)

Appendix: Spherical Trigonometry from square one

App 1. *Vectors* http://www.physics.uoguelph.ca/tutorials/vectors/vectors.html (accessed 2008-04-12)

App 2. *Spherical Trigonometry* http://www.math10.com/en/university-math/spherical-triangle/spherical-triangle. html (accessed 2008-04-12)

App 3. *Radian* http://www.clarku.edu/~djoyce/trig/angle.html (accessed 2008-04-12)

App 4. *Vector triple expansion* http://planetphysics.org/encyclopedia/VectorTripleProduct.html (accessed 2008-04-13)

Index

ABC azimuth table 62-3, 84-5, 101-10, 165
- formula 62-3, 110-11, 164-5, 198, 201, 251
- interpolation 102-5
- method 201-2
- tables and interpolation 105-10

Admiralty mile 7-8
albedo – Moon 181
altitude 30-2, 35-6, 45-52
- apparent 39, 45
- calculated 57-62, 64-6, 96-100
  - comparison with observed 56-7, 62-8
  - equation for calculator 165
- observed 36-7, 40, 45, 49-57
- planet 93-95, 222-3
- sextant 38-40, 45
- Sun at meridian passage 33-4, 147-9
- star altitude 94, 251

angle of dip 39-40, 45, 93-4, 149-51, 159, 164, 166
angle – linkage to time 26-9
- cut (of) 149, 248, 251
- parallax 41-2, 189
- reflection -law governing 78

Antarctic/circle 21, 164, 190, 218
aphelion 120-1, 151
apogee – Moon 180, 187
apparent altitude – see altitude
apparent magnitude 208, 231, 235, 237
- Moon 237
- planets 208-209
- Pogson's ratio 235
- stars 235, 237
- Sun 208, 235, 237

apparent solar time 26, 28, 104-6,
- effect of Earth's orbit on 110-13
- effect of obliquity on 114-8

apparent Sun 27, 93, 115-6, 121
Arctic circle 20-1, 218, 245-6
arc to time 27-9, 90, 145, 160, 218, 222
Aries – see first point of Aries
assumed position 57, 62, 64-6, 68-9, 167, 259, 263

astro-nautical almanac 9-13, 17, 29, 33, 43, 59,115,131,137,154,184
atomic clock 115, 134
axial obliquity 112, 123, 135, 147
azimuth 17-8, 36, 54-9, 62-3, 65-7,
- calculated 57, 62, 137, 178
- equation for calculator 110
- tables 62-3, 84-5, 101, 105-9
- interpolation 105-9
- zenith distance (and) 56

azimuth angle 36, 54-6, 62, 96, 99, 109
- PZX triangle (and) 54-6

BBC time signals 134
Big-bang theory 3
Brahe, Tycho 117
Bruno, Giordano 232

Calculated altitude –see altitude - calculated
calculated azimuth 57, 62, 137, 178
calculated zenith distance 100, 165, 224
celestial equator 2, 9, 11, 13-4, 32, 48, 51-2
celestial horizon 32-4, 37-8, 49-50, 52, 153-4, 158
celestial latitude (declination) 8-9, 32, 233
celestial longitude (Greenwich hour angle) 8, 10-11, 13, 26, 53, 95, 114, 131
celestial movement 15
celestial poles 2, 10, 12-4, 26, 32, 43, 271
celestial sphere 1-4, 8-18, 22, 26, 30-2, 34
- equatorial view 16, 18, 31, 154
- polar view 14, 140,144, 153, 184

chosen position 36, 57, 60-1, 66-9, 72
chronometer 30, 115, 134, 149, 180-1
circle – great 5, 7, 47-50, 54, 56-7, 288
- small 4-6
- position 30, 45-7, 56, 61-2, 64, 66-8
- cocked hat 70, 178

collimation error 81-2
conjunction –Moon and Sun 186
- planets 214

constellations - 12-3, 232, 235, 239, 241
- Aquila 241
- Auriga 266, 268
- Cassiopeia 266
- Cygnus 241, 266
- Great Bear (Ursa Major) 3, 241-3, 245, 265
- Little Bear (Ursa Minor) 3, 265
- Lyra 241
- Orion 240-1, 244
- Pegasus 266-7
- Taurus 241

correction tables 40, 80, 85-91, 184
- altitude
 Sun/star/planet 86, 94, 223
 Moon 42, 196-7
- minute increment 137, 254
- Norie's 100

cosine - see mathematics
cross product 284-7, 291-2

Day length 21
declination 9-10, 13-4
- apparent solar time (and) 123-7
- correction 95-6
- ephemeris table 86-8
- Moon 194-6
- planets 217, 222, 226
- stars 234-5, 251-2
- Sun 144-6, 164-7
 - vertical/horizontal elements 123-7

departure 294
dip – angle 39-40, 45, 93-4, 149-51, 159, 164, 166
dead reckoning (d.r.) position 57, 67-8, 164
distance off 56, 78
dot product 283-5, 287, 290-2

Earth -axial tilt 19-22, 35
- seasons 20-1, 95, 123-5, 146, 241
- circumference 5, 7, 8, 48
 - nautical mile 7-8
eccentricity 118-9
- Earth 131, 135

- Moon 181-2
- Planets 207-8

Ecliptic 11-12, 22-3, 26, 34, 137, 235
ellipse – 117-9, 207
ephemeris table – see table
equations – local hour angle 13-15, 140-144
- meridian altitude 31-33, 154-160
- solving the spherical triangle 57-60

equation of time 29, 31, 115, 127, 131, 134-5, 160, 163, 215
- graph 116
- Greenwich Mean Time 26, 28, 103, 105-6, 118-20, 123, 125,

equator 2
- celestial equator 2, 9, 11, 13-4, 32, 48, 51-2

equatorial view 16, 18, 31, 154
equinoctial points 21
equinox 21, 23, 25-6, 125-6, 130, 145, 154, 156

errors – sextant - collimation 81-2
- index 79
- perpendicularity 79-80
- side 79-80

First point of Aries 11-14, 34, 233
- GHA Aries 12, 86-8, 252, 260, 262,
- LHA Aries 99, 248, 259-60, 262
- Polaris (and} 271, 276

formula
- ABC azimuth 62-3, 110-11, 164-5, 198, 201, 251
- calculated azimuth 69, 110, 164, 253
- cosine 62-3, 165, 167, 198, 293
 - calculator modific. 165, 167, 200, 253, 293
- haversine 100, 165
- Lagrange 292
- versine 100, 165

Geocentric view of the Sun 22-7, 34, 114, 124-5, 128, 207, 238-9
geographical position 17-8, 35, 45, 48-9, 56, 61, 64, 67, 201, 226, 256
gravitation 112, 119-

– index

20, 131, 181, 187, 207, 232
gravitational constant 119
great circle 5, 7, 47-50, 54, 56-7, 288
Greenwich 6, 9
Greenwich Hour Angle 10-1, 13-5, 26, 29-30, 91, 93
   - Aries 12, 14-5, 86-8, 252, 260, 262
   - correction to 95
   - Moon 194-6
   - planets 217
   - PZX triangle (and) 53, 56
   - Star 13, 252
   - Sun 30, 131, 137-44
Greenwich Meridian 10-1, 13-4, 26, 31, 53
Greenwich Hour Angle of Aries 12, 14-5, 86-8, 252, 260, 262
Greenwich Mean Time (GMT/UT) 29-31, 33, 114-5, 128-9, 132, 134-5 118-20, 123, 125
   - local apparent noon 26, 29-30, 35, 153
   - time zones
112, 132-4

Hadley, John - sextant development 74
Halley, Edmund – Saros cycle 191
haversines – see mathematics
Hawking, Stephen 111
height of eye 17, 39, 93-4, 103, 159, 163,
   - graph 40
   - interpolation 102-5
heliocentric view of the Earth 22, 24, 26, 29, 114, 206
Hilaire, Capt. M St 57
horizon mirror 74-6
   - shades 78
   -side error 79-80
   - index error 80-1
horizon natural 16-7, 20-1, 24
   - celestial 32-4, 37-8, 49-50, 52, 153-4, 158
   - sensible 37-9
   - visible 38-39
hour angle – see Greenwich/Local Hour Angle

Inclination -Moon 182
   - planets 207-8
index arc 74-5
index error 80-1
index mirror 75
   -perpendicularity 79
   -shades 78
intercept –72, 96, 105, 137, 164
   - method 57
   - plotting 66-9
interpolation 102-5

Jupiter – 15, 206-7,
   - arc of visibility 210
   - conjunction 212-4
   - eccentricity 207
   - ephemeris 88
   - inclination 208
   - orbit 118

Kepler, Johannes - 117-8, 135, 207

Latitude 4-9, 14, 21-2, 25-6,
   - noon sight- see meridian passage
laws of planetary motion 117-8
limb – upper/lower - Sun 150-1
   -Moon 196-8
local apparent noon 26, 29-30, 35, 153
Local Hour Angle (LHA) 13-15
   - from GHA and longitude 140-44
   - latitude estimation 34
   - Moon 198-201
   - planet 222-26
   - PZX triangle 53-6
   - stars 251-256
   - Sun 164-8
local time – conversion to GMT/UT 34, 114, 132-5, 159
logarithm 101-2
   - magnitude scale 208, 235
longitude 4-8, 12-4, 26
   -GMT/UT linkage 30-1
   -link to hour angle 140-144
Longitude in time 29, 90
lunars – Neville Maskelyne 30, 180-1

303

Macdonald, Angus 76
magnitude – see apparent magnitude
Mars 15, 206-7,
    - arc of visibility 210
    - conjunction 212-4
    - eccentricity 207
    - ephemeris 88
    - inclination 207
    - orbit 118, 213
    - parallax 42, 44, 93, 222, 224

mathematics
    -cosine 41, 58, 100-2, 253, 277,
        280-1, 283-4, 287, 292-3
        -formula 62-3, 165-7, 198, 200
        253, 255
        - ratio 280
        - rule 281
    - degrees/minutes to decimals 90
    - describing 3D space 283-287
    - dot and cross products 283-87,
        291-2
    - haversine 100-1
        - formula 165
    - safe practice in subtraction 156-7
    - sine 100, 102, 253, 277-292
        - sine ratio 278-9
        - sine rule 279
        - sine wave (orbital speed) 123
    - spherical. triangle 47-52, 56-60, 96,
        100-2
    - vectors 120, 277, 283-292
        - dot and cross products 283-87,
        291-2
    - versine 100-2,
        - formula 165
mean solar time 29, 114-6, 123, 132, 134
meridian passage of Sun 29-32, 48, 52, 71,
    86-8, 132-35, 145-9, 153-7,
    159-163,178
    equations for 33-4, 154-60
    -linked variables at 153-5
meridians of longitude 5, 7, 26, 31, 34, 57,
    132, 146
micrometer – sextant 74, 76-7, 79-82, 148
mile - Admiralty 7-8

    - nautical 7-8, 39, 56, 64-5, 67,
        69, 150
    - statute 7-8
mirror - optical coatings 75-67-8 - horizon
  – see horizon mirror
    - index – see index mirror
month - sidereal 182
    - synodic 182
Moon 179-204
    - albedo 181
    - altitude correction 196-8
    - apogee 187
    - apparent size 187-8
    - augmentation 187-8
    - correction tables 95-6
    - declination 194-6
    - difference table (merid.
        pass.) 220-1
    - eccentricity 181-2
    - eclipses - Moon 189-91
        - Sun 189-91
    - first quarter 184-5
    - full Moon 185-6
    - general information 180-1
    - Greenwich Hour angle corrn.
        194-6
    - gibbous phase 185
    - horizontal parallax 188-9, 196-7
    - libration 181
    - lunars 30, 180-1
    - lunation 181
    - meridian passage 193-4
    - Moonrise and Moonset 191
    - new Moon 184, 186
    - orbit, length of 181-2
    - parallax 42, 188-9
    - perigee 187
    - phases 184-6
    - precession 180
    - proximity 15, 180
    - Saros cycle 191
    - sidereal month 182
    - sighting 196-203
    - synchronous rotation 181
    - synodic month 182
    - time of day (and) 186-7

– index

- waxing 185
Nautical almanac –see astronautical almanac
  nautical mile 7-8, 39, 56, 64-5, 67, 69, 150
navigation – with the Moon 179-204
    - with the planets 205-30
    - with the Sun 136-178
    - with the stars 265-276
Newton, Isaac 74, 119, 131
night sky 40, 231, 238, 241
noon sight – see meridian passage
nutation 130-1, 137
    - Polaris (and) 259, 271

Obliquity 20, 24-5
    - effect on solar time 116, 123-4, 127-30, 137, 147
    - Moon, effect on 193-4
Observer's view 16, 25
    - Moon 185-6
    - stars 238
opposition – planets 213-4
orbit - angular effect of speed changes 116-117, 120-3
    - Earth's orbit 120-3
    - eccentricity 118-9
        Moon 181-2
        Planets 119-20, 207

Parallax - 40-2, 93-5
    - Moon 188-9, 196-7
    - planets, 42, 222
    - Sun 159
parallel of latitude 5-6, 239-243
perigee – Moon 187
perihelion – Earth 120-2, 151
perpendicularity – index mirror 79-81
    - horizon mirror – side error 79-80
planets – 14, 205-220
    - altitude correction table 222-34
    - annual arc of position 210
    - approximate SHA 2014-16
    - apparent magnitude - 208
    - axial tilt 208
    - celestial sphere - movement 210
    - conjunctions in 2011 - 214
    - conjunction of Venus 211-2
    - declination 217-8
    - diagram 215-7
    - eccentricity 207-8
    - elongation of Venus 211-2
    - general information 206-7
    - Greenwich hour angle 217
    - inclination 207-8
    - magnitude/brightness 208
    - Mercury 213
    - naming 206
    - navigationally useful 207
    - opposition 213-4
    - orbits 117-20, 206
    - parallax 42, 222
    - planetary motion 119-20
    - sidereal hour angle 210
    - twilight 218-9
    - sight reduction 222-9
Pogson, Norman 235
plotting sheet 294
plotting of sight –
    - limitations 70-1
    - summary 68-9
polar view 14, 140, 144, 153, 184
Polaris 16, 52, 73, 88, 213, 265-276
    - altitude proforma 275
    - apparent movement 265
    - brightness reference star 237
    - circumpolar stars 245-6
    - constellation Ursa Minor 265
    - ephemeris table 88, 271-2, 274
    - finding in the sky 266-8
    - Kochab 243, 259, 262-3, 265
    - latitude estimation 270-5
        - mathematics 270
    - precession 271
polar distance 51
position
    - assumed 57, 62, 64-6, 68-9, 167
    - chosen 57, 62, 90, 156
    - dead reckoning 57, 67-8, 164
    - estimated 57, 68, 164, 172-3, 178

position circle 45-6, 61
position line 36, 39, 57, 67-8, 70-2, 84
    - transferred 71-2, 172-8
positioning on the celestial sphere 8-12
precession – Earth 130-1, 137
    - equinoxes 13
    - Moon 180, 187
    - Polaris 259, 271
Prime Meridian 6, 10, 26, 30, 115, 218
proforma – 160, 162, 170
    - Moon 204
    - planet 230
    - Polaris 275
    - star 258
    - Sun meridian altitude 161
    - Sun 171
Proxima centauri 15-16
PZX triangle 47-56
    - extended 52
    - orientation and LHA 55
    - summary PZX 56

Quadrant – seaman's 73-4
Quadrantal notation 101, 109-10, 169

Radian 277, 289-90
Rayleigh scattering 113
refraction 40, 43, 45, 72, 93, 95, 150-1

Seasons 20-1, 95, 123-5, 146, 241
semi-diameter 40, 42-5, 88, 93,159
    - Moon 42-4, 187, 196
    - planet 44, 222, 224
    - Sun 42-44, 145, 150-1, 159, 164
    - star 251, 271
sensible horizon – see horizon
sextant -73-83
    - acquiring new or second hand 83
    - coated mirrors 75-6
    - development of 73-4
    - distance off 77-8
    - errors 78-82
    - MacDonald, Angus 76
    - Hadley, John 74
    - index error 80-1
    - micrometer 74, 76-7, 79-82, 148

    - optics 73-4
    - perpendicularity
        - index mirror 75, 79
    - shades 78
    - swinging for horizon setting 82-3
    - side error – horizon mirror 79-80
sidereal Hour Angle 11-14, 34, 88, 102, 233, 235, 241, 248
    - planets 210, 215
sight reduction
    - Moon 196
    - multiple sights 70-1
    - multiple star sight 259-264
    - noon sight for latitude 152-60
    - planet 222-228
    - Polaris for altitude 273-5
    - star 251-7
    - single sight limitations 70-1
    - sight reduction tables 96-99, 165
    - Sun meridian 152-60
    - Sun non-meridian 164-70,
    - transferred position lines 71-2, 172
    - twilight 20-1, 86-8, 146, 186-207, 218-22
sine – see mathematics
Sirius 13, 16, 208, 213, 240-41, 260
    - apparent magnitude 235-7
small circle 4-6
solar day 114
solar time 114-5
    - apparent 114-5
    - mean 29, 114-6, 123, 132, 134
    - obliquity effect on 123-7
    - orbital speed effect on 120-3
    - visualizing effect on 128-30
solar system
    - geocentric/heliocentric 22-6
solstice –summer/winter 23, 25-6, 125-6, 130, 144-5
spherical triangle 47-52, 56-60, 96, 100-2
spherical trigonometry 57-60, 277-292
Stars 11-13, 231-263
    - altitude correction table 94, 251
    - brightness – see magnitude
    - celestial sphere 1-4, 8-18, 22, 26,
        - equatorial view 16, 18, 31,

- *index*

    - polar view 14, 140,144, 153,
- changing night sky 238-9
- circumpolar 245-6
- declination 234-5, 251-2
- geocentric parallax 40-42
- general 232-3
- Greenwich Hour Angle 252
- leading and trailing 241
- list 102, 248
- magnitude 235-7
- meridian passage 247
- movement 245-7
- named
    - *Alioth* 243-4
    - *Alkaid* 243-4
    - *Altair* 241-2
    - *Aldebaran* 214, 240-1, 261
    - *Arcturus* 2243-4, 246, 260
    - *Betelgeuse* 240-1, 260
    - *Capella* 241, 260, 266, 268
    - *Castor and Pollux* 240-1
    - *Deneb* 241-2, 266
    - *Dubhe* 241-2, 266
    - *Gienah* 242, 266
    - *Kochab* 243, 259, 262-3, 265
    - *Markab* 266-7
    - *Menkalinan* 266, 268
    - *Merak* 2241, 243, 244, 266
    - *Mizar* 244
    - *Polaris* 16, 52, 73, 88, 213, 265-276
    - *Procyon* 241, 260
    - *Proxima centauri* 15-6
    - *Rigel* 240-1
    - *Scheat* 266-7
    - Sirius 13, 16, 208, 213, 240-41, 260
        - apparent magnitude 235-7
    - *Theta-Aurigae* 266-68
    - *Vega* 237, 241-
- navigationally useful 233-5
- non-circumpolar 247
- non-visible 245
- peak height 247
- pattern recognition 241-246

- semi-diameter 251, 271
- sidereal hour angle 11-14, 34, 88, 102, 233, 235, 241, 248
- sight planning 248-251
- sight reduction 251-8
- star-finder 249-50
- summer triangle - 241-2, 266

Sun and time 112-35

Sun - altitude correction table 44, 93-5 150-1
- apparent movement 23, ,27, 121-8
- apparent solar time –see solar time- apparent
- declination 24-6, 29, 32-4, 86-8, 93-96, 123-30, 144-6, 164-7
- annual changes graph 127
- forenoon/afternoon sights 71, 111, 149, 172
- general information 113
- GHA correction 137-44
- latitude measurement 31-3, 147, -161
- local noon (at) 30, 34, 145, 164, 172, 269
- meridian passage 29-32, 48, 52, 71, 86-8, 132-35, 145-9, 153-7, 159-163, 178
- motion across sky 146-7
- movement on celestial sphere 15-6 22-25, 123-7
- navigating with the - 136-78
- non-meridian sights timing 149-50
- non-noon Sun sight 164-70
- noon Sun sight 152-60
- parabolic motion 146-7, 163
- parallax 40-2, 95, 159, 164, 167, 171
- power/energy density 19-20, 113
- rise and fall 24-5, 146-7
- semi-diameter 42-44, 145, 150-1, 159, 164
- sextant shades – use of 78
- Sun-sight reduction 150-1
    - meridian 159-161
    - non-meridian 164-171

307

- Sun-run-Sun 71-2, 149, 171-8
- Sun and time 112-35
- sunrise-sunset table 218-222
- upper/lower limb sights 150-151
sun dial - 131
syzygy 185, 190

Tables 84-111
- ABC 62-3, 69, 100, 105-11, 251, 256
- altitude
    - calculated 95-100, 148
    - Moon 196-7
    - planets 93-4, 222-4
    - Sun 93-4,
    - stars 93-4, 222-51,
- arc to time conversion 27-9, 90, 145, 160, 218, 222
- azimuth 62-3, 84-5, 101, 105-9
- calculated altitude 96-100
    - Marine tables 173-176
    - Rapid Air tables - 260-264
- corrections 40, 80, 85-91, 184
    - altitude
        - Sun/star/planet 86, 94, 223
        - Moon 42, 196-7
        - minute increment 137, 254
    -Norie's 100
- declination 86-89, 95-96
    - Moon 192, 194-6
    - planets 217, 219
    - star 234, 252
    - Sun 138-9, 144-6
-ephemerides 86-9
    - Aries 86-7, 234
    - GHA 86-8
    - Moon 192, 194-5
    - planets 217-20
    - Polaris 271-2
    - stars 234, 252
    - Sun 137-9
- height of eye/dip 93-4
- increments 91-2
- interpolation moonrise/set 220-2
- mathematical 101-2

- sight reduction 96-100
- sunrise-sunset/twilight 204-8
- versines 101-2
time – arc to time 27-9, 90, 145, 160, 218, 222
- GMT/UT – see Greenwich Mean Time
- linkage to angle 26-9
- mean/apparent difference 114, 125
time checks 134
time zones 112, 132-4
- world map 133

Universal coordinated time (UTC) 112, 114-5
universal theory of gravitation 112, 119-20, 131, 181, 187, 207, 232

Vectors 120, 277, 283-292
- dot and cross products 283-87, 291-2
Venus 205-30
- annual arc of position 210
- apparent magnitude 208-9
- conjunction 212
- elongation 211-2
- inclination 207-8
- orbital features 206-7, 211-2
Vernier, Pierre 77
- vernier scale 77
versines 100-2, 165
- longhand method calc. alt. 101-2
visible sea 16
visible sky 16-7, 38-9

Weems and Plath – star-finder 249-50, 259

Zenith 1, 17-8, 26, 32-3, 36-8, 48
zenith distance 18, 33, 52, 56, 100, 153
Zodiacal signs 3, 235